**Mathematics for Industry**

Volume 23

**Aims & Scope**

The meaning of "Mathematics for Industry" (sometimes abbreviated as MI or MfI) is different from that of "Mathematics in Industry" (or of "Industrial Mathematics"). The latter is restrictive: it tends to be identified with the actual mathematics that specifically arises in the daily management and operation of manufacturing. The former, however, denotes a new research field in mathematics that may serve as a foundation for creating future technologies. This concept was born from the integration and reorganization of pure and applied mathematics in the present day into a fluid and versatile form capable of stimulating awareness of the importance of mathematics in industry, as well as responding to the needs of industrial technologies. The history of this integration and reorganization indicates that this basic idea will someday find increasing utility. Mathematics can be a key technology in modern society.

The series aims to promote this trend by (1) providing comprehensive content on applications of mathematics, especially to industry technologies via various types of scientific research, (2) introducing basic, useful, necessary and crucial knowledge for several applications through concrete subjects, and (3) introducing new research results and developments for applications of mathematics in the real world. These points may provide the basis for opening a new mathematics-oriented technological world and even new research fields of mathematics.

More information about this series at http://www.springer.com/series/13254

Bernard Bonnard · Monique Chyba
Editors

# Recent Advances in Celestial and Space Mechanics

Springer

*Editors*
Bernard Bonnard
Faculté des Sciences Mirande
University of Burgundy
Dijon
France

Monique Chyba
Department of Mathematics
University of Hawaii, Manoa
Honolulu, HI
USA

ISSN 2198-350X ISSN 2198-3518 (electronic)
Mathematics for Industry
ISBN 978-3-319-27462-1 ISBN 978-3-319-27464-5 (eBook)
DOI 10.1007/978-3-319-27464-5

Library of Congress Control Number: 2016931203



Printed on acid-free paper

This Springer imprint is published by SpringerNature
The registered company is Springer International Publishing AG Switzerland

# Preface

The objective of this book is to present recent advances in celestial and space mechanics written by contributors from both academic institutions and space agencies. It is organized into seven complementary chapters, two of which are devoted to celestial mechanics and five are dedicated to orbital mechanics. The central line of the work presented here emphasizes the development and application of computational techniques in both areas. This edited volume is aimed at an audience interested in celestial mechanics with a focus on the N-body problem and astrodynamics. It is appropriate for advanced graduate students, as well as for potential and current researchers in the field.

Below, we further detail each contribution:

The first two chapters belong to the domain of celestial mechanics, and deal with the integrability and stability problem of the N-body problem:

1. T. Combot focuses on integrability questions of the N-body problem. The computations are based on recent advances in the application of Galois differential theory to the integrability problem for a mechanical system whose potential V is a meromorphic potential of order$-1$. In the computations, the classical notion of central configuration/Darboux point plays a key role. In particular, the author proves the non-integrability for the colinear three- and four-body problem.
2. In the chapter by D.J. Scheeres, general conditions for the existence of stable, minimum energy configurations are derived and investigated for the full N-body problem. The found relative equilibria include central configurations but also the so-called resting configurations. The results are applied to the analysis of the so-called spherical, equal mass full body problem. A detailed analysis is conducted for N = 2, 3 and a preliminary one for N = 4.

Space mission design and control is the focus of the remaining five chapters:

3. A. Farres and A. Jorba demonstrate station keeping of a spacecraft equipped with a solar sail. Using the sail orientation as control, the authors take advantage of dynamical properties of the restricted three-body problem for a solar sail and

the dynamical substitutes of the N-body problem for a solar sail in order to maintain the spacecraft in a closed neighborhood of either unstable equilibria or periodic orbits. Finally, the authors present the principles and results of numerical computations realized to simulate the NASA's Sun-jammer mission.
4. M. Chyba, T. Haberkorn, and R. Jedicke consider a non-coplanar rendezvous problem from a parking orbit around the EM-L2 Lagrange point to Asteroid 2006RH 120, using the mass minimization cost function. The model is the circular restricted four-body problem where the Sun is considered as a perturbation of the Earth-Moon circular restricted three-body problem and the control structure is assumed to be bang–bang with at most three trust arcs. The solutions are computed numerically using a shooting algorithm supplemented by direct methods and continuation methods to address initialization issues.
5. R. Epenoy introduces a three-step methodology to compute low-thrust minimum-energy transfers between Lyapunov orbits in the planar circular restricted three-body problem. The approach used by the author is based on a mixture of different methods including simplex methods, single shooting methods, continuation methods, variational equations, and gradient descent. The methodology is illustrated with a numerical application for transfers between Lyapnuov obits around EM-L1 and EM-L2 Lagrange points.
6. The chapter by M. Chyba, G. Patterson, and G. Picot deals with the optimal control of the elliptic restricted three-body problem. This nonautonomous system is analyzed and local controllability along smooth enough trajectories is proved. Numerical computations with this model on three different missions are provided. Key to these simulations is a continuation on the eccentricity to solve the shooting problem associated with normal extremals.
7. The chapter by J.-B. Caillau and A. Farres is dedicated to the study of local optima in time-minimum control for the restricted circular three-body problem. Their investigation is based on the use of numerical homotopic methods. The paper presents a first attempt to provide a global optimal solution for a problem of such complexity for some important physical transfers, e.g., from a geostationary orbit to the EM-L1 Lagrange point.

The editors would like to thank Dr. Masato Wakayama, Editor-in-Chief of Mathematics for Industry—Springer, for inviting them to generate this collection of chapters and his constant support, as well as Mieke van der Fluit, senior publishing assistant, for her help, patience, and assistance with the publishing logistics associated to this special issue.

Dijon, France — Bernard Bonnard
Honolulu, HI, USA — Monique Chyba
October 2015

# Contents

# Contributors

**Jean-Baptiste Caillau** UBFC & CNRS/INRIA, Math. Institute, Université de Bourgogne, Dijon Cedex, France

**Monique Chyba** Department of Mathematics, University of Hawaii at Manoa, Honolulu, HI, USA

**Thierry Combot** IMB, Universié de Bourgogne, Dijon Cedex, France

**Richard Epenoy** Centre National d'Etudes Spatiales, Toulouse, France

**Ariadna Farrés** Departament de Matemàtica Aplicada i Anàlisi, Universitat de Barcelona, Barcelona, Spain

**Thomas Haberkorn** MAPMO-Fédération Denis Poisson, University of Orléans, Orléans, France

**Robert Jedicke** Institute for Astronomy, University of Hawaii, Honolulu, HI, USA

**Àngel Jorba** Departament de Matemàtica Aplicada i Anàlisi, Universitat de Barcelona, Barcelona, Spain

**Geoff Patterson** Department of Mathematics, University of Hawaii at Manoa, Honolulu, HI, USA

**Gautier Picot** Department of Mathematics, University of Hawaii at Manoa, Honolulu, HI, USA

**Daniel J. Scheeres** Department of Aerospace Engineering Sciences, The University of Colorado, Boulder, USA

# Integrability and Non Integrability of Some $n$ Body Problems

**Thierry Combot**

**Abstract** We prove the non integrability of the colinear 3 and 4 body problem, for any positive masses. To deal with resistant cases, we present strong integrability criterions for 3 dimensional homogeneous potentials of degree $-1$, and prove that such cases cannot appear in the 4 body problem. Following the same strategy, we present a simple proof of non integrability for the planar $n$ body problem. Eventually, we present some integrable cases of the $n$ body problem restricted to some invariant vector spaces.

**Keywords** Morales-Ramis theory · Homogeneous potential · Central configurations · Differential Galois theory · Integrable systems

## 1 Introduction

In this article, we will consider the $n$ body problem whose Hamiltonian is given by

$$H_{n,d} = T_{n,d}(p) + V_{n,d}(q) = \sum_{i=1}^{n} \frac{\|p_i\|^2}{2m_i} + \sum_{1\leq i<j\leq n} \frac{m_i m_j}{\|q_i - q_j\|}$$

The quadratic form $T$ correspond to kinetic energy, $V$ is the potential, which is a homogeneous function of degree $-1$ in $q$. The coordinates $q_1, \dots, q_n$ correspond respectively to the coordinates of the bodies $m_1, \dots, m_n$.

Already since Poincare and Bruns [1, 2], it is known that the $n$-body problem is for $n \geq 3$ not integrable in general. Bruns in [1] proved the non-existence of additional algebraic first integrals, later generalized by Julliard-Tosel [3], and more recent work like [4–6] prove the meromorphic non-integrability or non existence of meromorphic first integrals in some cases. All these proofs strongly suggest that the $n$-body problem is never integrable for $n \geq 3$, even in particular cases (as proven for example for the

T. Combot (✉)
IMB, Universié de Bourgogne, 9 Avenue Alain Savary, 21078 Dijon Cedex, France
e-mail: thierry.combot@u-bourgogne.fr



B. Bonnard and M. Chyba (eds.), *Recent Advances in Celestial and Space Mechanics*, Mathematics for Industry 23, DOI 10.1007/978-3-319-27464-5_1

isosceles 3-body problem in [7]). The colinear problem (in dimension 1) is a priori more difficult than the non-integrability proof of the $n$ body problem in the plane and higher dimension, because it needs fewer additional first integrals to be integrable. Recall that as the energy and the impulsion of the center of mass are first integrals, in dimension 1 we only need $n-2$ additional first integrals for integrability. We will see that even if the problem is not so easy as the planar case, it can be completely studied in the case $n = 3, 4$ through the bounding of eigenvalues of the Hessian of $V$ at central configurations (see Definition 1). A similar trick allows to obtain a simple proof of the non integrability of the planar case with positive masses. In the opposite direction, the $n$-body problem also possesses explicit algebraic orbits, linked to central configurations [8]. Restricting the $n$-body problem to a vector space associated to a central configuration leads in particular to an integrable problem, although very simple. Still, as we will see, there are also less trivial invariant vector spaces of the $n$-body problem on which the potential is integrable.

In the integrability analysis of the $n$ body problem, and in the more general case of homogeneous potential, the notion of central configuration/Darboux point plays a key role.

**Definition 1** We consider the potential $V_{n,d}$ of the $n$ body problem. We will say that $c \in \mathbb{C}^{nd}$ is a central configuration if there exists $g \in \mathbb{C}^d$, $\alpha \in \mathbb{C}$ such that

$$\frac{\partial}{\partial q_i} V(c_1 - g, \ldots, c_n - g) = \alpha(c_1 - g, \ldots, c_n - g) \qquad i = 1 \ldots n$$

The scalar $\alpha$ is called the multiplier. We say that the central configuration is proper if $\alpha \neq 0$ (the case $\alpha = 0$ is called an absolute equilibrium). In the more general setting of $V$ a homogeneous potential of degree $-1$, we call $c$ a Darboux point if moreover $g = 0$.

We add this constant $g$ in our Definition for the $n$ body problem as the potential is in this case invariant by translation, and thus we do not (always) want to require that the center of mass be at 0. Our non-integrability proofs will be based on variational equations of the corresponding differential system near these central configurations. The main theorem behind these non-integrability proofs is the following

**Theorem 1** (Morales et al. [9]) *Let $V$ be a meromorphic homogeneous potential of degree $-1$ and $c$ a Darboux point. If $V$ is meromorphically integrable, then the identity component of the Galois group of the variational equation near the homothetic orbit associated to $c$ is abelian at any order. Moreover, the identity component of the Galois group of the first order variational equation is abelian if and only if*

$$Sp(\nabla^2 V(c)) \subset \left\{ \tfrac{1}{2}(k-1)(k+2),\ k \in \mathbb{N} \right\}$$

Note also that in dimension 1, $V_{n,1}$ is a rational potential (thus univaluated on $\mathbb{C}^n$), but is not in higher dimension. In the complex domain, the potential $V_{n,d},\ d \geq 2$ is properly defined on an algebraic variety $S$. An extension of Theorem 1 has been done in [10], and proves that in the $n$ body problem, the necessary condition for integrability on the Galois group of variational equations still holds.

Such a Theorem can be either used for each central configuration separately, or simultaneously using some algebraic properties. In the case of the $n$ body problem, a direct computation of central configurations is often too difficult. The colinear case with $n = 3, 4$ is still tractable, and we prove moreover that a complete computation of central configurations is not necessary, only upper bounds on eigenvalues of the Hessian matrix of $V_{n,1}$ at Darboux points is necessary.

Using the real algebraic geometry software RAGlib [11], we prove such an upper bound for $n = 3, 4$ and we conjecture that a similar upper bound always hold for any $n$. The software RAGlib is a Maple package, and the command we will mostly use is **HasRealRoots**. This command take in input a system of polynomials with rational coefficients, and (possibly) a set of polynomial inequalities. The answer is true/false, saying if the system has (at least) one real solution. This also allows to prove upper bounds for a (multivariate) rational function $f$ by just looking for solutions of the equation $f = B$ where $B$ is a (numerically guessed) upper bound. Note that the real conditions on the masses will be heavily used: in particular some polynomial integrability conditions cannot be satisfied in the real but would be in the complex.

We then prove very strong non-integrability Theorem that rules out any potential which satisfies these bounds. In the planar case, we also prove a similar upper bound, which holds moreover for any $n$. This allows to prove the non-integrability of the planar $n$-body problem. The main theorems of this article are the following

**Theorem 2** *For any* $(m_1, m_2, m_3) \in {\mathbb{R}_+^*}^3$, *the potential* $V_{3,1}$ *is not meromorphically integrable. Moreover, if* $m_1 + m_2 + m_3 = 1$, *the variational equations near the unique real central configuration have an Abelian Galois group (over the base field* $\mathbb{C}(t)$*) up to an order*

- *greater than* 1 *if and only if there exist* $\rho \in \mathbb{R}_+^*$ *and* $k \in \{5, 9, 14\}$ *such that*

$$\begin{aligned} m_1 &= \frac{(\rho+1)(-8\rho^5+k\rho^5-12\rho^4+3k\rho^4-8\rho^3+3k\rho^3+3k\rho^2+3k\rho+k)}{k(1+2\rho^3+\rho^4+2\rho+\rho^2)^2} \\ m_2 &= -\frac{(-8\rho^4+k\rho^4-28\rho^3+2k\rho^3+k\rho^2-40\rho^2-28\rho+2k\rho-8+k)\rho^2}{k(1+2\rho^3+\rho^4+2\rho+\rho^2)^2} \\ m_3 &= \frac{(\rho+1)(k\rho^5+3k\rho^4+3k\rho^3-8\rho^2+3k\rho^2-12\rho+3k\rho-8+k)\rho^2}{k(1+2\rho^3+\rho^4+2\rho+\rho^2)^2} \end{aligned} \qquad (E_k)$$

- *equal to* 2 *if and only if moreover* $m_1 = m_3$ *or* $(m_1, m_2, m_3) \in E_9$.

**Theorem 3** *For any* $m_1, m_2, m_3, m_4 > 0,\ m_1 + m_2 + m_3 + m_4 = 1$, *the potential* $V_{4,1}$ *is not integrable. Moreover, near the unique real central configuration, there are at most* 14 *one dimensional irreductible algebraic curves in the space of masses for which the variational equations have virtually Abelian Galois groups at least up to order* 1. *At least one of them, and at most* 10 *of them correspond to masses for* *which the second order variational equations have a virtually Abelian Galois group. None of them have a variational equation whose Galois group is virtually Abelian at order 5.*

**Theorem 4** *For any n-tuplet of positive masses, the planar n body problem is not meromorphically integrable.*

**Theorem 5** *The planar 5 body problem with masses $m = (-1/4, 1, 1, 1, 1)$ restricted to the vector space*

$$W = \{q \in \mathbb{R}^{10},\ q_{1,1} = q_{1,2} = q_{2,1} + q_{4,1} = q_{2,2} + q_{4,2} = q_{3,1} + q_{5,1} = q_{3,2} + q_{5,2} = 0\}$$

*is integrable in the Liouville sense.*

*The spatial $n+3$ body problem with masses $m = (m_1, \ldots, m_n, -\alpha, 4\alpha, 4\alpha)$ restricted to the vector space*

$$W = \Big\{q \in \mathbb{R}^{3(n+3)},\ \ q_{n+1,1} = q_{n+1,2} = q_{n+1,3} = q_{n+2,1} = q_{n+2,2} = q_{n+3,1}$$
$$= q_{n+3,2} = q_{n+2,3} + q_{n+3,3} = 0,\ \ q_{i,3}|_{i=1\ldots n} = 0,\ \ q_i|_{i=1\ldots n} = \beta R_\theta c,\ \ \beta, \theta \in \mathbb{R}\Big\}$$

*where c is a central configuration of n bodies with masses $(m_1, \ldots, m_n)$ in the plane on the unit circle with center of mass at 0, $R_\theta$ being a rotation in this plane and $\alpha$ chosen such that the configuration c with the central mass $-\alpha$, is an absolute equilibrium is integrable in the Liouville sense.*

The Theorem 2 implies the non integrability of the colinear 3 body problem, which was already done in [12] using the systematic approach using all central configurations and a relation between the eigenvalues of Hessian matrices. This approach is hard to apply to more complicated systems as its cost is exponential in the number of central configurations. This is due to the fact that all central configurations are analyzed, even if only a few of them would probably be enough to conclude to non integrability. Also, the physical assumption that the masses are real positive is not used. In the next section, we thus make a more precise analysis of variational equations near the unique real central configuration, whose existence and uniqueness is a result of Moulton [13]:

**Theorem 6** (Moulton [13]) *For any fixed positive masses $m_1, \ldots, m_n$ with a fixed order of the masses, the colinear n body problem admits exactly one real central configuration.*

Remark that also in the not trivially integrable example we found, central configurations seem to play a key role. In particular, they all contain continuums of central configurations (the first case contains the famous 5 body central configuration of Roberts [14]). According to a conjecture of Smale, proved for $n = 4, 5$ in [8, 15], such continuums are not possible with positive masses.

# 2 The Colinear 3 Body Problem

## 2.1 Central Configurations

**Proposition 1** (Euler) *We pose $c = (-1, 0, \rho)$ with $\rho \in \mathbb{C}\backslash\{0, -1\}$. If $c$ is a central configuration of the colinear 3 body problem (corresponding to the potential $V_{3,1}$), then the following equation is satisfied*

$$\begin{aligned}(m_2 + m_3) + (2m_2 + 3m_3)\rho + (3m_3 + m_2)\rho^2 \\ -(3m_1 + m_2)\rho^3 - (3m_1 + 2m_2)\rho^4 - (m_1 + m_2)\rho^5 = 0\end{aligned} \tag{1}$$

In the colinear 3 body problem, we can always translate a central configuration because the potential is invariant by translation. Moreover, due to this definition, the set of central configurations is also invariant by dilatation, so for any central configuration $q \in \mathbb{C}^3$, after translation and dilatation, we can always write it $q = (-1, 0, \rho)$ with $\rho \in \mathbb{C}\backslash\{0, -1\}$. The biggest problem that authors on the subject (see [4]) seem to have encountered is the fact that we have a polynomial of degree 5, which is not very easy to use. We will see that the complexity of central configuration equations is not a problem at all if we consider the problem differently.

The Theorem 6 of Moulton suggests that we should work in the opposite way. We fix $\rho > 0$ and we seek the masses such that $c = (-1, 0, \rho)$ is a central configuration. We are then sure that if we consider all possible $\rho$ we will then consider all positive masses (because for each triplet of masses, there is at least one $\rho$ that is convenient). More precisely, we have

**Proposition 2** *The set of masses $m_1, m_2, m_3$ such that $m_1 + m_2 + m_3 = 1$ and $c = (-1, 0, \rho)$ with*

$$\rho \in \mathbb{C}\backslash\{\rho,\ \ \rho(\rho+1)(1 + 2\rho + \rho^2 + 2\rho^3 + \rho^4) = 0\} \tag{2}$$

*is a central configuration, is an affine subspace of dimension 1 parametrized by*

$$\begin{aligned}m_1 &= s \\ m_2 &= -\frac{3s\rho^3 + 3s\rho^4 + s\rho^5 + s - 1 + 3\rho s - 3\rho + 3\rho^2 s - 3\rho^2}{\rho(1 + 2\rho + \rho^2 + 2\rho^3 + \rho^4)} \\ m_3 &= \frac{2\rho s + \rho^2 s + 2s\rho^3 + s\rho^4 + s - 1 - 2\rho - \rho^2 + \rho^3 + 2\rho^4 + \rho^5}{\rho(1 + 2\rho + \rho^2 + 2\rho^3 + \rho^4)}\end{aligned} \tag{3}$$

*Conversely, for each triplet of masses $(m_1, m_2, m_3) \in {\mathbb{R}_+^*}^3$, $m_1 + m_2 + m_3 = 1$, there exists a central configuration of the form $(-1, 0, \rho)$ with condition (2) and $\rho \in \mathbb{R}_+^*$. Eventually, for $\rho \in \mathbb{R},\ \rho \geq 1$, the $m_1, m_2, m_3$ are positive if and only if*

$$s \in \left]0, \frac{1 + 3\rho + 3\rho^2}{(1 + 2\rho + \rho^2 + 2\rho^3 + \rho^4)(1 + \rho)}\right[$$

*Proof* Using equation of Proposition 1, we get the following equations

$$\begin{pmatrix} 3\rho^3 - 3\rho^4 - \rho^5 & 1 + 2\rho + \rho^2 - \rho^3 - 2\rho^4 - \rho^5 & 1 + 3\rho + 3\rho^2 \\ 1 & 1 & 1 \end{pmatrix} \begin{pmatrix} m_1 \\ m_2 \\ m_3 \end{pmatrix} = \begin{pmatrix} 0 \\ 1 \end{pmatrix}$$

This is an affine equation and so the space of solutions is an affine subspace. Taking $\rho \in \mathbb{C}\backslash\{\rho, \;\; \rho(1 + 2\rho + \rho^2 + 2\rho^3 + \rho^4)\}$, the matrix has always maximal rank, and so the space of solution is of dimension 1, which we parametrize by $s$. Conversely, the Euler equation (1), thanks to Moulton's result for $n = 3$, has always exactly one real positive solution.

Finally, let us look at the case $\rho \in \mathbb{R}, \;\; \rho \geq 1$. We want the masses to be positive, and according to our parametrization, the masses are affine functions in $s$. An affine function changes sign at most once. Solving $m_i = 0$, we get

$$m_1 = 0 \Rightarrow s = 0$$

$$m_2 = 0 \Rightarrow s = \frac{1 + 3\rho + 3\rho^2}{3\rho^3 + 3\rho^4 + \rho^5 + 1 + 3\rho + 3\rho^2}$$

$$m_3 = 0 \Rightarrow s = \frac{1 + 2\rho + \rho^2 - \rho^3 - 2\rho^4 - \rho^5}{2\rho + \rho^2 + 2\rho^3 + \rho^4 + 1}$$

The last equality gives us for $\rho \geq 1$ $s \leq 0$ which is impossible because $m_1 \geq 0$. So $m_3$ does not change sign for any $s > 0$ and is positive. The positivity of $m_2$ gives us the constraint. □

Let us remark that the constraint $\rho \geq 1$ is in fact not a constraint, because using dilatation and the symmetry which consists of reversing to reverse the order of **all the masses**, we exchange $\rho$ by $1/\rho$. After this first proposition, we can study the integrability of the colinear 3 body problem for real positive masses.

In the following, we will note $W(c) \in M_3(\mathbb{C})$ the $3 \times 3$ matrix such that

$$W(c)_{i,j} = \frac{1}{m_i} \frac{\partial^2}{\partial q_i \partial q_j} V_3(c) \tag{4}$$

where $V_3$ is the potential of the colinear 3 body problem and $c \in \mathbb{C}^3$.

## 2.2 Non-integrability

In this subsection, we will prove Theorem 2.

**Lemma 7** *For any $\rho \in \mathbb{R}, \;\; \rho \geq 1$, there exists, among the masses $(m_1, m_2, m_3) \in {\mathbb{R}_+^*}^3$ such that $m_1 + m_2 + m_3 = 1$ and $c = (-1, 0, \rho)$ is a central configuration for* *the triplet of masses $(m_1, m_2, m_3)$, at most 3 triplets of masses for which the Galois group of first order variational equation has a Galois group whose identity component is Abelian.*

*Proof* The matrix $W$ for the central configuration of the form $c = (-\gamma + g, g, \rho\gamma + g)$ is given by

$$\frac{2}{\gamma^3}\begin{pmatrix} \frac{m_2+3m_2\rho+3m_2\rho^2+m_2\rho^3+m_3}{(1+\rho)^3} & -m_2 & -\frac{m_3}{(1+\rho)^3} \\ -m_1 & \frac{m_1\rho^3+m_3}{\rho^3} & -\frac{m_3}{\rho^3} \\ -\frac{m_1}{(1+\rho)^3} & -\frac{m_2}{\rho^3} & \frac{m_1\rho^3+m_2+3m_2\rho+3m_2\rho^2+m_2\rho^3}{(1+\rho)^3\rho^3} \end{pmatrix}$$

We need to choose $\gamma, g$ such that the multiplier of the central configuration is $-1$ and the center of mass is at 0 (because we want an orbit of the form $c.\phi(t)$). We first compute the spectrum of $W$ which gives

$$\left[0, \frac{4(2\rho^2+3\rho+2)}{(3\rho^3+3\rho^4+\rho^5+1+3\rho+3\rho^2)\gamma^3}, -\frac{2(s\rho^4+2s\rho^3-\rho^2+\rho^2 s-2\rho+2\rho s-1+s)}{\rho^3(1+2\rho+\rho^2)\gamma^3}\right]$$

where the masses $m_1, m_2, m_3$ are parametrized by $s$ according to the formula (3). The constraint that the multiplier of $c$ should be equal to $-1$ gives

$$\gamma^3 = -\frac{(s\rho^4+2s\rho^3-\rho^2+\rho^2 s-2\rho+2\rho s-1+s)}{\rho^3(1+2\rho+\rho^2)}$$

and so we get

$$Sp(W(c)) = \left\{0, 2, -\frac{4(1+\rho)\rho^3(2\rho^2+3\rho+2)}{(s\rho^4+2s\rho^3-\rho^2+\rho^2 s-2\rho+2\rho s-1+s)(1+2\rho+\rho^2+2\rho^3+\rho^4)}\right\}$$

Let us note $G(s, \rho)$ this last eigenvalue, which is a fractional linear function in $s$. The singularity in $s$ of $G$ is at

$$s = \frac{1+2\rho+\rho^2}{1+2\rho+\rho^2+2\rho^3+\rho^4}$$

This value of $s$ corresponds to the case where the central configuration is in fact an absolute equilibrium. Indeed, we then have the multiplier of the central configuration equal to zero. This special case produces the following set of masses

$$(m_1, m_2, m_3) = \left(\frac{(\rho+1)^2}{1+2\rho+\rho^2+2\rho^3+\rho^4}, \frac{-\rho^2}{1+2\rho+\rho^2+2\rho^3+\rho^4}, \frac{(\rho+1)^2\rho^2}{1+2\rho+\rho^2+2\rho^3+\rho^4}\right)$$

The mass $m_2$ is always non-positive, and so this case is impossible. Now in the general case, we solve the equation

$$G(s, \rho) \in \left\{\tfrac{1}{2}(i-1)(i+2) \;\; i \in \mathbb{N}\right\}$$

and we obtain the following solutions

$$\begin{aligned} m_1 &= \frac{(\rho+1)(-8\rho^5+k\rho^5-12\rho^4+3k\rho^4-8\rho^3+3k\rho^3+3k\rho^2+3k\rho+k)}{k(1+2\rho^3+\rho^4+2\rho+\rho^2)^2} \\ m_2 &= -\frac{(-8\rho^4+k\rho^4-28\rho^3+2k\rho^3+k\rho^2-40\rho^2-28\rho+2k\rho-8+k)\rho^2}{k(1+2\rho^3+\rho^4+2\rho+\rho^2)^2} \\ m_3 &= \frac{(\rho+1)(k\rho^5+3k\rho^4+3k\rho^3-8\rho^2+3k\rho^2-12\rho+3k\rho-8+k)\rho^2}{k(1+2\rho^3+\rho^4+2\rho+\rho^2)^2} \end{aligned} \qquad (E_k)$$

with $k \in \{\frac{1}{2}(i-1)(i+2)\ \ k \in \mathbb{N}\}$. These solutions are not valid for $k=0$, but we have that if $G(s,\rho)=0$ then

$$(1+\rho)(2\rho^2+3\rho+2)=0$$

which is excluded because $\rho \in \mathbb{R}_+^*$.

Let us look now what happen if we restrict ourselves to positive masses. We take $\rho \geq 1$ and we look at the sign of the masses given by the curves $(E_k)$. We already know according to Proposition 2 that the interval $I(\rho)$ to consider for $s$ is the following

$$I(\rho) = \left[0, \frac{1+3\rho+3\rho^2}{(1+2\rho+\rho^2+2\rho^3+\rho^4)(1+\rho)}\right]$$

and noting that $(1+2\rho+\rho^2) > (1+3\rho+3\rho^2)/(1+\rho)$ for $\rho \geq 1$, the singularity of $G(s,\rho)$ is never in $I(\rho)$, and so for $\rho \geq 1$, $G(.,\rho)$ increases on $I(\rho)$.

Then $G(.,\rho)$ is a bijection of $I(\rho)$ on

$$G(I(\rho),\rho) = \left]\frac{4(1+\rho)\rho^3(2\rho^2+3\rho+2)}{(1+\rho^2+2\rho)(1+2\rho+\rho^2+2\rho^3+\rho^4)}, \frac{4(2\rho^2+3\rho+2)(1+\rho)^2}{1+2\rho+\rho^2+2\rho^3+\rho^4}\right[$$

Studying these functions, we prove that the interval $G(I(\rho),\rho)$ is decreasing when $\rho \geq 1$ increases. Knowing that $G(I(1),1) = ]2,16[$, the only possible eigenvalues corresponding to a Galois group with an Abelian identity component are $5, 9, 14$. □

Let us now remark that the potential $V_{3,1}$ of the colinear 3 body problem can be reduced. Indeed, this potential is invariant by translation, and by making the symplectic variable change $p_i \longrightarrow \sqrt{m_i}p_i,\ q_i \longrightarrow q_i/\sqrt{m_i}$, the kinetic part in the Hamiltonian becomes the standard kinetic energy $(p_1^2+p_2^2+p_3^2)/2$. So the set of potential $V_{3,1}$ with parameters $(m_1,m_2,m_3) \in {\mathbb{R}_+^*}^3$, $m_1+m_2+m_3=1$ is a set of homogeneous potentials of degree $-1$ in the plane.

**Corollary 1** *The colinear* 3*-body problem with positive masses is not meromorphically integrable.*

*Proof* We proved that only the eigenvalues 5, 9, 14 are possible for integrability of the colinear 3-body problem. In [16], all potentials having these eigenvalues have been classified and they are not meromorphically integrable. □

Remark that in the limit case when two masses tend to 0, the potential $V_{3,1}$ after reduction is not singular and converges to a potential of the form $\alpha/q_1 + \alpha/q_2$, which has the eigenvalue 2 and is integrable.

## 2.3 Higher Variational Equations

Let us now compute exactly at which order the variational equations near the unique real Darboux point have a Galois group whose identity component is not Abelian. Indeed, using [16], we note that on the curves $E_5, E_{14}$, the potentials are integrable at most up to order 4, and on $E_9$ at most to order 6 (which reduces to 4 in our case, because the potential $V_3$ is real and integrable cases to order 5, 6 are complex).

### 2.3.1 At Order 2

To study the Galois group of second order variational equations, we apply Theorem 2 of [17]. We have however to take into account that the kinetic energy is $p_1^2/(2m_1) + p_2^2/(2m_2) + p_3/(2m_3)$ instead of $(p_1^2 + p_2^2 + p_3^2)/2$. This standard form of kinetic energy can be obtained by a symplectic change of variable. The Hessian matrix $\nabla^2 V(c)$ after this variable change is simply the matrix $W$ defined in (4) (Fig. 1).

**Lemma 8** *Let $\rho \in \mathbb{R}$, $\rho \geq 1$ be a real number, $k \in \{5, 9, 14\}$ and masses $(m_1, m_2, m_3) \in E_k$. The variational equations at order 2 near the homothetic orbit associated to c have a Galois group whose identity component is Abelian if and only if the masses belong to the set*

$$\left\{ \left( \frac{12}{35}, \frac{11}{35}, \frac{12}{35} \right), \left( \frac{24}{49}, \frac{1}{49}, \frac{24}{49} \right) \right\} \cup E_9$$

*Proof* We compute the third order derivatives of $V$ at $c$. Denoting by $X_2$ the eigenvector of eigenvalue 2 and $X_3$ the eigenvector of eigenvalue $k$, we have

$$D^3V(X_2, X_2, X_2) = D^3V(X_3, X_3, X_2) = -\frac{3\sqrt{2\rho^2 + 3\rho + 2}\sqrt{2k}}{(\rho+1)^2 g^{\frac{4}{3}} \sqrt{k - 2}\rho^{\frac{3}{2}}} \quad D^3V(X_3, X_2, X_2) = 0$$

$$D^3V(X_3, X_3, X_3) = \frac{-3\sqrt{2}\sqrt{2\,\rho^2 + 3\,\rho + 2}(\rho - 1)P(\rho)}{(1 + 2\rho^3 + \rho^4 + 2\rho + \rho^2)^3 \rho^{\frac{3}{2}} g^{\frac{4}{3}} (\rho+1)^2 \sqrt{k(k-2)m_1 m_2 m_3}}$$

where

$$g = \frac{-4(2\rho^2 + 3\rho + 2)}{\left(\rho^5 + 3\,\rho^4 + 3\,\rho^3 + 3\,\rho^2 + 3\,\rho + 1\right) k}$$

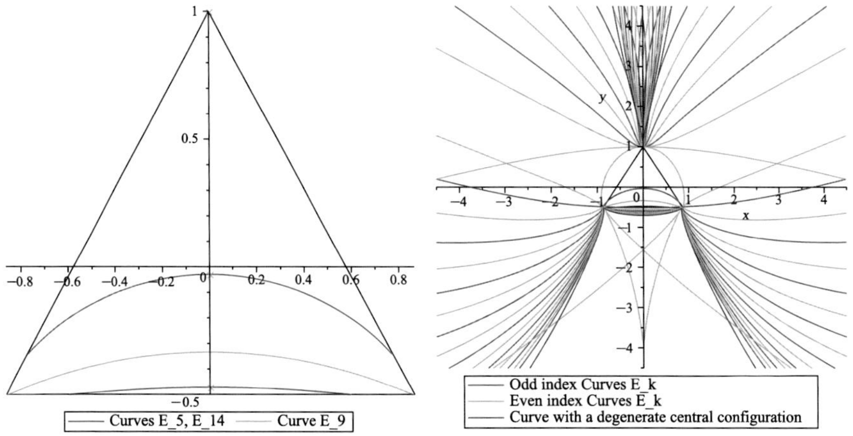


**Fig. 1** Graph of the masses having a first order variational equation with a Galois group whose identity component is Abelian. The masses are represented in barycentric coordinates. The masses inside the *black triangle* are positive. Drawing the curves outside the positive masses reveals that the curves $(E_k)$ accumulate on the curve $(E_\infty)$. They also intersect on the points $(m_1, m_2, m_3) = (1, 0, 0), (0, 1, 0), (0, 0, 1)$ which are the integrable cases (at the limit when the masses are going to zero through a limiting process)

$$P(\rho) = (k+10)\rho^6 + (5k+50)\rho^5 + (8k+120)\rho^4 + (7k+158)\rho^3 \\ + (8k+120)\rho^2 + (5k+50)\rho + k + 10$$

According to [17], the condition for integrability of the second order variational equations are that some of these third order derivative should vanish. Using the table of [17], the three first third order derivatives never lead to an integrability condition, but the last one does. In particular, for $k = 5, 14$, the integrability condition is $D^3V(X_3, X_3, X_3) = 0$, and there is none for $k = 9$.

The only real positive solution of equation $(\rho - 1)P(\rho) = 0$ for $k = 5, 14$ is $\rho = 1$. Putting this in the parametrization of $(E_5)$, $(E_{14})$, we obtain that the set of possible masses is given by

$$\left\{ \left( \frac{12}{35}, \frac{11}{35}, \frac{12}{35} \right), \left( \frac{24}{49}, \frac{1}{49}, \frac{24}{49} \right) \right\} \cup E_9$$

□

### 2.3.2 At Order 3

Let us now look at order 3. We will prove that $V_3$ is never integrable at order 3 near its unique real central configuration.

**Lemma 9** *The potential $V_3$ is never integrable at order 3 at its unique real central configuration.*

*Proof* We will directly use the main Theorem of [18]. A convenient variable change sends the potential $V_3$ to a planar homogeneous potential with standard kinetic energy, and a rotation dilatation puts the central configuration at $c = (1, 0)$. We then find that the third order integrability condition can be written

$$-\frac{256}{715}a^2 + \frac{13824}{5005}c = 0, \quad b = 0, \quad k = 5$$
$$-\frac{475136}{57057}a^2 - \frac{753664}{101745}b^2 + \frac{19759104}{323323}c = 0, \quad k = 9$$
$$-\frac{2755788800}{7436429}a^2 + \frac{19729612800}{7436429}c = 0, \quad b = 0, \quad k = 14$$

where the constants $a, b, c$ are

$$a = -\frac{3\sqrt{2\rho^2 + 3\rho + 2}\sqrt{2k}}{(\rho+1)^2 g^{\frac{4}{3}}\sqrt{k - 2}\rho^{\frac{3}{2}}},$$
$$b = \frac{-3\sqrt{2}\sqrt{2\,\rho^2 + 3\,\rho + 2}(\rho - 1)P(\rho)}{(1 + 2\rho^3 + \rho^4 + 2\rho + \rho^2)^3 \rho^{\frac{3}{2}} g^{\frac{4}{3}} (\rho+1)^2 \sqrt{k(k-2)m_1 m_2 m_3}}$$
$$c = F(\rho, k)$$

where $F$ is a rational fraction in $\rho, k$, and

$$g = \frac{-4(2\rho^2 + 3\rho + 2)}{\left(\rho^5 + 3\,\rho^4 + 3\,\rho^3 + 3\,\rho^2 + 3\,\rho + 1\right) k}$$

The constraint $b = 0$ for $k = 5, 14$ comes from order 2, and we already know the unique solution is $\rho = 1$. The other constraint gives

$$\frac{3024672}{1573}7^{\frac{2}{3}} \neq 0 \qquad \frac{2137106227200}{96577}7^{\frac{2}{3}} \neq 0$$

for $k = 5, 14$ respectively. For $k = 9$, the third order integrability constraint is

$$179523957 + 1436191656\,\rho + 5144769684\,\rho^2 + 11297844542\,\rho^3 + 17938383865\,\rho^4$$
$$+23104821764\,\rho^5 + 25814403801\,\rho^6 + 26361946842\,\rho^7 + 25814403801\,\rho^8 + 23104821764\,\rho^9$$
$$+17938383865\,\rho^{10} + 11297844542\,\rho^{11} + 5144769684\,\rho^{12} + 1436191656\,\rho^{13} + 179523957\,\rho^{14} = 0$$

This polynomial has no real positive root, and so the constraint is never satisfied. □

*Remark 1* One could compute the third order integrability condition for any curve $(E_k)$, and even test if this condition could be satisfied thanks to the holonomic approach of third order variational equations in [18]. Here the restriction $(m_1, m_2, m_3) \in {\mathbb{R}_+^*}^3$ is only for physical reasons, but a more complete study is possible. Still note that this constraint has allowed us to easily bound the eigenvalues, and then to study integrability near the unique real central configuration. If one would allow negative masses, or even complex masses, some results are no longer valid. Especially, there are complex masses which possess a non degenerate central configuration which is integrable at order 3.

## 3 The 4 Body Problem

The previous approach for non integrability proofs can be extended for more complicated systems, as the 4-body problem, for which a direct approach would be impossible due to the high number of central configurations. The difficulty of the problem of finding these central configurations is famous [8], thus we will try to require the least possible information on them. The most important quantity is the set of possible eigenvalues of Hessian matrices at the unique real central configuration. In particular, if this set is finite, then the classification approach of [16] is possible.

### 3.1 Eigenvalue Bounding

Following the method presented in [16], we will first try to prove a bound on eigenvalues of the Hessian matrices at Darboux points of $V_{4,1}$. In [16], the potential are planar, and so we need to operate a little differently. Instead of trying to bound directly these eigenvalues (whose expression could be complicated as they appear as roots of the characteristic polynomial), we simply bound the trace of the Hessian matrix. Indeed, the eigenvalues of the Hessian matrix are of the form $\{0, 2, \lambda_1, \lambda_2\}$, and so bounding the trace gives a bound on $\lambda_1 + \lambda_2$. Moreover, thanks to Theorem 1, we already know that for integrability we must have $\lambda_1, \lambda_2 \geq -1$, and thus we get also a bound on $\lambda_1, \lambda_2$ (Figs. 2 and 3).

**Theorem 10** *We consider the colinear* 4 *body problem with positive masses, whose potential is given by* $V_{4,1}$. *Let* $c$ *be the real central configuration (existence and uniqueness up to translation due to Theorem 6) with multiplier* $-1$. *Let* $W \in M_4(\mathbb{C})$ *be the matrix*

$$W_{i,j} = \frac{1}{m_i}\frac{\partial^2}{\partial q_i \partial q_j} V \quad i, j = 1 \ldots 4$$

*Then* $tr(W) < 70$.

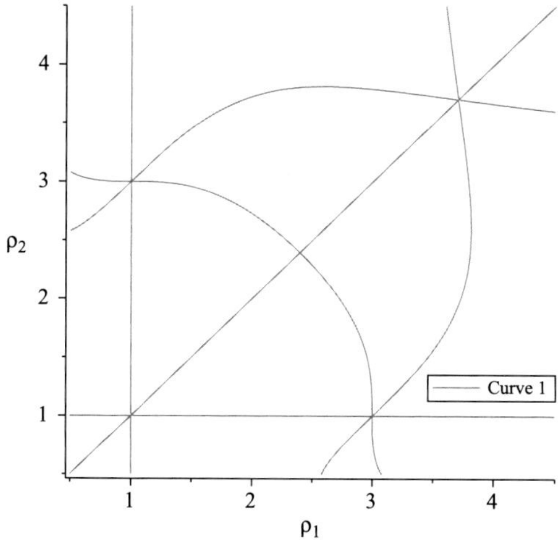


**Fig. 2** Diagram of the bifurcations between the index $i$ of the $M_i$ that realize the maximum of $tr(W)$. The index $i(\rho_1, \rho_2)$ can only change on one of these curves. Moreover there exists a zone (near $\rho_1, \rho_2 = 1$) where the set $S_{\rho_1,\rho_2}$ is empty. Numerical analysis gives a maximum around 69.74

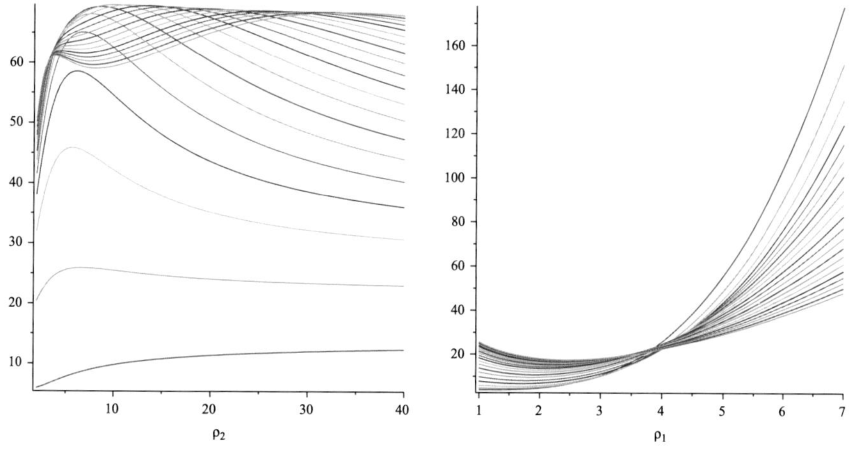


**Fig. 3** Graph of the functions $M_i, i = 1 \dots 4$. We see that $M_2, M_3, M_4$ are bounded but not $M_1$. This is why we proved that the curve $M_1$ has only to be considered for $\rho_1 \leq 5$, allowing us to bound the function

This value is not the optimal one which has a complicated algebraic expression. Still considering a better bound than this one is not useful as it will not allow us to reduce the number of exceptional cases we will have to deal with.

*Proof* We first remark that after translation, dilatation and changing the order of **all the masses**, a central configuration of $V_4$ can always be written in the form $c = (-\rho_1, -1, 1, \rho_2)$ with $\rho_1 \geq \rho_2 > 1$. Moreover, thanks to Moulton Theorem 6, we also know that for any fixed positive masses, there always exists a unique central configuration. So we will first fix our central configuration $c = (-\rho_1, -1, 1, \rho_2)$, and then compute the masses for which $c$ is a central configuration. Moreover, we will assume that $m_1 + m_2 + m_3 + m_4 = 1$ because multiplying all the masses by a constant does not change the trace of the matrix $W$.

The equation of central configurations is a linear system in the masses, with 3 equations for 4 unknowns. The solution is of the form

$$(m_1, m_2, m_3, m_4) = (J_1(\rho_1, \rho_2, m_3), J_2(\rho_1, \rho_2, m_3), J_3(\rho_1, \rho_2, m_3), J_4(\rho_1, \rho_2, m_3))$$

where $J_i$ are rational in $\rho_1, \rho_2$ and affine in $m_3$ (and $J_3(\rho_1, \rho_2, m_3) = m_3$). Now we compute the trace of matrix $W$, and we obtain that $tr(W)$ is also rational in $\rho_1, \rho_2$ and affine in $m_3$.

**Lemma 11** *The functions $J_i$ have no singularities for $\rho_1 \geq \rho_2 > 1$, and their coefficient in $m_3$ does not vanish for $\rho_1 \geq \rho_2 > 1$.*

*Proof* We simply build a polynomial whose factors are the denominators of the functions $J_i$ and numerators of the coefficient in $m_3$ of the functions $J_i$. This polynomial has no real solutions for $\rho_1 \geq \rho_2 > 1$. □

So we can handle safely these $J_i$, and solve equations of the form $J_i = 0$ in $m_3$ without dealing with singular cases. We need to prove

$$\max_{J_i>0,\ i=1\ldots4,\ \rho_1\geq\rho_2>1} tr(W) < 70$$

Let us now remark that for fixed $\rho_1 \geq \rho_2 > 1$, the function $tr(W)$ in $m_3$ on the set

$$S_{\rho_1,\rho_2} = \{m_3 \in \mathbb{R}, J_i(\rho_1, \rho_2, m_3) > 0,\ i = 1\ldots4\}$$

has its maximum on the boundary of $S_{\rho_1,\rho_2}$ (because $tr(W)$ is affine in $m_3$). So for fixed $\rho_1 \geq \rho_2 > 1$, the maximum on the possible $m_3$ has 4 possible values

$$M_i(\rho_1, \rho_2) = tr(W)(\rho_1, \rho_2, J_i(\rho_1, \rho_2, \cdot)^{-1}(0)) \quad i = 1\ldots4$$

Let us now prove the following Lemma

**Lemma 12** *The following bounds hold*

$$M_2(\rho_1, \rho_2) \leq 69.9 \quad M_3(\rho_1, \rho_2) \leq 69.9 \quad M_4(\rho_1, \rho_2) \leq 69.9 \qquad \forall\, \rho_1 \geq \rho_2 > 1$$

$$M_1(\rho_1, \rho_2) \leq 69.9 \qquad \forall\, \rho_1 \geq \rho_2 > 1,\quad \rho_1 \leq 5$$

*Proof* These inequalities are automatically proved using RAGlib. □

**Lemma 13** *If* $\rho_1 \geq 5$, $\rho_1 \geq \rho_2 > 1$, *then*

$$\max_{m_3 \in S_{\rho_1,\rho_2}} tr(W) \in \{M_2(\rho_1, \rho_2), M_3(\rho_1, \rho_2), M_4(\rho_1, \rho_2)\}$$

*Proof* We set $\rho_1 \geq 5$ with $\rho_1 \geq \rho_2 > 1$. Assume now that $S_{\rho_1,\rho_2} \neq \emptyset$ and $M_1(\rho_1, \rho_2)$ is the maximum of $tr(W)$ on $S_{\rho_1,\rho_2}$. Then the corresponding masses $(m_1, m_2, m_3, m_4)$ should be all non-negative (recall that the maximum could be reached at the boundary of the domain of positive masses, so for non-negative masses). Solving equation $J_1(\rho_1, \rho_2, m_3) = 0$ in $m_3$, we get a rational fraction $D$ in $\rho_1, \rho_2$. We now prove using RAGlib that

$$D(\rho_1, \rho_2) \leq 0 \qquad \forall \rho_1 \geq 5, \rho_1 \geq \rho_2 > 1$$

So the only possibility left for having all non-negative masses is that $m_3 = D = 0$. This implies that $M_1(\rho_1, \rho_2) = M_3(\rho_1, \rho_2)$ and so the Lemma follows. □

Using Lemma 13, we know that if $\rho_1 \geq 5$, $\rho_1 \geq \rho_2 > 1$, the maximum $M_2$, $M_3$ or $M_4$. These are bounded by 69.9 thanks to Lemma 12. For $5 \geq \rho_1 \geq \rho_2 > 1$, the maximum of $tr(W)$ can be any of the $M_i$, but due to Lemma 12, all of these are then bounded by 69.9. So

$$\max_{J_i > 0,\ i=1\ldots4,\ \rho_1 \geq \rho_2 > 1} tr(W) < 70$$ □

## 3.2 Symmetric Central Configurations

For symmetric central configurations, several cases are possible which are not possible in the non-symmetric case. So we will analyze in this part the case where the real central configuration is of the form $(-\rho, -1, 1, \rho)$.

**Lemma 14** *The function* $tr(W)$ *has no singularities for* $\rho_1 > \rho_2 > 1$, *and its coefficient in* $m_3$ *does not vanish for* $\rho_1 > \rho_2 > 1$.

This Lemma is immediately proved by RAGlib. For $\rho_1 = \rho_2$, the coefficient in $m_3$ of $tr(W)$ vanishes, making it a special case. On the other hand, this produces an additional symmetry that reduce the number of parameters by 1 and greatly simplify the formulas

**Theorem 15** (Pacella [19]) *We consider the colinear* 4 *body problem potential* $V_{4,1}$ *with positive masses and the central configuration* $c$ *with multiplier* $-1$ *(existence and unicity up to translation due to 6). Noting* $W \in M_4(\mathbb{C})$ *with*

$$W_{i,j} = \frac{1}{m_i} \frac{\partial^2}{\partial q_i \partial q_j} V$$

*the spectrum of* $W$ *is of the form* $Sp(W) = \{0, 2, \lambda_1, \lambda_2\}$ *with* $\lambda_1, \lambda_2 > 2$.

This already allows to reduce somewhat the possible set of eigenvalues. We will now check if some curves (in the space of masses) corresponding to a pair of eigenvalues $\lambda_1, \lambda_2$ are non-empty for real positive masses.

**Lemma 16** *If the potential $V_{4,1}$ with positive masses possesses a real central configuration of the form $(-\rho, -1, 1, \rho)$, then the spectrum of the Hessian matrix $W$ at the real central configuration with multiplier $-1$ has the form $Sp(W) = \{0, 2, \lambda_1, \lambda_2\}$ with*

$$\{\lambda_1, \lambda_2\} \in \{\{5, 9\}, \{5, 14\}, \{9, 27\}, \{14, 44\}\}$$

*Proof* Using Pacella Theorem, we obtain a better minoration $\lambda_1, \lambda_2 > 2$. Knowing that $2 + \lambda_1 + \lambda_2 < 70$, we get the following possibilities

$$\begin{gathered}\{5,5\},\{5,9\},\{5,14\},\{5,20\},\{5,27\},\{5,35\},\{5,44\},\{5,54\},\{9,9\},\\ \{9,14\},\{9,20\},\{9,27\},\{9,35\},\{9,44\},\{9,54\},\{14,14\},\{14,20\},\{14,27\},\\ \{14,35\},\{14,44\},\{20,20\},\{20,27\},\{20,35\},\{20,44\},\{27,27\},\{27,35\}\end{gathered} \quad (5)$$

We first compute the characteristic polynomial of matrix $W$. Using the same notations as before, the characteristic polynomial has rational coefficients in $\rho_1, \rho_2, m_3$. Factoring it, we take out the $z(z-2)$ factor (corresponding to eigenvalues 0, 2) and we then get a degree 2 polynomial $P$ in $z$. The coefficient in $z$ corresponds to the trace of $W$, and is affine in $m_3$. We now put $\rho_1 = \rho_2$ in the expression of the characteristic polynomial. The coefficient corresponding to the trace only depends on $\rho_2$. The equation $P(z) = (z-\lambda_1)(z-\lambda_2)$ in $\rho_2, m_3$ gives rise to two equations in $\rho_2, m_3$, and we have moreover the constraint of positivity of the masses $m_i$ which can be written as a function of $\rho_2, m_3$ with the functions $J_i$. This polynomial system of equations and inequalities has real solutions only for $\lambda_1, \lambda_2$ given by the Lemma. □

## 3.3 Reduction of Exceptional Curves

In this part, we will always assume that the real central configuration $(-\rho_1, -1, 1, \rho_2)$ is such that $\rho_1 > \rho_2$.

**Lemma 17** *If the potential $V_4$ with positive masses is meromorphically integrable, then the real central configuration $c$ with multiplier $-1$ has a Hessian matrix $W$ with spectrum of the form $Sp(W) = \{0, 2, \lambda_1, \lambda_2\}$ with*

$$\begin{gathered}\{\lambda_1, \lambda_2\} \in \{\{5,5\},\{5,9\},\{5,14\},\{5,20\},\{5,27\},\{5,35\},\\ \{5,44\},\{5,54\},\{9,20\},\{9,27\},\{9,35\},\{9,44\},\{9,54\},\{14,44\}\}\end{gathered}$$

*Proof* Using Pacella Theorem, we obtain a better minoration $\lambda_1, \lambda_2 > 2$. Knowing that $2 + \lambda_1 + \lambda_2 < 70$, we get the possibilities (5). So we only need to eliminate the cases

$$\{9, 9\}, \{9, 14\}, \{14, 14\}, \{14, 20\}, \{14, 27\}, \{14, 35\},$$
$$\{20, 20\}, \{20, 27\}, \{20, 35\}, \{20, 44\}, \{27, 27\}, \{27, 35\}$$

We first compute the characteristic polynomial of matrix $W$. Using the same notations as before, the characteristic polynomial has rational coefficients in $\rho_1, \rho_2, m_3$. Factoring it, we take out the $z(z-2)$ factor (corresponding to eigenvalues 0, 2) and we then get a degree 2 polynomial $P$ in $z$. The coefficient in $z$ corresponds to the trace of $W$, and is affine in $m_3$. We then solve the equation $tr(W)(\rho_1, \rho_2, m_3) = 2 + \lambda_1 + \lambda_2$ in $m_3$ (using Lemma 14, this always produces exactly one solution) and put this solution in $P$. So the only equation we have to study is of the form

$$Z_0(\rho_1, \rho_2) = P_{\rho_1,\rho_2}(0) - \lambda_1\lambda_2 = 0 \qquad \rho_1 > \rho_2 > 1 \tag{6}$$

Using RAGlib, we prove that for $\lambda_1, \lambda_2$ in the upper 12 cases, this equation has no solutions. This proves the Lemma. □

*Remark 2* Remark that all the remaining curves are non empty for $\rho_1 \geq \rho_2 > 1$, but this does not imply they are non empty for positive masses (contrary to the previous part where we have taken into account the positivity of the masses). Numerical evidence suggest that for positive masses, the only possible eigenvalues $\{\lambda_1, \lambda_2\}$ are

$$\{5, 9\}, \{5, 14\}, \{5, 20\}, \{5, 27\}, \{9, 20\}, \{9, 27\}, \{9, 35\}, \{9, 44\}, \{14, 44\}$$

but taking into account this additional constraint seems too complicated.

## 3.4 Second Order Variational Equations

Using the integrability table of [17], integrability at second order requires that some of the third order derivatives of the potential vanish. Considering only the eigenvalues $\lambda_1, \lambda_2$ (the other ones do not lead to any additional integrability condition) we obtain the following number of conditions (i.e. the number of third order derivatives that should vanish)

| | |
|---|---|
| $\{5, 5\}, \{5, 14\}, \{5, 27\}, \{14, 44\}$ | 4 conditions |
| $\{5, 44\}, \{5, 20\}, \{5, 35\}, \{5, 54\}$ | 3 conditions |
| $\{5, 9\}, \{9, 27\}, \{9, 44\}$ | 2 conditions |
| $\{9, 35\}, \{9, 54\}$ | 1 condition |
| $\{9, 20\}$ | 0 condition |

The main drawback is that we need a priori to compute the eigenvalues of the Hessian matrix, and due to the parameters, this is quite difficult in our problem. In particular, testing the constraint implies solving 2-variables polynomials of degree 172 and this seems too large to rule out real solutions (if there are none at all). Still in some cases, we can avoid this computation

**Proposition 3** *Let $V$ be a meromorphic homogeneous potential of degree $-1$ in dimension $n$, $c$ a Darboux point of $V$ with multiplier $-1$, and $E$ a stable subspace of $\nabla^2 V(c)$. Assume that $\nabla^2 V(c)$ is diagonalizable and*

$$\exists B \subset \mathbb{N}, \textit{ with } \max(B) \leq 2\min(B) + 1, \ Sp\left(\nabla^2 V(c)\big|_E\right) \subset \{k(2k+3), \ k \in B\} \tag{7}$$

*If the second order variational equation near the homothetic orbit associated to $c$ has a Galois group whose identity component is Abelian then*

$$D^3 V(c) \cdot (X, Y, Z) = 0 \qquad \forall X, Y, Z \in E$$

*Proof* Using the integrability table of [17], we see that the condition on eigenvalues (7) implies that the table $A$ for such eigenvalues will only have zeros. So denoting $X_1, \ldots, X_p$ the eigenvectors associated to eigenvalues $\lambda_i$ $i = 1 \ldots p$ of $\nabla^2 V(c)$, we obtain the integrability condition

$$D^3 V(c) \cdot (X_i, X_j, X_k) = 0 \ \ \forall i, j, k = 1 \ldots p$$

These $p$ eigenvectors span the invariant subspace $E$, and so by multilinearity, this gives the Proposition. □

We try to avoid computing the eigenvectors associated to eigenvalues $\{\lambda_1, \lambda_2\}$ for the Hessian matrix of the real central configuration of $V_4$. In the cases $\{\lambda_1, \lambda_2\} \in \{\{5, 5\}, \{5, 14\}, \{5, 27\}, \{14, 44\}\}$, the hypotheses of Proposition 3 are satisfied using for $E$ the stable subspace generated by the eigenvectors associated to $\lambda_1, \lambda_2$. And it appears that this subspace is much easier to compute. Remark also that when the two eigenvalues are equal, then finding the eigenvectors is not necessary as any vector in the corresponding eigenspace is an eigenvector.

**Lemma 18** *We consider $V_4$ the potential of the colinear 4 body problem with positive masses, $c$ the real central configuration with multiplier $-1$, and $W \in M_4(\mathbb{C})$ the matrix such that*

$$W_{i,j} = \frac{1}{m_i} \frac{\partial^2}{\partial q_i \partial q_j} V$$

*If $Sp(W) = \{0, 2, 5, 5\}, \{0, 2, 5, 14\}, \{0, 2, 5, 27\}, \{0, 2, 14, 44\}$, then the potential $V_4$ is not meromorphically integrable.*

*Proof* We want to consider the stable subspace $E$ of $W$ corresponding to eigenvalues $\lambda_1, \lambda_2$. We already know an eigenvector of eigenvalue 0, $v = (1, 1, 1, 1)$, and an eigenvector of eigenvalue 2, the vector $c = (-\rho_1, -1, 1, \rho_2)$. As the matrix is symmetric, the eigenspaces are orthogonal, and thus we have $E = \text{Span}(v, c)^{\perp}$. We obtain

$$E = \text{Span}((2, -1 - \rho_1, \rho_1 - 1, 0), (0, \rho_2 - 1, -1 - \rho_2, 2))$$

denoting $w_1, w_2$ these two basis vectors of $E$.

Let us first consider the non-symmetric case. As Lemma 14 applies, we can consider the polynomial $Z_0 \in \mathbb{R}[\rho_1, \rho_2]$ given by Eq. (6), and

$$Z_1 = D^3V(c)(w_1, w_1, w_1), \quad Z_2 = D^3V(c)(w_1, w_1, w_2)$$

$$Z_3 = D^3V(c)(w_1, w_2, w_2), \quad Z_4 = D^3V(c)(w_2, w_2, w_2)$$

We obtain a system of 5 equations in two variables (the polynomials $Z_i$ being of degree 58), and we prove that this system has no solutions for $\rho_1 > \rho_2 > 1$. Thus the second order variational equation has not a Galois group with an Abelian identity component.

The symmetric case. Only the cases $Sp(W) = \{0, 2, 5, 14\}, \{0, 2, 14, 44\}$ are possible. We have $\rho_1 = \rho_2$, and then the condition to have these eigenvalues are of the form of two polynomials in $\rho_2, m_3$. The polynomials $Z_i$ above are still defined, and are polynomials in $\rho_2, m_3$. This system of 6 equations has no real solutions for $\rho_2 > 1, m_3 > 0$, and thus the second order variational equation does not have a Galois group with an Abelian identity component. Thus the potential $V_4$ is not meromorphically integrable in these cases. □

## 4 Higher Variational Equations

*Proof of Theorem* 3 The still open cases are

$$\{5, 44\}, \{5, 20\}, \{5, 35\}, \{5, 54\}, \{5, 9\}, \{9, 27\}, \{9, 44\}, \{9, 35\}, \{9, 54\}, \{9, 20\} \tag{8}$$

The case $\{9, 20\}$ is particularly interesting (and difficult) as there are no integrability conditions at order 2, and numerical evidence suggest that this case is really possible for positive masses. So this curve gives masses for which all integrability conditions near the unique (up to translation) real Darboux point up to order 2 are satisfied.

In the same manner as in [16], we will compute for these remaining sets of eigenvalues higher variational equations. We only need to study real 3 dimensional homogeneous potentials of degree $-1$. Asuming there exists a real Darboux point $c$, after rotation we can assume that $c = (1, 0, 0)$ (and the potential is still real). Then the series expansion of $V$ at $c$ will be of the form

$$V(1+q_1,q_2,q_3) = q_1^{-1}\left(1+\frac{1}{2}\left(\lambda_1\frac{q_2^2}{q_1^2}+\lambda_2\frac{q_3^2}{q_1^2}\right)+\sum_{i=3}^{\infty}\sum_{j=0}^{i}u_{i,j}\frac{q_2^{i-j}q_3^j}{q_1^i}\right) \tag{9}$$

As in [16], the main part of the algorithm consists of finding solutions in $\mathbb{C}(t)\left[\operatorname{arctanh}\left(\frac{1}{t}\right)\right]$ of a large system of linear differential equations, which are the $k$-th variational equations. These $k$-th variational equations are put in block triangular form to make computation faster. Only the last equation is solved through the variation of parameters technique and then its monodromy analyzed through commutativity condition of monodromy in [17].

Instead of computing a basis of solutions, we only compute several solutions, that through empirical evidence, will lead to the strongest integrability conditions. The output of the algorithm is a set of polynomial conditions on higher-order derivatives of the potential $V$ at the Darboux point $c$, so here it will be polynomial conditions on the $u_{i,j}$. As presented in [16], if a non degeneracy type condition is satisfied (see [16] Definition 4.1), we will be able to express higher-order derivatives in function of lower order ones. In our cases, this will always be the case for variational equations of order $\geq 3$ (but we are lucky, because it seems that if eigenvalues are spaced enough, degeneracy at any order is possible). This allows us in particular to express all derivatives of order $\geq 4$ as functions of $u_{3,0}, u_{3,1}, u_{3,2}, u_{3,3}$. The possible series expansions are written in Appendix A. Variational equations up to order 4 have been analyzed. Still, at order 4, some combinations of eigenvalues are still possible, and thus looking at order 5 is necessary. However, a speed-up is possible in certain cases:

## 4.1 An Invariant Subspace of the 5-th Order Variational Equation

**Lemma 19** *Let $V$ be a real meromorphic homogeneous potential of degree $-1$ in dimension 3. Assume that $V$ has a series expansion of the form*

$$V(1+q_1,q_2,q_3) = q_1^{-1}\left(1+\frac{1}{2}\left(\lambda_1\frac{q_2^2}{q_1^2}+\lambda_2\frac{q_3^2}{q_1^2}\right)+\sum_{i=3}^{\infty}\sum_{j=0}^{i}u_{i,j}\frac{q_2^{i-j}q_3^j}{q_1^i}\right)$$

*with $u_{3,1} = u_{4,1} = u_{5,1} = 0$ and $\lambda_1 \in \{5, 9, 14, 20\}$. Then $V$ is not meromorphically integrable.*

*Proof* The dynamical system associated to $V$ is of the form $\ddot{q} = \nabla V(q)$. Let us compute $\partial_{q_3} V$

$$\partial_{q_3} V = q_1^{-1} \left( \lambda_2 \frac{q_3}{q_1^2} + \sum_{i=3}^{5} \sum_{j=2}^{i} j u_{i,j} \frac{q_2^{i-j} q_3^{j-1}}{q_1^i} + \sum_{i=6}^{\infty} \sum_{j=1}^{i} j u_{i,j} \frac{q_2^{i-j} q_3^{j-1}}{q_1^i} \right)$$

Thus we get that the series expansion of $\partial_{q_3} V$ at order 5 for $q_3 = 0$ is

$$\partial_{q_3} V = u_{6,1} \frac{q_2^5}{q_1^7} + O\left( (q_2, q_3)^6 / q_1^8 \right)$$

As we see, there is only one term left, and it is of order 5. Let us now look at the 5-th order variational equation.

This variational equation will have an invariant subspace $\mathcal{W}$ corresponding to the 5-th order variational equation of $\tilde{V}$, the restriction of $V$ to the plane $q_3 = 0$. Let us now look the variational equation on $\mathcal{W}$. The potential $\tilde{V}$ is a 2-dimensional homogeneous potential of degree $-1$, and it has a Darboux point at $(1, 0)$. The eigenvalues of the Hessian matrix of $\tilde{V}$ at this point are $\{2, \lambda_1\}$. Now using [16], we know that for any choice of **real** $\tilde{V}$ with $\lambda_1 \in \{5, 9, 14, 20\}$, the 5-th order variational equation does not have a virtually Abelian Galois group. Thus the 5-th order variational equation of $V$ does not have a virtually Abelian Galois group, and thus $V$ is not meromorphically integrable. □

*Remark 3* Note that physically, the condition $u_{3,1} = u_{4,1} = u_{5,1} = 0$ implies that the plane $q_3 = 0$ is invariant at order 4. At order 5, it is no longer invariant, however the derivatives in time of $q_1, q_2$ do not depend on $q_3$.

We now use Lemma 19. Looking at the series expansions in Appendix A we have computed, we see that for all of them except the last one, we have either $u_{3,1} = u_{4,1} = u_{5,1} = 0$ or $u_{3,2} = u_{4,3} = u_{5,4} = 0$ (or both). In the first case we can apply directly Lemma 19. In the second case, we just have to exchange $q_2, q_3$, and the hypotheses of Lemma 19 are satisfied. So except for the case $(\lambda_1, \lambda_2) = (9, 20)$, the hypotheses are satisfied and thus there is no real meromorphic homogeneous potential of degree $-1$ in dimension 3 with $c = (1, 0, 0)$ as a Darboux point of $V$ with multiplier $-1$ and these pairs of eigenvalues are meromorphically integrable.

## 4.2 *The Case* {9, 20}

In the last subsection, we tried to avoid computing the Galois group of the 5-th order variational equation as it is computationally expensive. At order 4, the ideal $\mathcal{I}_4$ is zero-dimensional, but still has real solutions (given in Appendix A). As we can see in Appendix A, the previous Lemma does not apply for these eigenvalues. Thus it is necessary to compute completely the 5-th order variational equation. The coefficients of the series expansion at order 4 are polynomials in $u_{3,0}, u_{3,1}, u_{3,2}, u_{3,3}$ modulo the ideal $\mathcal{I}_4$. As the algorithm never needs to inverse an element of this ring (which contains zero divisors), it also works at order 5. The output (after one week of computation) is the ideal $\mathcal{I}_5$, which happens to be improper. Thus $\mathcal{I}_5 =< 1 >$, and so no solutions (even complex) at all. We deduce then

**Lemma 20** *Let $V$ be a real meromorphic homogeneous potential of degree $-1$ in dimension 3. Assume that $c = (1, 0, 0)$ is a Darboux point of $V$ with multiplier $-1$ and $Sp(\nabla^2 V(c)) = \{2, 9, 20\}$. Then $V$ is not meromorphically integrable.* □

## 5 The Planar $n$-Body Problem

Let us prove in this section Theorem 4. The main tool will be the following Theorem

**Theorem 21** (Pacella [19] Theorem 3.1) *We consider the colinear $n$ body problem with positive masses and $c$ a configuration with multiplier $-1$, given by potential $V_{n,2}$. Noting $W \in M_n(\mathbb{C})$ with*

$$W_{i,j} = \frac{1}{m_i} \frac{\partial^2}{\partial q_i \partial q_j} V_{n,2}$$

*the spectrum of $W$ is of the form $Sp(W) = \{0, 2, \lambda_1, \dots, \lambda_{n-2}\}$ with $\lambda_i > 2,\ i = 1 \dots n-2$.*

*Proof* For the $n$ body problem in the plane, the Hessian matrix to compute is of the form $W \in M_{2n}(\mathbb{C})$

$$W_{i,j} = \frac{1}{m_i} \frac{\partial^2}{\partial q_i \partial q_j} V_{n,2}$$

with the notation $m_{i+n} = m_i$. Computing this matrix at a colinear central configuration, we obtain a matrix of the form

$$W = \begin{pmatrix} A & 0 \\ 0 & -\frac{1}{2}A \end{pmatrix}$$

Due to Pacella Theorem, we have moreover that $Sp(A) = \{0, 2, \lambda_1, \dots, \lambda_{n-2}\}$ with $\lambda_i > 2,\ i = 1 \dots n-2$. Then the spectrum of $W$ is of the form

$$\{0, 2, \lambda_1, \dots, \lambda_{n-2}, 0, -1, -\tfrac{1}{2}\lambda_1, \dots, -\tfrac{1}{2}\lambda_{n-2}\}$$

According to the integrability condition of the Morales-Ramis Theorem 1, all allowed eigenvalues for integrability are greater or equal to $-1$. These conditions cannot be satisfied as $-\frac{1}{2}\lambda_i < -1,\ i = 1 \dots n-2$. Thus the planar $n$ body problem with positive masses is not meromorphically integrable. □

# 6 Integrable *n*-Body Problems

In this section, we will progress in the opposite way. Instead of trying to prove non integrability, we come from already known integrable cases, and we try to determine if after some transformations, they correspond to particular cases of the $n$ body problem.

**Proposition 4** *The potential $V$ in $n$ variables $q_1, \dots, q_n$*

$$V(q) = \sum_{l=1}^{p} a_l \left( \sum_{j=j_l+1}^{j_{l+1}} q_j^2 \right)^{-1/2} \tag{10}$$

*with $0 = j_1 < j_2 < \dots < j_{p+1} \le n,\ a_l \in \mathbb{C}$ is integrable in the Liouville sense. For any complex orthogonal matrix $R \in \mathbb{O}_n(\mathbb{C})$, the potential $V(Rq)$ is integrable in the Liouville sense.*

Here the kinetic part is assumed to be $T(p) = \|p\|^2/2$ and so the potential $V$ is associated to a Hamiltonian system with $H(p, q) = T(p) + V(q)$.

*Proof* The potential $V$ of equation (10) is a decoupled linear combination

$$V(q) = \sum_{l=1}^{p} a_l V_l(q_{j_l+1}, \dots, q_{j_{l+1}}), \qquad V_l(q_{j_l+1}, \dots, q_{j_{l+1}}) = \left( \sum_{j=j_l+1}^{j_{l+1}} q_j^2 \right)^{-1/2}$$

These potentials are invariant by the rotation group $\mathbb{O}_{j_{l+1}-j_l}(\mathbb{C})$ and so are integrable. Thus the potential $V$ is integrable. As integrability is preserved by any orthogonal transformation, the potential $V(Rq)$ will also be integrable in the Liouville sense. □

Although these potentials seem to have a quite simple expression, the orthogonal transformation $R$ can mix the variables (the decomposition of $V$ is not necessarily conserved). However, the potential can always be written

$$V(q) = \sum_{l=1}^{p} a_l Q_l(q)^{-1/2} \tag{11}$$

with $Q_i$ quadratic forms. And as an orthogonal transformation conserves the rank of these quadratic forms, we have moreover $\sum_{l=1}^{p} \operatorname{rank} Q_l \le n$.

The Hamiltonian of the $n$ body problem in dimension $d$ can be written

$$H(p,q) = \sum_{i=1}^{n} \frac{\|p_i\|^2}{2m_i} + \sum_{i<j} m_i m_j \left( \sum_{k=1}^{d} (q_{i,k} - q_{j,k})^2 \right)^{-1/2}$$

The kinetic part of $H_{n,d}$ is not $\|p\|^2/2$ (as in Proposition 4). To transform the kinetic part to the standard one, we only have to make the variable change $q_{i,k} \mapsto q_{i,k}/\sqrt{m_i}$. The potential now becomes

$$\tilde{V}_{n,d} = \sum_{i<j} m_i m_j \left( \sum_{k=1}^{d} (q_{i,k}/\sqrt{m_i} - q_{j,k}/\sqrt{m_j})^2 \right)^{-1/2}$$

This expression is similar to Eq. (11), but there could be too many quadratic forms. After reduction by translation, the potential becomes a $(n-1)d$-dimensional potential. There are $n(n-1)/2$ independent quadratic forms, and to be of the form (11), we need $n(n-1)d/2 \leq (n-1)d$, which implies $n \leq 2$. So in general, the potential $\tilde{V}_{n,d}$ is not of the form (10), but it could be for some restricted cases.

**Definition 2** We say that a vector space $W \subset \mathbb{R}^{nd}$ is an invariant vector space if

$$\forall q \in W, \quad \nabla V_{n,d} \in W$$

This definition generalizes central configurations, which correspond to the case $\dim W = 1$. Needless to say, as it is more difficult to find the invariant vector spaces than to find central configurations, we will not try to be exhaustive in this search. Let us remark that we already know some invariant vector spaces such as the isosceles 3-body problem and the collinear 3-body problem (which is an invariant vector space of the planar 3 body problem). Several others can be found using symmetries.

Let us now establish some rules for finding vector spaces $W$ and masses $m$ such that $V|_W$ is of the form (10) up to an orthogonal transformation. A necessary condition is that it can be written under the form (11). So in the expression of $V|_W$ we should try to have the lowest possible number of independent quadratic forms (corresponding to mutual distances) with the lowest possible rank.

Remark that if we allow negative masses, some terms in the sum in $V_{n,d}$ could cancel each other, thus reducing greatly the number of quadratic forms. So it seems that finding examples will be easier when negative masses are allowed. And indeed, all interesting examples we will find require a negative mass. Let us now prove Theorem 5.

## 6.1 An Integrable 5 Body Problem

*Proof* The vector space $W$ is of dimension $10 - 6 = 4$. The mass $-1/4$ is at the origin which is the center of mass of the system. On the vector space $W$, the 2-nd and the 4-th body are symmetric with respect to the origin, as well as the 3-th and 5-th bodies. Due to these symmetries, the vector space $W$ is invariant. We can thus restrict our potential to $W$. Now computing the potential $V$ on $W$ we find $V_{5,2}\big|_W =$

$$\left((q_{2,1} - q_{3,1})^2 + (q_{2,2} - q_{3,2})^2\right)^{-1/2} + \left((q_{2,1} + q_{3,1})^2 + (q_{2,2} + q_{3,2})^2\right)^{-1/2} \quad (12)$$

There are only two quadratic forms each with rank two for each. As $2 + 2 = 4 = \dim W$, we are in the form (11). We can now try to put this potential in the form (10). This is done by the following orthogonal transformation

$$R = \frac{1}{\sqrt{2}} \begin{pmatrix} 1 & -1 & 0 & 0 \\ 0 & 0 & 1 & -1 \\ 1 & 1 & 0 & 0 \\ 0 & 0 & 1 & 1 \end{pmatrix}$$

acting on $q_{2,1}, q_{3,1}, q_{2,2}, q_{3,2}$ in this order. Thus $V_{5,2}\big|_W$ is integrable. □

On $W$, the bodies are always on the edges of a parallelogram whose center is the origin (where lies the mass $-1$). Looking at the forces acting on the bodies, we see that they are not attracted by the center at all (because the repulsion of the central mass $-1/4$ exactly compensates the attraction of the opposite mass 1 at twice the distance). The masses are then only attracted by their neighbours. Looking at the expression of the potential (12), we see that the force acting on the center of vertices of the parallelogram (which are $(\pm q_{2,1} \pm q_{3,1}, \pm q_{2,2} \pm q_{3,2})$) is toward the center. Thus the motion of these centers are conics with focus at the origin.

Thus the motion of a body of mass 1 is the composition of two conic motions. The body has a conic motion whose focus is the center of mass of two bodies of mass 1, and this center of mass has a conic motion with focus at the origin. If the two conics are ellipses with rational period ratio, this leads to (algebraic) periodic orbits of the bodies (Fig. 4).

## 6.2 An Integrable $n + 3$ Body Problem

*Proof* The space $W$ is of dimension 3. The forces between the $n$ cocyclic masses and the central mass exactly compensate. The forces between the 3 last masses also compensate (as this is also an absolute equilibrium). So the only forces between the bodies are between the last two masses and the cocyclic masses. But due to symmetry

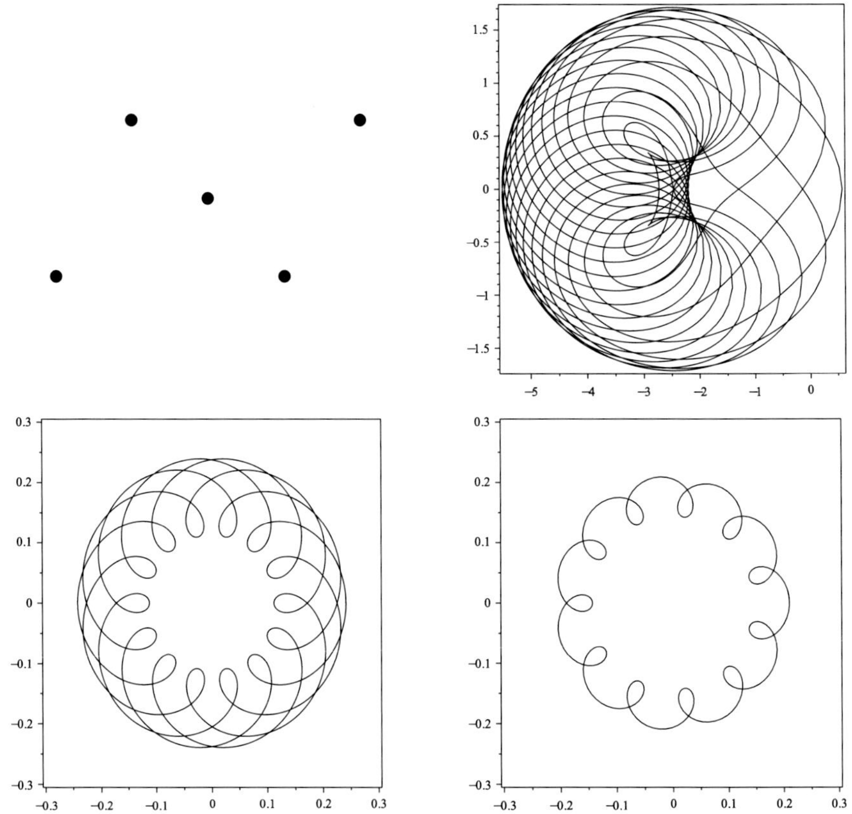


**Fig. 4** The configuration of the 5 bodies and examples of motions of a body of mass 1 with ellipses with rational period ratio

and the fact that the masses are cocyclic, this force only involves one distance. Thus the potential is of the form

$$V = \gamma \left(q_{1,1}^2 + q_{1,2}^2 + q_{n+2,3}^2\right)^{-1/2}$$

This potential corresponds to a central force, and thus is integrable. □

Let us look at an example. The most known cyclic central configuration is the regular polygon. We have $m_1 = \cdots = m_n = 1$. The central mass (chosen to produce an absolute equilibrium) and the potential are then

$$-\alpha = -\frac{1}{2}\sum_{k=1}^{n-1} \sin\left(\frac{k\pi}{n}\right)^{-1} \qquad V = 4n\alpha \left(q_{1,1}^2 + q_{1,2}^2 + q_{n+2,3}^2\right)^{-1/2}$$

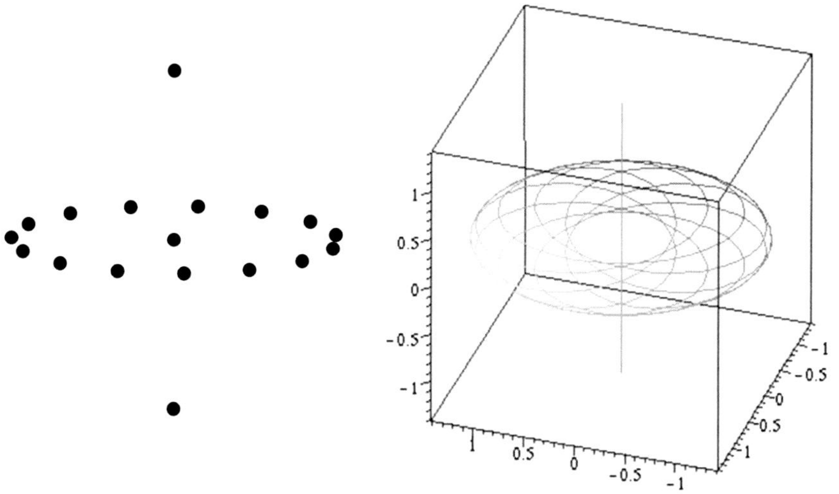


**Fig. 5** A configuration of the $n + 3$ bodies with a regular polygon and an example of motion of the bodies with an ellipse with non-zero inclination

respectively. The motion is the following: the $n$ bodies describe conics in the plane, and the two symmetrical last bodies move along the vertical line. Note that the motion of the bodies on the vertical line is not determined by the motion of the bodies in the plane (this is not a rigid motion as in the case of central configurations). This vertical motion depends on the "inclination" of the conic orbit chosen for the above potential (Fig. 5).

## Appendix A. Integrable Series Expansions at Order 4

The results are written in the following way. We give a series expansion of the form (9) of $V$, such that the $k$-th order variational equation of $V$ near $c$ has a virtually Abelian Galois group if and only if $(u_{3,0}, u_{3,1}, u_{3,2}, u_{3,3}) \in \mathcal{I}_k^{-1}(0)$. The sequence of ideals $\mathcal{I}_k$ is growing, and we compute these conditions up to order 4. For the eigenvalues in (8), they are given below. Remark that the Hilbert dimension of the ideals $\mathcal{I}_4$ greatly depend on eigenvalues, and that sometimes exceptional possible solutions appear in the 4-th order variational equation. In particular, the restriction of these series expansions to the planes in $(q_1, q_2)$ and $(q_1, q_3)$ does not always lead to integrable series expansion at order 4 on these planes.

$$q_1^{-1}\left(1+\tfrac{5}{2}\tfrac{q_2^2}{q_1^2}+\tfrac{44}{2}\tfrac{q_3^2}{q_1^2}+u_{3,1}\tfrac{q_2^2q_3}{q_1^3}+\left(\tfrac{7}{459}u_{3,1}^2+\tfrac{175}{24}\right)\tfrac{q_2^4}{q_1^4}+\left(-\tfrac{734}{7775}u_{3,1}^2+\tfrac{130779}{622}\right)\tfrac{q_2^2q_3^2}{q_1^4}+\tfrac{682538736}{1082611}\tfrac{q_3^4}{q_1^4}+O\left(\tfrac{(q_1,q_2)^6}{q_1^6}\right)\right)\quad \mathcal{I}_2=\mathcal{I}_3=<u_{3,0},u_{3,2},u_{3,3}>,\ \mathcal{I}_4=<u_{3,0},u_{3,1},u_{3,2},u_{3,3}>$$

$$q_1^{-1}\left(1+\tfrac{5}{2}\tfrac{q_2^2}{q_1^2}+\tfrac{20}{2}\tfrac{q_3^2}{q_1^2}+u_{3,3}\tfrac{q_3^3}{q_1^3}+\tfrac{175}{24}\tfrac{q_2^4}{q_1^4}+\tfrac{13575}{142}\tfrac{q_2^2q_3^2}{q_1^4}+\left(\tfrac{105347}{840700}u_{3,3}^2+\tfrac{153200}{1201}\right)\tfrac{q_3^4}{q_1^4}+O\left(\tfrac{(q_1,q_2)^6}{q_1^6}\right)\right)\quad \mathcal{I}_2=\mathcal{I}_3=<u_{3,0},u_{3,1},u_{3,2}>\quad \mathcal{I}_4=<u_{3,0},u_{3,1},u_{3,2},u_{3,3}>$$

$$q_1^{-1}\left(1+\tfrac{5}{2}\tfrac{q_2^2}{q_1^2}+\tfrac{35}{2}\tfrac{q_3^2}{q_1^2}+u_{3,3}\tfrac{q_3^3}{q_1^3}+\tfrac{175}{24}\tfrac{q_2^4}{q_1^4}+\tfrac{13575}{142}\tfrac{q_2^2q_3^2}{q_1^4}+\left(\tfrac{108970043}{1516693800}u_{3,3}^2+\tfrac{8028474825}{20222584}\right)\tfrac{q_3^4}{q_1^4}+O\left(\tfrac{(q_1,q_2)^6}{q_1^6}\right)\right)\quad \mathcal{I}_2=\mathcal{I}_3=<u_{3,0},u_{3,1},u_{3,2}>\quad \mathcal{I}_4=<u_{3,0},u_{3,1},u_{3,2},u_{3,3}>$$

$$q_1^{-1}\left(1+\tfrac{5}{2}\tfrac{q_2^2}{q_1^2}+\tfrac{54}{2}\tfrac{q_3^2}{q_1^2}+u_{3,3}\tfrac{q_3^3}{q_1^3}+\tfrac{175}{24}\tfrac{q_2^4}{q_1^4}+\tfrac{15646365}{60604}\tfrac{q_2^2q_3^2}{q_1^4}+\left(\tfrac{6634990046129}{14245925733270}u_{3,3}^2+\tfrac{93293112339}{97889959}\right)\tfrac{q_3^4}{q_1^4}+\tfrac{119177438935}{9191020828}\tfrac{q_2^2q_3^3}{q_1^5}+\right.$$
$$\left.\left(\tfrac{1116401622922069362821}{500825552245950208634834 4}u_{3,3}^3+\tfrac{17921365151237750718}{239428834674960487}u_{3,3}\right)\tfrac{q_3^5}{q_1^5}+O\left(\tfrac{(q_1,q_2)^6}{q_1^6}\right)\right)\quad \mathcal{I}_2=\mathcal{I}_3=\mathcal{I}_4=<u_{3,0},u_{3,1},u_{3,2}>$$

$$q_1^{-1}\left(1+\tfrac{5}{2}\tfrac{q_2^2}{q_1^2}+\tfrac{9}{2}\tfrac{q_3^2}{q_1^2}+u_{3,1}\tfrac{q_2^2q_3}{q_1^3}+u_{3,3}\tfrac{q_3^3}{q_1^3}+\left(\tfrac{7}{324}u_{3,1}^2+\tfrac{175}{24}\right)\tfrac{q_2^4}{q_1^4}+\left(\tfrac{19}{123}u_{3,1}^2+\tfrac{95}{246}u_{3,1}u_{3,3}+\tfrac{7965}{164}\right)\tfrac{q_2^2q_3^2}{q_1^4}+\left(\tfrac{3289}{12060}u_{3,3}^2+\tfrac{13311}{536}\right)\tfrac{q_3^4}{q_1^4}+\left(\tfrac{3497470290125}{504750025968}u_{3,1}-\tfrac{664048660625}{504750025968}u_{3,3}\right)\tfrac{q_2^4q_3}{q_1^5}+\right.$$
$$\left.\left(\tfrac{2220239578205}{252375012984}u_{3,1}-\tfrac{921441864205}{252375012984}u_{3,3}\right)\tfrac{q_2^2q_3^3}{q_1^5}+\left(-\tfrac{187329008582632851}{386510093636424240}u_{3,1}+\tfrac{775252202045235367}{193270046818212120}u_{3,3}\right)\tfrac{q_3^5}{q_1^5}+O\left(\tfrac{(q_1,q_2)^6}{q_1^6}\right)\right)$$

$$\mathcal{I}_2=\mathcal{I}_3=<u_{3,0},u_{3,2},\ \mathcal{I}_4=\left\langle u_{3,0},u_{3,2},u_{3,3}^5+\tfrac{189924461418873225}{1610583723485101}u_{3,3}^3-\tfrac{312185150911564368 75}{6442334893940404}u_{3,3},u_{3,3}^3+\tfrac{322885767353079225}{3221167446970202}u_{3,3}+\tfrac{103594358565795300}{1610583723485101}u_{3,1}\right\rangle$$

$$q_1^{-1}\left(1+\tfrac{9}{2}\tfrac{q_2^2}{q_1^2}+\tfrac{27}{2}\tfrac{q_3^2}{q_1^2}+u_{3,0}\tfrac{q_2^3}{q_1^3}+u_{3,2}\tfrac{q_2q_3^2}{q_1^3}+\left(\tfrac{3289}{12060}u_{3,0}^2+\tfrac{13311}{536}\right)\tfrac{q_2^4}{q_1^4}+\left(\tfrac{42887}{2155950}u_{3,2}^2+\tfrac{47791}{143730}u_{3,0}u_{3,2}+\tfrac{376407}{1597}\right)\tfrac{q_2^2q_3^2}{q_1^4}+\left(\tfrac{11446993}{324669060}u_{3,2}^2+\tfrac{376298379}{1603304}\right)\tfrac{q_3^4}{q_1^4}+\right.$$
$$\left(\tfrac{20757059609765573267416645528141821064521799326228648986 9}{84019386579526787481989378262356547408870350000637040 00}u_{3,0}-\tfrac{16787633864550203502743667842716175416793657678815071}{33607754631810714992795727304942618963548140002524160}u_{3,2}\right)\tfrac{q_2^5}{q_1^5}+\left(\tfrac{41420797622285419653283151280696451 41}{56466151342246656632909541204212 60}u_{3,0}-\right.$$
$$\left.\left.\tfrac{16869769031486182277210016047384863}{56466151342246656632909541204212 60}u_{3,2}\right)\tfrac{q_2^3q_3^2}{q_1^5}+\left(\tfrac{20533796486740053950290290586390781 97}{13899360330399177017331579373344 64}u_{3,0}+\tfrac{12138261325480316078257448116190049 3}{69496801651995885086657896866723 20}u_{3,2}\right)\tfrac{q_2q_3^4}{q_1^5}+O\left(\tfrac{(q_1,q_2)^6}{q_1^6}\right)\right)$$

$$\mathcal{I}_2=\mathcal{I}_3=<u_{3,1},u_{3,3}>\quad \mathcal{I}_4=\left\langle u_{3,1},u_{3,3},u_{3,0}^5-\tfrac{183692999458343787808210893960821881229602107524169}{2037783629727328800869143913079543129453945056 1232}u_{3,0}^3+\cdot\tfrac{5871817960967665190055792996619628724237316321335937 5}{40755672594546596017382878261590862549078901122464}u_{3,0},\right.$$
$$\left.u_{3,0}^3-\tfrac{355417444337263595295695896022205344612417387175741 9}{2037783629727328800869143913079543129453945056 1232}u_{3,0}+\cdot\tfrac{1350984520019956288893495927826867723730667397880 25}{2037783629727328800869143913079543129453945056 1232}u_{3,2}\right\rangle$$

$$q_1^{-1}\left(1+\tfrac{9}{2}\tfrac{q_2^2}{q_1^2}+\tfrac{35}{2}\tfrac{q_3^2}{q_1^2}+u_{3,0}\tfrac{q_2^3}{q_1^3}+u_{3,2}\tfrac{q_2q_3^2}{q_1^3}+u_{3,3}\tfrac{q_3^3}{q_1^3}+\tfrac{3289}{12060}u_{3,0}^2\tfrac{q_2^4}{q_1^4}+\left(\tfrac{299}{19730}u_{3,2}^2+\tfrac{6539}{19730}u_{3,0}u_{3,2}+\tfrac{1202985}{3946}\right)\tfrac{q_2^2q_3^2}{q_1^4}+\tfrac{2171}{27090}u_{3,2}u_{3,3}\tfrac{q_2q_3^3}{q_1^4}+\left(\tfrac{1084733}{30333876}u_{3,2}^2+\tfrac{108970043}{1516693800}u_{3,3}^2+\right.\right.$$
$$\left.\tfrac{8028474825}{20222584}\right)\tfrac{q_3^4}{q_1^4}+\left(\tfrac{99171}{8576}u_{3,0}^3+\tfrac{363467}{4824000}u_{3,0}\right)\tfrac{q_2^5}{q_1^5}+\left(\tfrac{1250075444814383346 1}{1601323583570180591940 0}u_{3,2}^3-\tfrac{106381137094520737}{9000423588396965634000}u_{3,3}^2u_{3,2}+\tfrac{236392873}{2319952050}u_{3,0}^2u_{3,2}+\tfrac{1575063218080739091500 63}{1771114719097939422958 4}u_{3,2}+\right.$$
$$\left.\tfrac{203530243}{2667496}u_{3,0}\right)\tfrac{q_2^3q_3^2}{q_1^5}+\left(\tfrac{11081754182194243668456646 7}{15886067589169436518805105300 00}u_{3,3}^3-\tfrac{38522925037517849096607 9}{12309652055185871405368}u_{3,3}\right)\tfrac{q_2^2q_3^3}{q_1^5}+\left(\tfrac{2133042536605191925}{570808716287973 12}u_{3,2}+\tfrac{6522159327691787}{179804745630711532 8}u_{3,2}^3+\right.$$
$$\left.\left.\tfrac{904165813438076266 61}{12462124968549644724000}u_{3,2}u_{3,3}^2\right)\tfrac{q_2q_3^4}{q_1^5}+\left(\tfrac{179117095742904615655002021}{175812605478192208347168 46}u_{3,3}+\tfrac{4508155134073481644830319664918 9}{8334836094207397117218388767450000}u_{3,3}^3\right)\tfrac{q_3^5}{q_1^5}+O\left(\tfrac{(q_1,q_2)^6}{q_1^6}\right)\right)$$

$$\mathcal{I}_2=\mathcal{I}_3=<u_{3,1}>\quad \mathcal{I}_4=\left\langle u_{3,3}^3-\tfrac{2692736387071909369924 25}{697103584127013953981 4}u_{3,2}^2u_{3,3}-\tfrac{1951743012788974613845395937 5}{6041564395767454267838 8}u_{3,3},\tfrac{2960366553926052591241012}{2960366553926052591241 01}u_{3,0}u_{3,3}^3-\tfrac{2001534598250263118013 01}{5920733107852105182482 02}u_{3,2}u_{3,3}^3-\right.$$
$$\tfrac{956354076266597560784244009375}{2960366553926052591241 01}u_{3,0}u_{3,3}+\tfrac{1634510379739805414109277481 25}{2960366553926052591241 01}u_{3,2}u_{3,3},-\tfrac{52777633186609423650515300}{461892599596214565695 31}u_{3,0}u_{3,2}u_{3,3}+u_{3,3}^3-\tfrac{3269020759479610828218554962 50}{2001534598250263118013 01}u_{3,3},$$
$$\left.u_{3,2}u_{3,3}^2+\tfrac{76926697336726202 0}{815441353231641}u_{3,0}u_{3,2}^2-\tfrac{364191525409188890}{516446190380039 3}u_{3,2}^3+\tfrac{5093536748382230147212 5}{67138004749405109}u_{3,2}\right\rangle$$

$$q_1^{-1}\left(1+\tfrac{9}{2}\tfrac{q_2^2}{q_1^2}+\tfrac{54}{2}\tfrac{q_3^2}{q_1^2}+u_{3,0}\tfrac{q_2^3}{q_1^3}+u_{3,2}\tfrac{q_2q_3^2}{q_1^3}+u_{3,3}\tfrac{q_3^3}{q_1^3}+\left(\tfrac{13311}{536}+\tfrac{3289}{12060}u_{3,0}^2\right)\tfrac{q_2^4}{q_1^4}+\left(\tfrac{8131447}{843807240}u_{3,2}^2+\tfrac{1469946609}{3125212}+\tfrac{46560929}{140634540}u_{3,0}u_{3,2}\right)\tfrac{q_2^2q_3^2}{q_1^4}+\tfrac{26821}{525690}u_{3,2}u_{3,3}\tfrac{q_2q_3^3}{q_1^4}+\left(\tfrac{3843346549}{105721155720}u_{3,2}^2+\right.\right.$$
$$\left.\tfrac{93293112339}{97889959}+\tfrac{6634990046129}{14245925733270}u_{3,3}^2\right)\tfrac{q_3^4}{q_1^4}+\left(\tfrac{99171}{8576}u_{3,0}+\tfrac{363467}{4824000}u_{3,0}^3\right)\tfrac{q_2^5}{q_1^5}+\left(\tfrac{118579147678287}{965637379396}u_{3,0}+\tfrac{5483698568191354811285544825005306 73}{5939392725426156545743050602228909 6}u3,2+\tfrac{21459422475171895871107095123965213}{70607855310476622219169250816049649200}u_{3,2}^3+\right.$$
$$\left.\tfrac{1543078654120439}{14556983494394700}u_{3,0}^2u_{3,2}-\tfrac{6532333876427547851207247 8}{18007002218330647802772268712859}u_{3,2}u_{3,3}^2\right)\tfrac{q_2^3q_3^2}{q_1^5}+$$
$$\left(-\tfrac{563994532368308540620474922409653 25}{501489449847342564242033164136492}u_{3,3}+\tfrac{16450587697848383603452787103914 84}{499295433504260440523474269043394847 5}u_{3,3}^3\right)\tfrac{q_2^2q_3^3}{q_1^5}+\left(\tfrac{631084173799659528109335 3}{1083266165141870049136 04}u_{3,2}+\tfrac{10277293173559961713898 51}{4387227968824573699000962 00}u3,2^3+\right.$$
$$\left.\left.\tfrac{30151163445704395316027384 98}{10096714493437453619876483672 25}u_{3,2}u_{3,3}^2\right)\tfrac{q_2q_3^4}{q_1^5}+\left(-\tfrac{4008495460000401399979252028072066166113121}{29916876551082911336452430097420793501516}u_{3,3}+\tfrac{53741296159240513905164009584597889825626085272 1}{18880266902383600180381843412484720891442241221020 0}u_{3,3}^3\right)\tfrac{q_3^5}{q_1^5}+O\left(\tfrac{(q_1,q_2)^6}{q_1^6}\right)\right)$$

$$\mathcal{I}_2=\mathcal{I}_3=<u_{3,1}>\quad \mathcal{I}_4=\left\langle u_{3,3}^3-\tfrac{214050374934841338395989765180 21}{4495112199619735814187970929088}u_{3,2}^2u_{3,3}-\tfrac{38018197051893130425988535699449612 5}{1123778049904933953546992732272}u_{3,3},\tfrac{4035092449451502815699543327165595075}{1842996001844091683817068080926 08}u_{3,2}u_{3,3}-\right.$$
$$\tfrac{38018197051893130425988535699449612 5}{1123778049904933953546992732272}u_{3,0}u_{3,3}-\tfrac{51588237214057}{1157111566196588}u_{3,2}u_{3,3}^3+u3,0u_{3,3}^3,-\tfrac{87760653723284948742355803723886 1}{82167457058793670335631729657 12}u_{3,0}u_{3,2}u_{3,3}+u_{3,3}^3-$$
$$\left.\tfrac{4035092449451502815699543327165595075}{82167457058793670335631729657 12}u_{3,3},u_{3,2}u_{3,3}^2+\tfrac{582782943344153167092437 73}{327480205042360804589 60}u_{3,0}u_{3,2}^2-\tfrac{10465956758061712316821512 1}{12444247791609710574404 80}u_{3,2}^3+\tfrac{27466841687145693668090580055 5}{12444247791609710574404 8}u_{3,2}\right\rangle$$

$$q_1^{-1}\left(1+\frac{9}{2}\frac{q_2^2}{q_1^2}+\frac{44}{2}\frac{q_3^2}{q_1^2}+u_{3,0}\frac{q_2^3}{q_1^3}+u_{3,2}\frac{q_2q_3^2}{q_1^3}+\left(\frac{3289}{12060}u_{3,0}^2+\frac{13311}{536}\right)\frac{q_2^4}{q_1^4}+\left(\frac{48787}{4090760}u_{3,2}^2+\frac{2031899}{6136140}u_{3,0}u_{3,2}+\frac{78370281}{204538}\right)\frac{q_2^2q_3^2}{q_1^4}+\left(\frac{4690489}{129913320}u_{3,2}^2+\frac{682538736}{1082611}\right)\frac{q_3^4}{q_1^4}+\right.$$

$$\left(\frac{363467}{4824000}u_{3,0}^3+\frac{99171}{8576}u_{3,0}\right)\frac{q_2^5}{q_1^5}+\left(\frac{44243462785788303173707584 1}{938912143429199537485438760400}u_{3,2}^3+\frac{8173166259659}{78016448675700}u_{3,0}^2u_{3,2}+\frac{4779618603758363251509386524 40}{524225946747969741762703307 89}u_{3,2}+\frac{85215635549}{862536746}u_{3,0}\right)\frac{q_2^3q_3^2}{q_1^5}+$$

$$\left.\left(\frac{28213191044570641817}{97969267165144009266 00}u_{3,2}^3+\frac{15427883670599970273 11}{32656422388381336422}u_{3,2}\right)\frac{q_2q_3^4}{q_1^5}+O\left(\frac{(q_1,q_2)^6}{q_1^6}\right)\right)$$

$$\mathcal{I}_2=\mathcal{I}_3=<u_{3,1},u_{3,3}>\quad \mathcal{I}_4=\left\langle u_{3,1},u_{3,3},u_{3,2}^3-\frac{122281489890241714 8900}{70922785617530321}-\frac{36284913764868151 58}{2127683568525909 63}u_{3,0}u_{3,2}^2\right\rangle$$

$$q_1^{-1}\left(1+\frac{9}{2}\frac{q_2^2}{q_1^2}+\frac{20}{2}\frac{q_3^2}{q_1^2}+u_{3,0}\frac{q_2^3}{q_1^3}+u_{3,1}\frac{q_2^2q_3}{q_1^3}+u_{3,2}\frac{q_2q_3^2}{q_1^3}+u_{3,3}\frac{q_3^3}{q_1^3}+\left(\frac{13311}{536}+\frac{3289}{12060}u_{3,0}^2+\frac{11}{1809}u_{3,1}^2\right)\frac{q_2^4}{q_1^4}+\left(\frac{44}{135}u_{3,1}u_{3,0}+\frac{11}{270}u_{3,1}u_{3,2}\right)\frac{q_2^3q_3}{q_1^4}+\left(\frac{2959}{8772}u_{3,0}u_{3,2}+\frac{451}{2924}u_{3,1}u_{3,3}+\frac{11}{408}u_{3,2}^2+\right.\right.$$

$$\left.\frac{814}{6579}u_{3,1}^2+\frac{258255}{1462}\right)\frac{q_2^2q_3^2}{q_1^4}+\left(\frac{253}{2241}u_{3,1}u_{3,2}+\frac{121}{830}u_{3,2}u_{3,3}\right)\frac{q_2q_3^3}{q_1^4}+\left(\frac{2981}{86472}u_{3,2}^2+\frac{153200}{1201}+\frac{105347}{840700}u_{3,3}^2\right)\frac{q_3^4}{q_1^4}+\left(\frac{24121663415883}{456685207040}u_{3,0}-\frac{3676530375246235296032796 7}{21962011519983239630131 20}u_{3,2}+\right.$$

$$\frac{89405326300298824860496291}{177892293311864241004062720 00}u_{3,2}^3-\frac{41358147528355848356789}{822889690590546031104000}u_{3,0}u_{3,2}^2+\frac{1346853263812693}{13769058992256000}u_{3,0}^2u_{3,2}-\frac{922257609158044398539993}{45571173903965222232522240000}u_{3,2}u_{3,3}^2+\frac{363467}{4824000}u_{3,0}^3+\frac{230463041841}{4566852070400}u_{3,0}u_{3,1}u_{3,3}-$$

$$\left.\frac{321053308153987608599 53}{15177743182003404573696 00}u_{3,1}u_{3,3}u_{3,2}\right)\frac{q_2^5}{q_1^5}+\left(\frac{1006362154607592365}{76272045092872776}u_{3,1}+\frac{151684318285194825}{5403817053617488}u_{3,3}-\frac{119025726972819079}{1126507350433189824 0}u_{3,2}^2u_{3,3}+\frac{236491188371309540 71}{36329862051470371824 0}u_{3,0}u_{3,2}u_{3,3}+\right.$$

$$\left.\frac{5798437746842933}{10723874942903904936}u_{3,1}^2u_{3,3}+\frac{386205492458418970315 9}{858624920428505988542 40}u_{3,1}u_{3,3}^2+\frac{240823623910428286 3}{136740522442502312812 8}u_{3,1}u_{3,2}^2+\frac{2556040771182469}{28463888934742363 2}u_{3,1}u_{3,0}u_{3,2}\right)\frac{q_2^4q_3}{q_1^5}+\left(-\frac{7295524609850}{130123321413}u_{3,1}+\frac{7533233096661}{105601052404}u_{3,3}-\right.$$

$$\left.\frac{114781940299}{12230075371440}u_{3,2}^2u_{3,3}+\frac{204029796932797}{1314733102429800}u_{3,0}u_{3,2}u_{3,3}-\frac{833363511404}{8316082876815}u_{3,1}^2u_{3,3}+\frac{2463793308929051}{26633641426812840}u_{3,1}u_{3,3}^2+\frac{46351753531771}{1166425453146132}u_{3,1}u_{3,2}^2-\frac{76656425825}{485605933866}u_{3,1}u_{3,0}u_{3,2}\right)\frac{q_2^3q_3^2}{q_1^5}+$$

$$\left(-\frac{7295524609850}{130123321413}u_{3,1}+\frac{7533233096661}{105601052404}u_{3,3}-\frac{114781940299}{12230075371440}u_{3,2}^2u_{3,3}+\frac{204029796932797}{1314733102429800}u_{3,0}u_{3,2}u_{3,3}-\frac{833363511404}{8316082876815}u_{3,1}^2u_{3,3}+\frac{2463793308929051}{26633641426812840}u_{3,1}u_{3,3}^2+\frac{46351753531771}{1166425453146132}u_{3,1}u_{3,2}^2-\right.$$

$$\left.\frac{76656425825}{485605933866}u_{3,1}u_{3,0}u_{3,2}\right)\frac{q_2^2q_3^3}{q_1^5}+\left(\frac{50525703076300}{161496415089}u_{3,2}-\frac{4235472447613}{52324838488836}u_{3,2}^3+\frac{3093278443}{4840858404}u_{3,0}u_{3,2}^2+\frac{4492719610495891}{187658834333418000}u_{3,2}u_{3,3}^2+\frac{13050298589}{34722700404}u_{3,1}u_{3,3}u_{3,2}\right)\frac{q_2q_3^4}{q_1^5}+$$

$$\left.\left(\frac{46809030}{1737847}u_{3,3}+\frac{1362858557}{173089561200}u_{3,2}^2u_{3,3}+\frac{68250852673}{4257725150000}u_{3,3}^3+\frac{133423}{51883632}u_{3,1}u_{3,2}^2\right)\frac{q_3^5}{q_1^5}+O\left(\frac{(q_1,q_2)^6}{q_1^6}\right)\right)$$

$$\mathcal{I}_2=\mathcal{I}_3=<0>\quad \mathcal{I}_4=\left\langle\frac{21492255493447480629464816241}{9682635067611564055515910 0}u_{3,1}-\frac{16197873805354317734 7}{584082355435906 96}u_{3,0}+\frac{368276978761658545444304917078892 9}{401764027768828799405354582800 0}u_{3,2}-\frac{153908497065685013278741413}{506763343612357343217200}u_{3,3}+\frac{83773728539}{2056568535750}u_{3,1}^3+\right.$$

$$\frac{3602135236781333}{1177796800424025 00}u_{3,1}^2u_{3,2}-\frac{2084161691351276274465739127892 23}{756811308122677505856598167600000}u_{3,2}^3+\frac{7730882033}{41131370715}u_{3,0}u_{3,1}^2+\frac{43978788551225893017676270488 43}{150536063693566161309250260000 0}u_{3,0}u_{3,2}^2+\frac{11182502409314151996181325 7}{97817110511222463923320000 0}u_{3,2}^2u_{3,3}-$$

$$\frac{19202818906143244099}{292041177717953480 0}u_{3,0}^2u_{3,2}+\frac{43128463577770140210243091327 3}{570219122859315840469296005000000}u_{3,2}u_{3,3}^2-\frac{74061391347985716079524677731}{10515339379956414871756900000}u_{3,0}u_{3,2}u_{3,3}-\frac{19813596076797265 89}{584082355435906960}u_{3,0}u_{3,1}u_{3,3}+$$

$$\frac{161591603756363504127259996323 47}{138827703184066570985197462000 00}u_{3,1}u_{3,3}u_{3,2}+\frac{11541178916148758193157317}{3651677034853751443771000 0}u_{3,1}^2u_{3,3}-\frac{3628412075057407138506005535897}{7455629002060904322747250700000}u_{3,1}u_{3,3}^2-\frac{16997850549154609014088319603}{22427684947304925114636852000 0}u_{3,1}u_{3,2}^2+$$

$$\frac{209831610644740256848667017}{3345789802713404731922650 00}u_{3,1}u_{3,0}u_{3,2}+u_{3,1}u_{3,0}^2,\ \frac{110235640405544276665397614 29}{619688644327140099553018 24}u_{3,1}-\frac{80989369026771588673 5}{467265884348725568}u_{3,0}+\frac{9500305040852292249125132917333559}{1645625457741122762364332371148 8}u_{3,2}-$$

$$\frac{91088276984695530979304697}{324328539911908699659008}u_{3,3}+\frac{12565459497}{438734620960}u_{3,1}^3+\frac{18240141713679571}{964851138907361280}u_{3,1}^2u_{3,2}-\frac{5376282199401071407015583505428 03}{309989911807048706398862609448960 0}u_{3,2}^3+\frac{7730882033}{65810193144}u_{3,0}u_{3,1}^2+\frac{1131799927960860804112866915655 3}{616595716888846996722689064960 0}u_{3,0}u_{3,2}^2+$$

$$\frac{66181848063068086388215533}{626029507271823769109248000}u_{3,2}^2u_{3,3}-\frac{19202818906143244099}{467265884348725568 0}u_{3,0}^2u_{3,2}+\frac{1133842119622475310043992533923}{2335617527231757682562236436480000}u_{3,2}u_{3,3}^2-\frac{43832047337179489300081998 39}{672981720317210551792441600 0}u_{3,0}u_{3,2}u_{3,3}-$$

$$\frac{19813596076797265 89}{934531768697451136}u_{3,0}u_{3,1}u_{3,3}+\frac{41676784231893787208363119215647}{5686382722419366747553688043520 0}u_{3,1}u_{3,3}u_{3,2}+\frac{100452525304217534839407441}{397302461392088157082284800}u_{3,1}^2u_{3,3}-\frac{2147417526703479037810755093693}{4771602561318978766558240448000}u_{3,1}u_{3,3}^2-$$

$$\frac{88408532612579993390951826 07}{14353718366275152073367585280 0}u_{3,1}u_{3,2}^2+\frac{3270228085780662089082754 69}{642391642120973708529148800}u_{3,1}u_{3,0}u_{3,2}+u_{3,1}u_{3,0}^2,\ \frac{8119705174366086529811522577}{48413175338057820277579550}u_{3,1}-\frac{16197873805354317734 7}{1168164710871813 92}u_{3,0}+$$

$$\frac{148703375823006369826778256064516 79}{321411222215063039524283666240 00}u_{3,2}-\frac{69826208977705124409476661}{2533816718061786716086 00}u_{3,3}+\frac{8860106111}{342761422625}u_{3,1}^3+\frac{28427099781479791}{1884474880678440000}u_{3,1}^2u_{3,2}-\frac{8415162668370113067830824469307 23}{6054490464981420046852785340800000}u_{3,2}^3+\frac{7730882033}{82262741430}u_{3,0}u_{3,1}^2+$$

$$\frac{1770580274836493303130809264059 3}{1204288509548529290474002080000 0}u_{3,0}u_{3,2}^2+\frac{507335049729722153776903 29}{489085552556112319616600000}u_{3,2}^2u_{3,3}-\frac{19202818906143244099}{584082355435906960 0}u_{3,0}^2u_{3,2}+\frac{178235672009457430199004941452 3}{45617529828745267237543680400000 00}u_{3,2}u_{3,3}^2-$$

$$\frac{33600654207135772471672151 07}{525766968997820743587845000 0}u_{3,0}u_{3,2}u_{3,3}-\frac{19813596076797265 89}{116816471087181392 0}u_{3,0}u_{3,1}u_{3,3}+\frac{652315139453621904043358449848 47}{111062162547253256788157969600000}u_{3,1}u_{3,3}u_{3,2}+\frac{7395050030314814242823703 3}{310392547962568872720535000}u_{3,1}^2u_{3,3}-$$

$$\frac{16461616131697982511442638744 09}{372781450103045216137362535000 0}u_{3,1}u_{3,3}^2-\frac{65395299955379290936582540 91}{112138424736524625573184260000 0}u_{3,1}u_{3,2}^2+\frac{1209054945894037098282477 61}{2509342352035053548941987 50}u_{3,1}u_{3,0}u_{3,2}+u_{3,1}u_{3,0}^2,\ \frac{73712383302787538188228465761}{387305402704462562220636400}u_{3,1}-$$

$$\frac{48593621416062953204 1}{233632942174362784}u_{3,0}+\frac{1777477588284361679676057626609403 33}{257128977772050431619426932992000}u_{3,2}-\frac{582369803672098936042126773}{2027053374449429372868800}u_{3,3}+\frac{790993099357}{246788224290 00}u_{3,1}^3+\frac{381194564379568 73}{16750887828252800 00}u_{3,1}^2u_{3,2}-$$

$$\frac{11176596945068479662681172649221 69}{538176930220570670831358696960000 0}u_{3,2}^3+\frac{7730882033}{54841827620}u_{3,0}u_{3,1}^2+\frac{23544244037889590806056576606779}{10704786751542482581991129600000}u_{3,0}u_{3,2}^2+\frac{42313139669577287169891029 7}{3912684420448898556932800000}u_{3,2}^2u_{3,3}-\frac{5760845671842973229 7}{116816471087181392 00}u_{3,0}^2u_{3,2}+$$

$$\frac{21102189756216048155978964167121}{364940238629962137900349443200000 00}u_{3,2}u_{3,3}^2-\frac{28023870521327646500034450051}{42061357519825659487027600000}u_{3,0}u_{3,2}u_{3,3}-\frac{5944078823039179767}{233632942174362784 0}u_{3,0}u_{3,1}u_{3,3}+\frac{7798038051302623410848709075408 69}{888497300378026054305263756800000}u_{3,1}u_{3,3}u_{3,2}+$$

$$\left.\frac{67210240403890614680456346 9}{248314038370055098176428000 0}u_{3,1}^2u_{3,3}-\frac{13729440986554737069516573973737}{29822516008243617290989002800000}u_{3,1}u_{3,3}^2-\frac{17654267268330405238165715528 9}{269132219367659101375642224000 0}u_{3,1}u_{3,2}^2+\frac{6532828008174407727920942563}{12044843289768257034921540000}u_{3,1}u_{3,0}u_{3,2}+u_{3,1}u_{3,0}^2\right\rangle$$

# References


1. Bruns H (1887) Acta Math **11**
2. Poincaré H (1890) Sur le problème des trois corps et les équations de la dynamique. Acta Math. 13:A3–A270
3. Julliard Tosel E (1999) Non-intégrabilité algébrique et méromorphe de problèmes de n corps
4. Morales-Ruiz J, Simon S (2009) On the meromorphic non-integrability of some $N$-body problems. Discret Contin Dyn Syst (DCDS-A) 24:1225–1273
5. Boucher D (2000) Sur la non-intégrabilité du probleme plan des trois corps de masses égalesa un le long de la solution de lagrange. CR Acad Sci Paris 331:391–394
6. Tsygvintsev A (2007) On some exceptional cases in the integrability of the three-body problem. Celest Mech Dyn Astron 99:23–29
7. Nomikos D, Papageorgiou V (2009) Non-integrability of the anisotropic stormer problem and the isosceles three-body problem. Physica D: Nonlinear Phenom 238:273–289
8. Albouy A, Kaloshin V (2012) Finiteness of central configurations of five bodies in the plane. Ann Math 176:535–588
9. Morales-Ruiz J, Ramis J, Simó C (2007) Integrability of Hamiltonian systems and differential Galois groups of higher variational equations. Annales scientifiques de l'Ecole normale supérieure 40:845–884
10. Combot T (2013) A note on algebraic potentials and Morales-Ramis theory. Celest Mech Dyn Astron 1–22. doi:10.1007/s10569-013-9470-2
11. Din MSE (2003) Raglib: a library for real algebraic geometry. http://www-calfor.lip6.fr/safey/RAGLib/
12. Bostan MS-E-D.A, Combot T (2014) Computing necessary integrability conditions for planar parametrized homogeneous potentials. In: ISSAC 2014
13. Moulton F (1910) The straight line solutions of the problem of n bodies. Ann Math 12:1–17
14. Roberts GE (1999) A continuum of relative equilibria in the five-body problem. Physica D: Nonlinear Phenom 127:141–145
15. Hampton M, Moeckel R (2006) Finiteness of relative equilibria of the four-body problem. Inventiones Mathematicae 163:289–312
16. Combot T (2011) Integrable homogeneous potentials of degree $-1$ in the plane with small eigenvalues, arXiv:1110.6130
17. Combot T (2013) Integrability conditions at order 2 for homogeneous potentials of degree $-1$, Non-linearity, vol 26
18. Combot T, Koutschan C (2012) Third order integrability conditions for homogeneous potentials of degree $-1$. J Math Phys 53
19. Pacella F (1987) Central configurations of the n-body problem via equivariant morse theory. Arch Ration Mech Anal 97:59–74

# Relative Equilibria in the Full $N$-Body Problem with Applications to the Equal Mass Problem

**Daniel J. Scheeres**

**Abstract** General conditions for the existence of stable, minimum energy configurations in the full $N$-body problem are derived and investigated. Then the minimum energy and stable configurations for the spherical, equal mass full body problem are investigated for $N = 2, 3, 4$. This problem is defined as the dynamics of finite density spheres which interact gravitationally and through surface forces. This is a variation of the gravitational $N$-body problem in which the bodies are not allowed to come arbitrarily close to each other (due to their finite density), enabling the existence of resting configurations in addition to orbital motion. Previous work on this problem has outlined an efficient and simple way in which the stability of configurations in this problem can be defined. This methodology is reviewed and derived in a new approach and then applied to multiple body problems. In addition to this, new results on the Hill stability of these configurations are examined and derived. The study of these configurations is important for understanding the mechanics and morphological properties of small rubble pile asteroids. These results can also be generalized to other configurations of bodies that interact via field potentials and surface contact forces.

**Keywords** N-body problem · Relative equilibria · Stability · Bifurcation · Asteroids

## 1 Introduction

Celestial Mechanics systems have two fundamental conservation principles: conservation of momentum and conservation of (mechanical) energy. While conservation of momentum is always conserved for a closed system, mechanical energy is often not conserved and can be dissipated through non-conservative internal interactions. Thus, for any closed mechanical system it makes sense to seek out what their local and global minimum energy configurations are at a fixed level of angular momentum.

D.J. Scheeres (✉)
Department of Aerospace Engineering Sciences,
The University of Colorado, Boulder, USA
e-mail: scheeres@colorado.edu



B. Bonnard and M. Chyba (eds.), *Recent Advances in Celestial and Space Mechanics*, Mathematics for Industry 23, DOI 10.1007/978-3-319-27464-5_2

We can use the existence of these local and global minima to define what we consider stable states for a system, which provides a strict and robust limit for any mechanical system and is distinguished from particle motion systems in astrodynamics which often are only oscillatory stable and in fact do not minimize the total energy. The ubiquity and pervasiveness of energy dissipation in the solar system and its role in the long-term evolution of bodies of all sizes motivates these questions concerning the minimum energy states for celestial mechanics systems at a given value of angular momentum.

If this question is applied to the traditional $N$-body gravitational problem, however, it can be shown that there are no minimum energy configurations for $N \geq 3$ [9]. This is due to the point-mass nature of the traditional $N$-body problem, as when there are more than 3 bodies it is always possible to drive the total energy of the system to $-\infty$ while maintaining a constant angular momentum. This issue is considered in an earlier paper [9], where it is shown that replacing point mass bodies with finite density, rigid bodies enables the system $N$-body gravitational problem to have unique minimum energy configurations. This occurs as now bodies can rest on each other, meaning that their relative gravitational potential energies are bounded from below. That paper explores this phenomenon for celestial mechanics systems and finds all possible relative equilibria for the 2 and 3-body problem assuming the bodies have equal size and density and are spherical. Among other results, it is found that the so-called "full" 3-body problem has seven distinct relative equilibria, five additional beyond the familiar orbiting Lagrange and Euler configurations. Out of these, only 3 can be stable and only one of these three can be the minimum energy configuration for a given angular momentum. In [11] the problem is developed for the equal mass, spherical 4-body problem and a number of candidate relative equilibria are also found. An interesting result in both studies is that the relative stability of these minimum energy configurations change as the angular momentum of the system is modified.

In the current chapter a new derivation of some of the fundamental results in these papers is given, with the criteria for stable configurations generalized and extended to constraints on when the system can also be bound energetically. This chapter also provides more detailed evaluation of the energetic stability conditions for the 3-body orbiting conditions and relates the energy values of the existing relative equilibria to the Hill stability energy levels. This overall methodology is also applied to the 4-body problem, restricted so that the bodies are spheres of equal size and density. The computational issues related to the 4-body problem are significant, as even for the point-mass case the total number of relative equilibria are not known and are at best bounded [3]. Due to this we do not analyze purely orbital relative equilibria in the 4-body problem, motivated by a theorem from Moeckel [5] that these can never be minimum energy configurations for the point-mass $N$ body problem. Instead, we focus on resting configurations, or configurations that are separated into a mixture of two groups of bodies that orbit each other (motivated by the Hypothesis in [9]).

The original method used for this analysis relies on a novel application of the Cauchy Inequality to find a modified version of the Sundman Inequality. From this approach the minimum energy function can be derived, which is in spirit the same function as Smale's Amended potential [13, 14]. The minimum energy function is defined for a given level of angular momentum and only involves internal degrees of freedom in the system. Thus, whenever this function is stationary or positive definite with respect to all degrees of freedom, then the system is in a relative equilibrium. Further, if the function is positive definite about this relative equilibria then the given configuration is energetically stable, meaning that it is impossible for it to deviation from its current configuration without the addition of exogenous energy. It can be shown that at a given angular momentum there may be multiple stable relative equilibria, in this case there is always a minimum energy configuration which corresponds to the lowest energy level that the system can exist in.

In addition to reviewing and summarizing current results on the equal mass, spherical $N$-body problem for $N = 2, 3, 4$, this paper develops a new derivation of the amended potential that is based on Lagrangian mechanics, providing a dynamical foundation for the previous derivation which only used integrals of motion for the system. Specifically, we show that when the $N$-body problem is transformed into a rotating frame, with the rotation rate chosen based on the angular momentum of the system and its instantaneous moment of inertia, that the resultant Lagrangian can be reduced by Routh reduction to a system that only depends on the internal degrees of freedom of the system. This derivation is also extended to gravitationally interacting rigid bodies as well, sometimes termed the "Full-Body Problem."

The outline of the paper is as follows. First, a number of fundamental results and definitions are reviewed, leading to the statement of the Lagrange equations of motion for the Full $N$-body problem. The angular momentum magnitude is then removed through an application of Routhian reduction. Then the necessary and sufficient conditions for relative equilibria and their stability in the Full $N$-body problem is derived and proven. Following this derivation we consider the simplified case of equal mass, finite density spheres. We specifically review previously published results for $N = 2, 3, 4$ with additional insight into the energetics of these cases and conditions under which they will be Hill stable. This chapter is designed to provide a rigorous introduction to this general problem, and is intended as serving as a starting point for future analyses of the Full $N$-body problem.

## 2 Fundamental Quantities and Specification of the Full $N$-Body Problem

Consider the dynamics of a set of $N$ rigid bodies that interact with each other through gravitation and surface contact forces. We specify these bodies in general as mass distributions, denoted as $\mathscr{B}_i$, $i = 1, 2, \ldots, N$. Each body has its own center of mass location, velocity and body orientation and rotation. Thus in total there are $6N$ degrees of freedom for the system, although we assume that we trivially remove the three degrees of freedom associated with linear momentum by fixing the system barycenter at the origin. Following the motivation in [9], we also assume that each body has a finite density and mass, and thus has finite dimensions. We do not account for the flexure or movement of the mass distributions relative to themselves, only relative to each other. We note that these body, so defined, will place mutual gravitational forces and torques across each other, independent of contact.

## 2.1 Implicit Energy Dissipation Interaction Models

We qualify our rigid body assumption, in that while we do not allow the mass distributions to distort, we assume that they will cause infinitesimal distortions of mass distribution and hence tidal torques across each other. This will allow otherwise decoupled spherical bodies to place infinitesimal torques on each other, and will essentially enable the transfer of angular momentum between non-touching components and cause energy to be dissipated. When surfaces are in contact, we again assume that there will be some coefficient of friction between the components along with infinitesimal distortions of the bodies, again leading to transfer of angular momentum with energy dissipation. These physical effects serve dual purpose, in that they will tend to synchronize collections of bodies, either resting on each other or orbiting each other, and will also dissipate excess energy in the system. This is needed in order for a resting or orbiting collection of particles to dissipate relative motion between each other. Inclusion of this notional model ensures that configurations will, when reduced to their minimum energy state, all rotate at a common rate.

## 2.2 Density Distributions and Fundamental Quantities

To more easily discuss the following derivations for the finite density $N$ body problem we state our problem in an integral form. Consider an arbitrary collection of $N$ mass distributions, denoted as $\mathcal{B}_i$, $i = 1, 2, \ldots, N$, following the derivation in [7]. Each body $\mathcal{B}_i$ is defined by a differential mass distribution $dm_i$ that is assumed to be a finite density distribution, denoted as

$$dm_i = \rho_i(\mathbf{r})dV \tag{1}$$

$$m_i = \int_{\mathcal{B}_i} dm_i \tag{2}$$

where $m_i$ is the total mass of body $\mathcal{B}_i$, $\rho_i$ is the density of body $\mathcal{B}_i$ (possibly constant) and $dV$ is the differential volume element. If $\mathcal{B}_i$ is described by a point mass density distribution, the body itself is just defined as a single point $\mathbf{r}_i$. Instead, if the body is defined as a finite density distribution, $\mathcal{B}_i$ is defined as a compact set in $\mathbb{R}^3$ over which $\rho_i(\mathbf{r}) \neq 0$. In either case the $\mathcal{B}_i$ are defined as compact sets.

We assume that each differential mass element $dm_i(\mathbf{r})$ has a specified position and an associated velocity. For components within a given body $\mathcal{B}_i$ a rigid body assumption is made so that the entire body can be defined by the position and velocity of its center of mass, its attitude, and its angular velocity. Finally, we assume that these positions and velocities are defined relative to the system barycenter, which is chosen as the origin, or

$$\sum_{i=1}^{N} \int_{\mathcal{B}_i} \mathbf{r} dm_i(\mathbf{r}) = 0 \tag{3}$$

$$\sum_{i=1}^{N} \int_{\mathcal{B}_i} \dot{\mathbf{r}} dm_i(\mathbf{r}) = 0 \tag{4}$$

Given these definitions an integral form of the kinetic energy, total system inertia dyad, gravitational potential energy, and angular momentum vector can be stated as [7]

$$T = \frac{1}{2} \sum_{i=1}^{N} \int_{\mathcal{B}_i} (\dot{\mathbf{r}} \cdot \dot{\mathbf{r}})\, dm_i(\mathbf{r}) \tag{5}$$

$$\bar{\mathbf{I}} = - \sum_{i=1}^{N} \int_{\mathcal{B}_i} (\widetilde{\mathbf{r}} \cdot \widetilde{\mathbf{r}})\, dm_i(\mathbf{r}) \tag{6}$$

$$\mathcal{U} = -\mathcal{G} \sum_{i=1}^{N-1} \sum_{j=i+1}^{N} \int_{\mathcal{B}_i} \int_{\mathcal{B}_j} \frac{dm_i\, dm_j}{|\mathbf{r}_{ij}|} \tag{7}$$

$$\mathbf{H} = \sum_{i=1}^{N} \int_{\mathcal{B}_i} (\mathbf{r} \times \dot{\mathbf{r}})\, dm_i(\mathbf{r}) \tag{8}$$

where $\mathbf{r}_{ij} = \mathbf{r}_j - \mathbf{r}_i$, a bolded quantity with an overline, $\bar{\mathbf{I}}$, denotes a dyad and $\widetilde{\mathbf{r}}$ is the cross-product dyad associated with a vector $\mathbf{r}$. Note that the definition of $\mathcal{U}$ in Eq. 7 eliminates the self-potentials of these bodies from consideration. As the finite density mass distributions are assumed to be rigid bodies this elimination is reasonable. This notation can be further generalized by defining the single and joint general mass differentials

$$dm(\mathbf{r}) = \sum_{i=1}^{N} dm_i(\mathbf{r}) \tag{9}$$

$$dm(\mathbf{r})\, dm'(\mathbf{r}') = \sum_{i=1}^{N-1} \sum_{j=i+1}^{N} dm_i(\mathbf{r})\, dm_j'(\mathbf{r}') \tag{10}$$

and the total mass distribution $\mathscr{B} = \{\mathscr{B}_i, i = 1, 2, \ldots, N\}$. Then the above definitions can be reduced to integrals over $\mathscr{B}$:

$$T = \frac{1}{2} \int_{\mathscr{B}} (\dot{\mathbf{r}} \cdot \dot{\mathbf{r}})\, dm(\mathbf{r}) \tag{11}$$

$$\bar{\mathbf{I}} = -\int_{\mathscr{B}} (\tilde{\mathbf{r}} \cdot \tilde{\mathbf{r}})\, dm(\mathbf{r}) \tag{12}$$

$$\mathscr{U} = -\mathscr{G} \int_{\mathscr{B}} \int_{\mathscr{B}} \frac{dm\, dm'}{|\mathbf{r} - \mathbf{r}'|} \tag{13}$$

$$\mathbf{H} = \int_{\mathscr{B}} (\mathbf{r} \times \dot{\mathbf{r}})\, dm(\mathbf{r}) \tag{14}$$

$$\int_{\mathscr{B}} \mathbf{r} dm(\mathbf{r}) = 0 \tag{15}$$

$$\int_{\mathscr{B}} \dot{\mathbf{r}} dm(\mathbf{r}) = 0 \tag{16}$$

## 2.3 *Relative Reduction*

As stated for the discrete forms of these results, a key simplification results since the barycenter and barycentric velocity are both zero. Applying Lagrange's Identity we can express all of the above as relative integrals. Applying the standard result provides:

$$T = \frac{1}{4M} \int_{\mathscr{B}} \int_{\mathscr{B}} \left(\dot{\mathbf{r}} - \dot{\mathbf{r}}'\right) \cdot \left(\dot{\mathbf{r}} - \dot{\mathbf{r}}'\right) dm(\mathbf{r}) dm(\mathbf{r}') \tag{17}$$

$$\bar{\mathbf{I}} = -\overline{\mathbf{A}} \cdot \frac{1}{2M} \int_{\mathscr{B}} \left(\tilde{\mathbf{r}} - \tilde{\mathbf{r}}'\right) \cdot \left(\tilde{\mathbf{r}} - \tilde{\mathbf{r}}'\right) dm(\mathbf{r}) dm(\mathbf{r}') \cdot \overline{\mathbf{A}}^T \tag{18}$$

$$\mathbf{H} = \overline{\mathbf{A}} \cdot \frac{1}{2M} \int_{\mathscr{B}} \int_{\mathscr{B}} \left(\mathbf{r} - \mathbf{r}'\right) \times \left(\dot{\mathbf{r}} - \dot{\mathbf{r}}'\right) dm(\mathbf{r}) dm(\mathbf{r}') \tag{19}$$

$$0 = \frac{1}{2M} \int_{\mathscr{B}} \int_{\mathscr{B}} \left(\mathbf{r} - \mathbf{r}'\right) dm(\mathbf{r}) dm(\mathbf{r}') \tag{20}$$

$$0 = \frac{1}{2M} \int_{\mathscr{B}} \int_{\mathscr{B}} \left(\dot{\mathbf{r}} - \dot{\mathbf{r}}'\right) dm(\mathbf{r}) dm(\mathbf{r}') \tag{21}$$

We note a distinct shift in this statement, as we go from assuming that all vectors are expressed relative to an inertial frame to a more general idea that all vectors are expressed in a frame that is only defined relative to each other. This necessitates the introduction of a transformation dyad that takes this internal frame and rotates it into an inertial frame, which we denote as $\overline{\mathbf{A}}$. Thus the expressions for the angular momentum vector and the inertia dyadic must now be transformed into the inertial frame, as is indicated above. The scalar quantities do not need this, of course.

Thus all of these quantities are reduced to integrals of relative vectors over the mass distribution, and mimic the potential energy function. This change is significant, as it explicitly shows some of the fundamental symmetries of the problem.

## 2.4 Degrees of Freedom

Despite the integral form of the kinetic and potential energies, we recall that there are only finite degrees of freedom for the system. Specifically, for $N$ bodies there are $3N$ translational degrees of freedom and $3N$ rotational degrees of freedom for a total of $6N$ DOF. In our formulation we have already removed 3 DOF by setting the center of the system at the barycenter, reducing the total to $3(2N - 1)$. The degrees of freedom are split between three general classes, the relative positions of the bodies, the orientations of the bodies relative to each other, and the overall inertial orientation of the system.

It is instructive to review these degrees of freedom. For $N = 1$ there are no relative position or attitude degrees of freedom, and thus there is only the inertial orientation degrees of freedom for the system, yielding a total of 3 DOF in agreement with the general rule. For $N = 2$ we start with one central body with no degrees of freedom. The position of a second body relative to this has 3 DOF and its relative attitude has 3 DOF. Finally, we add the inertial orientation to get a total of 9 DOF. Each additional body then adds 6 DOF again, reproducing our general rule.

We distinguish between the internal, relative degrees of freedom and the inertial orientation degrees of freedom. For the current system we represent the internal degrees of freedom as $q_i$: $i = 1, 2, \ldots, 6(N - 1)$. These are specifically the relative positions of the centers of mass and the orientations of the rigid bodies relative to each other. For convenience we can imagine these to be Cartesian position vectors and Euler angles. We note that their time derivatives are expressed with respect to an inertial frame. The additional 3 DOF that orient the system relative to inertial space is represented as the rotation dyad $\overline{\mathbf{A}}$ which takes the relative frame into inertial space.

Note again that in our general statements, the final 3 DOF that orient the system relative to inertial space do not change any of our fundamental integral quantities except that of $\mathbf{H}$ and $\overline{\mathbf{I}}$. This is because each of these orientations acts on the entire system but do not change the relative orientations or speeds. This invariance is tied to the existence of the angular momentum integral. Despite this, since the speeds are all defined relative to an inertial frame there remains a fundamental connection between the inertial and relative frames.

## 2.5 Transformation into a Rotating System

Of specific interest to us is the overall rotation of the system due to the non-zero but constant angular momentum. A specific goal is to remove this integral of motion, commonly termed the elimination of the nodes. In our analysis we can remove one degree of freedom quite simply, and by doing so define the fundamental quantity that we use to discuss relative equilibria and their stability.

We define a very specific rotating frame from which we will measure motion. This is done by defining a system angular velocity which is a function of the angular momentum integral. Specifically, define an angular velocity vector

$$\omega = \frac{\mathbf{H}}{I_H} \tag{22}$$

$$= \dot{\theta}\hat{\mathbf{H}} \tag{23}$$

where $I_H = \hat{\mathbf{H}} \cdot \bar{\mathbf{I}} \cdot \hat{\mathbf{H}}$ is the moment of interia of the system about the angular momentum direction (which is fixed in space). We note that $I_H$ is a function of both the internal system and its orientation relative to $\hat{\mathbf{H}}$, but not to rotations around this unit vector which we denote by the angle $\theta$. Since the angular momentum is fixed in space the angular velocity $\omega = \dot{\theta}\hat{\mathbf{H}}$ is a true velocity and not a quasi-velocity and thus do not need to worry about this aspect of Lagrangian systems. We can then rewrite our system relative to this rotating frame, noting that the rotation rate $\dot{\theta}$ is not necessarily constant as the moment of inertia $I_H$ is not a constant in general.

Now reformulate the system kinetic energy in this rotating frame, with rotation vector defined by $\omega$. The main change is that the time derivatives will now be expressed relative to a rotating frame. Given an inertial velocity $\Delta\dot{\mathbf{r}}$, it can be expressed relative to a rotating frame as

$$\Delta\dot{\mathbf{r}} = \Delta\mathbf{v} + \omega \times \Delta\mathbf{r} \tag{24}$$

where $\mathbf{v}$ represents the speed relative to the rotating frame and $\mathbf{r}$ is the location of the mass elements in question. The dot product of this with itself, which is the kinetic energy integrand, then becomes

$$\Delta\dot{\mathbf{r}} \cdot \Delta\dot{\mathbf{r}} = (\Delta\mathbf{v} + \tilde{\omega} \cdot \Delta\mathbf{r}) \cdot (\Delta\mathbf{v} + \tilde{\omega} \cdot \Delta\mathbf{r}) \tag{25}$$

$$= \Delta\mathbf{v} \cdot \Delta\mathbf{v} + 2\omega \cdot \widetilde{\Delta\mathbf{r}} \cdot \Delta\mathbf{v} - \omega \cdot \widetilde{\Delta\mathbf{r}} \cdot \widetilde{\Delta\mathbf{r}} \cdot \omega \tag{26}$$

where we have used the properties of the cross product dyad and rearranged the terms.

Now consider the double integration over each of these three terms. The first term is the kinetic energy relative to the rotating frame

$$T_r = \frac{1}{4M}\int_{\mathscr{B}}\int_{\mathscr{B}} \Delta\mathbf{v} \cdot \Delta\mathbf{v} dm(\mathbf{r})dm(\mathbf{r}') \tag{27}$$

The final term takes on a simple form as well, once one recalls the definition of the inertia dyad $\bar{\mathbf{I}}$

$$\frac{1}{2}\omega \cdot \bar{\mathbf{I}} \cdot \omega = -\frac{1}{4M}\omega \cdot \int_{\mathscr{B}}\int_{\mathscr{B}} \widetilde{\Delta \mathbf{r}} \cdot \widetilde{\Delta \mathbf{r}} dm(\mathbf{r})dm(\mathbf{r}') \cdot \omega \tag{28}$$

From the definition of $\omega = \mathbf{H}/I_H = \dot{\theta}\hat{\mathbf{H}}$, we find that

$$\frac{1}{2}\omega \cdot \bar{\mathbf{I}} \cdot \omega = \frac{1}{2} I_H \dot{\theta}^2 \tag{29}$$

Finally consider the middle term. It can be shown that

$$\omega \cdot \mathbf{H}_r = \mathbf{0} \tag{30}$$

$$\mathbf{H}_r = \frac{1}{2M}\int_{\mathscr{B}}\int_{\mathscr{B}} \widetilde{\Delta \mathbf{r}} \cdot \Delta \mathbf{v} dm(\mathbf{r})dm(\mathbf{r}') \tag{31}$$

This serves as a constraint between the coordinates and velocities, and even if the value must be zero its partials with respect to individual terms may not be. Thus we can show this to be true, but must also retain this term in the Lagrange equations and only apply this after the equations are formulated. The proof is easy, and consists of rewriting $\Delta \mathbf{v} = \Delta \dot{\mathbf{r}} - \widetilde{\omega} \cdot \Delta \mathbf{r}$. Then the first term equals

$$\omega \cdot \frac{1}{2M}\int_{\mathscr{B}}\int_{\mathscr{B}} \widetilde{\Delta \mathbf{r}} \cdot \Delta \dot{\mathbf{r}} dm(\mathbf{r})dm(\mathbf{r}') = \omega \cdot \mathbf{H} \tag{32}$$

The second term equals

$$\omega \cdot \frac{1}{2M}\int_{\mathscr{B}}\int_{\mathscr{B}} \widetilde{\Delta \mathbf{r}} \cdot \widetilde{\Delta \mathbf{r}} \cdot \omega dm(\mathbf{r})dm(\mathbf{r}') = -\omega \cdot \bar{\mathbf{I}} \cdot \omega \tag{33}$$

By definition the first term equals $H^2/I_H$. The second term equals $-H^2/I_H$, and thus the sum is zero.

The result is that the kinetic energy becomes

$$T = T_r + \frac{1}{2} I_H \dot{\theta}^2 + \omega \cdot \mathbf{H}_r \tag{34}$$

### 2.5.1 Lagrangian Function and Dynamics

The Lagrangian of the original system is just $L = T - \mathscr{U}$. In this new coordinate system it is

$$L = T_r + \frac{1}{2} I_H \dot{\theta}^2 + \omega \cdot \mathbf{H}_r - \mathscr{U} \tag{35}$$

We note that the kinetic energy and gravitational potential are both independent of the angle $\theta$, and thus $\partial L / \partial \theta = 0$ leading to the momentum integral.

$$\frac{d}{dt} \frac{\partial L}{\partial \dot{\theta}} = 0 \tag{36}$$

$$I_H \dot{\theta} + \hat{\mathbf{H}} \cdot \mathbf{H}_r = H \tag{37}$$

where we note that the value of $H$ is just $I_H \dot{\theta}$ due to Eq. 30 but that we retain the functional form still as it is a constraint that must be applied after the equations of motion are derived. Then we find

$$\dot{\theta} = \frac{1}{I_H} \left[ H - \hat{\mathbf{H}} \cdot \mathbf{H}_r \right] \tag{38}$$

We can apply Routhian reduction to this system (see [1, 12] for a rigorous application of this approach), yielding

$$L_r = L - \dot{\theta} \frac{\partial L}{\partial \dot{\theta}} \tag{39}$$

$$= T_r + \frac{H}{I_H} \hat{\mathbf{H}} \cdot \mathbf{H}_r + \frac{1}{2 I_H} \left( \hat{\mathbf{H}} \cdot \mathbf{H}_r \right)^2 - \left( \frac{H^2}{2 I_H} + \mathscr{U} \right) \tag{40}$$

We note that the term $\frac{1}{2I_H} \left( \hat{\mathbf{H}} \cdot \mathbf{H}_r \right)^2$ will not arise in the Lagrange equations as its first partial will always equal zero and its total time derivative must equal zero by Eq. 30, thus we leave it out of the following discussion. The term $\frac{H}{I_H} \hat{\mathbf{H}} \cdot \mathbf{H}_r$ is important, however, and leads to the Coriolis and frame tie acceleration terms in the Lagrange equations.

This reduced Lagrangian has an integral of motion, equal to

$$E = \sum_i \dot{q}_i \frac{\partial L}{\partial \dot{q}_i} + \sum_i \dot{\theta}_i \frac{\partial L}{\partial \dot{\theta}_i} - L \tag{41}$$

$$= T_r + \frac{H^2}{2 I_H} + \mathscr{U} \tag{42}$$

where the terms from $\frac{H}{I_H} \hat{\mathbf{H}} \cdot \mathbf{H}_r$ will cancel out due to the skew-symmetric form between coordinates and rates. We note that this is just equal to the original energy of the system, given that $T = T_r + \frac{H^2}{2I_H}$, and thus is nominally conserved. We denote the new, amended potential as

$$\mathscr{E} = \frac{H^2}{2 I_H} + \mathscr{U} \tag{43}$$

There are several conclusions we can draw from this analysis. First, from the energy equation we see that

$$E - \mathscr{E} = T_r \tag{44}$$

$$\geq 0 \tag{45}$$

and thus we have

$$E \geq \mathscr{E} \tag{46}$$

with equality occurring when the relative kinetic energy is $T_r = 0$. We note that it is not clear whether this minimum in the kinetic energy is possible, as for the earlier system we evidently could not have $T = 0$ for non-zero angular momentum. Thus, we need to do more work to claim that $E = \mathscr{E}$ can exist.

### 2.5.2 Limits on the Energy

First we derive the amended potential through the use of Cauchy's Inequality. This approach (first presented in [9]) provides a separate rigorous bound on the system energy.

Recall that $\mathbf{H} = H\hat{\mathbf{H}} = \int_{\mathscr{B}} \mathbf{r} \times \dot{\mathbf{r}}\, dm$. Dotting both sides by the (constant) unit vector aligned with the angular momentum vector yields the equality

$$H = \int_{\mathscr{B}} \hat{\mathbf{H}} \cdot (\mathbf{r} \times \dot{\mathbf{r}})\, dm \tag{47}$$

$$= \int_{\mathscr{B}} \dot{\mathbf{r}} \cdot (\hat{\mathbf{H}} \times \mathbf{r})\, dm \tag{48}$$

Now apply the triangle inequality to the integral to find:

$$\int_{\mathscr{B}} \dot{\mathbf{r}} \cdot (\hat{\mathbf{H}} \times \mathbf{r})\, dm \leq \int_{\mathscr{B}} |\dot{\mathbf{r}}| \left|\hat{\mathbf{H}} \times \mathbf{r}\right| dm \tag{49}$$

Squaring the original term, equal to $H^2$, and applying the Cauchy-Schwarz inequality yields the main result.

$$H^2 \leq \left[\int_{\mathscr{B}} |\dot{\mathbf{r}}| \left|\hat{\mathbf{H}} \times \mathbf{r}\right| dm\right]^2 \tag{50}$$

$$\left[\int_{\mathscr{B}} |\dot{\mathbf{r}}| \left|\hat{\mathbf{H}} \times \mathbf{r}\right| dm\right]^2 \leq \left[\int_{\mathscr{B}} (\hat{\mathbf{H}} \times \mathbf{r}) \cdot (\hat{\mathbf{H}} \times \mathbf{r})\, dm\right]\left[\int_{\mathscr{B}} \dot{r}^2\, dm\right] \tag{51}$$

But $\int_{\mathscr{B}} \dot{r}^2 \, dm = 2T$ and $\int_{\mathscr{B}} (\hat{\mathbf{H}} \times \mathbf{r}) \cdot (\hat{\mathbf{H}} \times \mathbf{r}) \, dm = \hat{\mathbf{H}} \cdot \int_{\mathscr{B}} -\tilde{\mathbf{r}} \cdot \tilde{\mathbf{r}} \, dm \cdot \hat{\mathbf{H}} = \hat{\mathbf{H}} \cdot \bar{\mathbf{I}} \cdot \hat{\mathbf{H}} = I_H$. Thus, $H^2 \leq 2TI_H$.

If we then use $T = E - \mathscr{U}$ the Sundman inequality becomes $\frac{H^2}{2I_H} + \mathscr{U} = \mathscr{E} \leq E$, which is exactly the same conclusion as we had from our Lagrangian analysis. This modified Sundman Inequality provides an important, and sharp, lower bound on the system energy for a given angular momentum. We note that the traditional Sundman inequality, using the polar moment of inertia, yields an angular momentum terms $H^2/(2I_P) \leq H^2/(2I_H)$ and thus the limit on the energy is not as sharp and the amended potential from that approach may not be equal to the energy.

Finally, to establish that the case $E = \mathscr{E}$ can exist, we construct a configuration and system state where this occurs. Consider the system energy when the entire system is at least momentarily stationary and spinning at a single, constant rate. If this occurs, then the inertial speeds are all $\dot{\mathbf{r}} = \omega \times \mathbf{r}$ and all the rigid body rotation speeds are $\omega$. At this momentary condition we can show that

$$T = -\frac{1}{2}\omega \cdot \int_{\mathscr{B}} (\tilde{\mathbf{r}} \cdot \tilde{\mathbf{r}}) \, dm(\mathbf{r}) \cdot \omega \tag{52}$$

$$= \frac{1}{2}\omega \cdot \bar{\mathbf{I}} \cdot \omega \tag{53}$$

Applying the same situation to the angular momentum yields

$$\mathbf{H} = -\int_{\mathscr{B}} (\tilde{\mathbf{r}} \cdot \tilde{\mathbf{r}}) \, dm(\mathbf{r}) \cdot \omega \tag{54}$$

$$= \bar{\mathbf{I}} \cdot \omega \tag{55}$$

with the magnitude then being $H^2 = \omega \cdot \bar{\mathbf{I}} \cdot \bar{\mathbf{I}} \cdot \omega$.

Now, forming the total energy we find

$$E = \frac{1}{2}\omega \cdot \bar{\mathbf{I}} \cdot \omega + \mathscr{U} \tag{56}$$

This cannot be directly related to the angular momentum as of yet. To do that, assume that the $\omega$ is chosen along a principal moment of inertia at this instantaneous configuration. Then we can state that $H^2 = \omega^2 I_\omega^2$, where $I_\omega$ is a principal moment of inertia. Also, then $T = \frac{1}{2}\omega^2 I_\omega = \frac{1}{2}H^2/I_\omega$. Then the energy equals

$$E = \frac{1}{2}\frac{H^2}{I_\omega} + \mathscr{U} \tag{57}$$

which is identically equal to $\mathscr{E}$. Thus, by construction, we prove that the lower limit can be achieved, or $E = \mathscr{E}$, and thus the relative kinetic energy can be identically zero at an instant of time, $T_r = 0$.

We note that the condition $E = \mathscr{E}$ is not sufficient for the system to be in a relative equilibrium, as the forces acting within the system may not be balanced and thus may cause the system to evolve in time.

## 2.6 Additional Integrals of Motion

In this new rotating frame the angular momentum magnitude is explicitly removed, yielding a quadrature for the total system rotation once the internal motion is known

$$\theta - \theta_o = H^2 \int_{t_o}^{t} \frac{d\tau}{I_H(\mathbf{q}(\tau))} \tag{58}$$

Still, the additional two integrals of motion for conservation of angular momentum transverse to the angular momentum direction are not explicitly present. These can be transformed into the relative frame as follows. Consider the projection of the internal angular momentum into the inertial frame $\overline{\mathbf{A}} \cdot \mathbf{H}_r$. We know that $\mathbf{H}_r$ is already perpendicular to $\hat{\mathbf{H}}$ and that for our defined frame the rotation is always about this direction. Thus we can infer that $\overline{\mathbf{A}} \cdot \mathbf{H}_r = \mathbf{0}$.

# 3 Equilibrium and Stability Conditions

With these results we can determine conditions for relative equilibrium and conditions for stability. In fact, given the classical form of the Energy, split into a quadratic and a potential part, the derivation of stability conditions is simple. The only catch involves the presence of uni-lateral constraints which exist when the rigid bodies are in contact. We consider the cases separately. First we present some definitions.

**Definition 1** (*Relative Equilibrium*) A given configuration is said to be in "Relative Equilibrium" if its internal kinetic energy is null ($T_r = 0$), meaning that $\mathscr{E} = E$ at an instant, and that it remains in this state over at least a finite interval of time.

**Definition 2** (*Energetic Stability*) A given relative equilibrium is said to be "Energetically Stable" if any equi-energy deviation from that relative equilibrium requires a negative internal kinetic energy, $T_r < 0$, meaning that this motion is not allowed.

Note that energetic stability is different than Lyapunov or spectral stability, which are the usual notions of stability in astrodynamics (these distinctions are discussed in detail for the Full Body Problem in [8]). Energetic stability is stronger in general, as it is robust to any energy dissipation and in fact—if it applies—means that a given relative equilibrium configuration cannot shed any additional energy and thus is static without the injection of exogenous energy, a condition we refer to as being in a minimum energy state.

## 3.1 No Contact Case

When there are no contacts between the bodies, there are necessarily no active unilateral constraints and all of the degrees of freedom are unconstrained. We also note that the kinetic energy is quadratic in the generalized coordinate rates and has the form of a natural system ([2], p. 72). Then the condition for a relative equilibrium are that all of the $\dot{\mathbf{q}} = 0$ (yielding $T_r = 0$) and $\partial\mathscr{E}/\partial\mathbf{q} = \mathbf{0}$. Energetic stability of the configuration occurs when the Hessian of the amended potential is positive definite, or $\partial^2\mathscr{E}/\partial\mathbf{q}^2 > 0$, meaning that it has only positive eigenvalues. Neutral stability can occur when $\partial^2\mathscr{E}/\partial\mathbf{q}^2 \geq 0$, meaning that at least one eigenvalue is equal to zero. In this case it is possible for the system to drift at a constant rate relative to the equilibrium, ultimately destroying the configuration. If the configuration is not positive definite or semi-definite, then there exists at least one negative eigenvalue and the system can escape from the equilibrium configuration while conserving energy. Another way to consider this case is that the system can still dissipate energy, and thus can evolve to a lower energy state. We note that this is a stronger form of stability than is sometimes used in celestial mechanics and astrodynamics, where spectral stability of linearized motion can sometimes be stable (as in the Lagrange configurations of the 3-body problem that satisfy the Routh criterion).

It is a remarkable fact of celestial mechanics that in the point mass $n$-body case for $n \geq 3$, the Hessian of any relative equilibrium configuration has at least one negative eigenvalue and is unstable [5]. Thus for the point mass $n$-body problem all central configurations are always energetically unstable except for the 2-body problem. For the $n = 2$ body problem there is only a single relative equilibrium and it is positive definite and thus stable. If we consider the 3-body problem, we note that while the Lagrange configurations may be spectrally stable when they satisfy the Routh criterion, they are not at a minimum of the amended potential and thus if energy dissipation occurs they can progressively escape from these configurations. We note that for the finite density cases there are always stable configurations at any angular momentum [9].

In keeping with a variational approach, in the no-contact case (i.e., when there are no constraints on the coordinates), the equilibrium condition is

$$\delta\mathscr{E} = 0 \tag{59}$$

where $\delta\mathscr{E} = \sum_{i=1}^{n} \left(\partial\mathscr{E}/\partial q_i\right)\delta q_i$ which corresponds to $\partial\mathscr{E}/\partial q_i = 0$ for all $i$. The stability condition is

$$\delta^2\mathscr{E} > 0 \tag{60}$$

which corresponds to the Hessian $\left[\partial^2\mathscr{E}/\partial q_i \partial q_j\right]$ being positive definite.

## 3.2 Contact Case

The equilibrium and stability conditions must be modified if there are constraints which are activated. We assume, without loss of generality, that generalized coordinates are chosen to correspond to each contact constraint, such that in the vicinity of their being active the unilateral constraint can be restated as

$$\delta q_j \geq 0 \tag{61}$$

for $j = 1, 2, \ldots, m$ constraints. We note that these constrained generalized coordinates may either be relative positions or Euler angles between bodies. Assume we have the system at a configuration $\mathbf{q}$ with $m$ active constraints as just enumerated and $T_r = 0$. Further, assume that the $n - m$ unconstrained states satisfy $\mathscr{E}_{q_i} = 0$ for $i = m + 1, \ldots, n$. For this system to be at rest the principle of virtual work and energy states that the variation of the $m$ constrained states are such that the amended potential only increases, or

$$\delta\mathscr{E} \geq 0 \tag{62}$$

$$\delta\mathscr{E} = \sum_{j=1}^{m} \mathscr{E}_{q_j}\delta q_j \tag{63}$$

which, for our assumed constraints on the states, is the same as $\mathscr{E}_{q_j} \geq 0$ for $j = 1, 2, \ldots, m$. The derivation of this just notes that, as defined, if the amended potential can only increase in value then motion is not allowed, as this corresponds to a decrease in kinetic energy from its zero value, which of course is non-physical.

For stability, we require the $n - m$ unconstrained variables to satisfy the same positive definite condition as derived earlier. For the constrained states we only need to tighten the condition to $\delta\mathscr{E} > 0$ or $\mathscr{E}_{q_j} > 0$ for $j = 1, 2, \ldots, m$. This last assertion demands some more specific proof and motivates the following general theorem on necessary and sufficient conditions for a relative equilibrium.

**Theorem 1** *Consider a system with an amended potential $\mathscr{E}$ as defined above with $n$ degrees of freedom, $m$ of which are activated in such a way that only the variations $\delta q_j \geq 0$, $j = 1, 2, \ldots, m$ are allowed. The degrees of freedom $q_i$ for $m < i \leq n$ are unconstrained.*

*The Necessary and Sufficient conditions for the system to be in a relative equilibrium are that:*

1. $T_r = 0$
2. $\mathscr{E}_{q_i} = 0 \; \forall \; m < i \leq n$*The Necessary and Sufficient conditions for*
3. $\mathscr{E}_{q_j} \geq 0 \; \forall \; 1 \leq j \leq m$

*The Necessary and Sufficient conditions for a system in a relative equilibrium to be energetically stable are that:*

1. $\left[\frac{\partial^2 \mathscr{E}}{\partial q_i \partial q_k}\right] > 0 \; \forall \; m < i, k \leq n$
2. $\mathscr{E}_{q_j} > 0 \; \forall \; 1 \leq j \leq m$

*Proof* First, note that by construction the relative kinetic energy $T_r$ is of the form $T_r = \frac{1}{2}\sum_{i,j} a_{ij}(\mathbf{q})\dot{q}_i\dot{q}_j$ where $[a_{ij}] > 0$ (is positive definite as a matrix) and $\partial L/\partial \dot{q}_i = \partial T_r/\partial \dot{q}_i = \sum_j a_{ij}\dot{q}_j$ (c.f. [2]). Thus it can be easily shown that all $\dot{q}_i = 0$ if and only if $T_r = 0$. Also note that $d/dt(H/I_H\hat{\mathbf{H}} \cdot \mathbf{H}_r$ will be linear in the $\dot{q}_i$ and thus will also go to zero if $T_r = 0$. Furthermore, even though the $a_{ij}$ can be a function of the $\mathbf{q}$, if $T_r = 0$ then $\partial L/\partial \mathbf{q}|_{T_r=0} = -\partial \mathscr{E}/\partial \mathbf{q}$.

Next consider the relative equilibrium conditions for the unconstrained degrees of freedom. It is easily shown that if $\mathscr{E}_{q_i} = 0$ and $T_r = 0$ then $d/dt(\partial L/\partial \dot{q}_i) = 0$, meaning that all $\ddot{q}_i = 0$. Conversely, if all $\dot{q}_i = 0$ then for all $\ddot{q}_i = 0$ to hold we must have $\partial L/\partial q_i = 0$. However, for zero speeds we have $L = \mathscr{E}$, completing the proof for the conditions of the unconstrained relative equilibria.

Next we develop the stability conditions for the unconstrained coordinates. Note that conservation of energy implies that under all finite variations we must have $\Delta T_r + \Delta \mathscr{E} = 0$, where $\Delta \mathscr{E} = \mathscr{E}(\mathbf{q} + \delta\mathbf{q}) - \mathscr{E}(\mathbf{q})$. If the system is at a relative equilibrium at the test point $\mathbf{q}^*$ (and only these unconstrained coordinates are varied) we find $T_r = \Delta T_r = -\frac{1}{2}\delta\mathbf{q} \cdot \left.\frac{\partial^2 \mathscr{E}}{\partial \mathbf{q}^2}\right|^* \cdot \mathbf{q} + \ldots$, as the first order partials are all zero. If the amended potential $\mathscr{E}$ is positive definite at the relative equilibria, then the quantity on the right is $< 0$ and we have $T_r < 0$, which is impossible, meaning that finite variations cannot occur. Conversely if $\delta^2\mathscr{E} = 0$ for some specific variations, then the configuration can be deviated from the relative equilibrium, and it is considered to be indeterminate as higher orders must be considered to determine stability. If $\delta^2\mathscr{E} < 0$ for some variation, then $T_r > 0$ for this variation and the system is no longer at a relative equilibrium, implying instability.

For the constrained coordinates, first assume that all of the unconstrained coordinates are in a stable relative equilibrium. As all of the $m$ constraints are active, for zero speeds we have $E = \mathscr{E}$. Consider a variation consistent with the constraints, yielding $\delta E = \delta T_r + \delta\mathscr{E} = 0$, where $\delta\mathscr{E} = \sum_{j=1}^{m} \mathscr{E}_{q_j}\delta q_j$. If $\delta\mathscr{E} > 0$ for all possible variations, then by necessity $\delta T_r < 0$, which is impossible. Thus in this case no motion can occur and, in fact, the system is stable against all variations in both the constrained and unconstrained variations.

If, instead, $\delta\mathscr{E} = 0$ for all variations of at least a subset of the constrained coordinates, then $\delta T_r = 0$ and by extension $\dot{\mathbf{q}} = 0$ for this subset. Then, since $\delta\mathscr{E} = 0$ for all allowable variations of this subset, this implies that $\mathscr{E}_{q_j} = 0$ and by Lagrange's equations all $\ddot{q}_j = 0$, leaving the system in equilibrium. The stability of this configuration is considered to be indeterminate, however, as the higher order partials must be considered, incorporating constraints at a higher order as well.

Finally, consider the case when $\delta\mathscr{E} < 0$ for some specific variation $\delta\mathbf{q}'$, chosen such that the unconstrained variations are all zero and the constrained variations are allowable. Given the assumed restrictions on $\delta q_j \geq 0$, this means that some values of $\mathscr{E}_{q_j} < 0$. Then we can form the test variation $\delta T_r = -\delta \mathscr{E} > 0$, which is allowed. Furthermore, writing the Lagrange equations yields $d/dt(\partial T_r/\partial \mathbf{q}) = -\partial \mathscr{E}/\partial \mathbf{q}$. Taking the dot product of this with the allowable variation $\delta \mathbf{q}'$ yields

$$\frac{d}{dt}\left(\frac{\partial T_r}{\partial \dot{\mathbf{q}}}\right) \cdot \delta \mathbf{q}' > 0 \tag{64}$$

This can be integrated over time from $t = 0$ to $t = \Delta t$ to find

$$\left.\frac{\partial T_r}{\partial \dot{\mathbf{q}}}\right|_{\Delta t} \cdot \delta \mathbf{q}' > 0 \tag{65}$$

Thus, after a finite time in such a configuration we see that the velocities in the relative kinetic energy are non-zero, meaning that the system is no longer in a relative equilibrium. □

# 4 Fundamental Quantities in the Full $N$-Body Problem

Now that the dynamical analyses are complete, we can progress to the main problem. This requires us to develop explicit expressions for the amended potential, and eventually will result in applying our restrictions on the bodies being equal mass and size spheres.

## *4.1 Rigid Body Expressions*

We first restate the fundamental quantities of the total kinetic energy, gravitational potential energy, moments of inertia and angular momentum of this system in explicit coordinates. Since finite density distributions are assumed for each body we must also incorporate rotational kinetic energy, rigid body moments of inertia, angular velocities and explicit mutual potentials that are a function of body attitude [7].

In the following the $i$th rigid body's center of mass is located by the position $\mathbf{r}_i$ and has a velocity $\dot{\mathbf{r}}_i$ in the inertial frame and velocity $\mathbf{v}_i$ in the rotating frame. In addition to its mass $m_i$, the $i$th body has an inertia dyadic $\overline{\mathbf{I}}_i$, an angular velocity vector $\omega_i$ and an attitude dyadic that maps its body-fixed vectors into inertial space, $\overline{\mathbf{A}}_i$. For expressing the kinetic energy of the rigid bodies relative to the rotating frame it is important to define their angular velocity relative to the rotating frame. If their inertial angular velocity is $\omega_i$, then the relative angular velocity will be $\Omega_i = \omega_i - \omega$. Using this definition the basic quantities are then defined as [9]:

$$T = \frac{1}{2}\sum_{i=1}^{N}\left[m_i\left(\dot{\mathbf{r}}_i\cdot\dot{\mathbf{r}}_i\right) + \omega_i\cdot\bar{\mathbf{I}}_i\cdot\omega_i\right] \tag{66}$$

$$T_r = \frac{1}{2}\sum_{i=1}^{N}\left[m_i\left(\mathbf{v}_i\cdot\mathbf{v}_i\right) + \Omega_i\cdot\bar{\mathbf{I}}_i\cdot\Omega_i\right] \tag{67}$$

$$\mathcal{U} = \sum_{i=1}^{N-1}\sum_{j=i+1}^{N}\mathcal{U}_{ij}(\mathbf{r}_{ij},\overline{\mathbf{A}}_{ij}) \tag{68}$$

$$\bar{\mathbf{I}} = \sum_{i=1}^{N}\left[m_i\left(r_i^2\mathbf{U} - \mathbf{r}_i\mathbf{r}_i\right) + \bar{\mathbf{I}}_i\right] \tag{69}$$

$$I_H = \sum_{i=1}^{N}\left[m_i\mathbf{r}_i\cdot\left(\mathbf{U} - \hat{\mathbf{H}}\hat{\mathbf{H}}\right)\cdot\mathbf{r}_i + \hat{\mathbf{H}}\cdot\bar{\mathbf{I}}_i\cdot\hat{\mathbf{H}}\right] \tag{70}$$

$$\mathbf{H} = \sum_{i=1}^{N}\left[m_i\left(\mathbf{r}_i\times\dot{\mathbf{r}}_i\right) + \bar{\mathbf{I}}_i\cdot\Omega_i\right] \tag{71}$$

In the above we have implicitly assumed that all vector quantities are expressed in an inertial frame, thus the only place where the relative orientations $\overline{\mathbf{A}}_i$ occur are in the potential energy, $\mathcal{U}_{ij}$, which must specify the orientation between two different rigid bodies $i$ and $j$ as $\overline{\mathbf{A}}_{ij} = \overline{\mathbf{A}}_j^T\cdot\overline{\mathbf{A}}_i$, which transfers a vector from the body $i$ frame into the body $j$ frame.

The kinetic energy, moments of inertia and angular momentum can also be stated in relative form between the center of masses (assuming barycentric coordinates and applying Lagrange's Identity). In introducing these specifications we also note that we can decouple the relative orientations of the bodies from their inertial orientations. Thus we add the rotation dyad $\overline{\mathbf{A}}$ which represents the rotation of a single internal body-relative frame into inertial space. This quantity is only needed for those expressions which are stated in an inertial frame. We assume all relative vectors to be specified in the internal frame, including inertia dyads of each individual rigid body.

$$T = \frac{1}{2M}\sum_{i=1}^{N-1}\sum_{j=i+1}^{N}m_im_j\left(\dot{\mathbf{r}}_{ij}\cdot\dot{\mathbf{r}}_{ij}\right) + \frac{1}{2}\sum_{i=1}^{N}\omega_i\cdot\bar{\mathbf{I}}_i\cdot\omega_i \tag{72}$$

$$T_r = \frac{1}{2M}\sum_{i=1}^{N-1}\sum_{j=i+1}^{N}m_im_j\left(\mathbf{v}_{ij}\cdot\mathbf{v}_{ij}\right) + \frac{1}{2}\sum_{i=1}^{N}\Omega_i\cdot\bar{\mathbf{I}}_i\cdot\Omega_i \tag{73}$$

$$\mathbf{H} = \overline{\mathbf{A}}\cdot\left[\frac{1}{M}\sum_{i=1}^{N-1}\sum_{j=i+1}^{N}m_im_j\left(\mathbf{r}_{ij}\times\dot{\mathbf{r}}_{ij}\right) + \sum_{i=1}^{N}\bar{\mathbf{I}}_i\cdot\Omega_i\right] \tag{74}$$

$$\bar{\mathbf{I}} = \overline{\mathbf{A}}\cdot\left[\frac{1}{M}\sum_{i=1}^{N-1}\sum_{j=i+1}^{N}m_im_j\left(r_{ij}^2\mathbf{U} - \mathbf{r}_{ij}\mathbf{r}_{ij}\right) + \sum_{i=1}^{N}\bar{\mathbf{I}}_i\right]\cdot\overline{\mathbf{A}}^T \tag{75}$$

$$I_H = \frac{1}{M}\sum_{i=1}^{N-1}\sum_{j=i+1}^{N} m_i m_j \mathbf{r}_{ij} \cdot \left(\mathbf{U} - \hat{\mathbf{H}}_r \hat{\mathbf{H}}_r\right) \cdot \mathbf{r}_{ij} + \sum_{i=1}^{N} \hat{\mathbf{H}}_r \cdot \bar{\mathbf{I}}_i \cdot \hat{\mathbf{H}}_r \tag{76}$$

where $\hat{\mathbf{H}}_r$ denotes the unit vector $\hat{\mathbf{H}}$ specified in the body-relative frame.

## 4.2 Finite Density Sphere Restriction

This paper focuses on the sphere-restriction of the Full-Body problem, where all of the bodies have finite, constant densities and spherical shapes defined by a diameter $d_i$. This allows for considerable simplification of the mutual potentials, although the rotational kinetic energy, moments of inertia and angular momentum of the systems are still tracked. In this case the moment of inertia of a constant density sphere is $m_i d_i^2/10$ about any axis and the minimum distance between two bodies will be $d_{ij} = (d_i + d_j)/2$. The resultant quantities for these systems are

$$T = \frac{1}{2M}\sum_{i=1}^{N-1}\sum_{j=i+1}^{N} m_i m_j \left(\dot{\mathbf{r}}_{ij} \cdot \dot{\mathbf{r}}_{ij}\right) + \frac{1}{2}\sum_{i=1}^{N} \frac{m_i d_i^2}{10}\omega_i^2 \tag{77}$$

$$T_r = \frac{1}{2M}\sum_{i=1}^{N-1}\sum_{j=i+1}^{N} m_i m_j \left(\mathbf{v}_{ij} \cdot \mathbf{v}_{ij}\right) + \frac{1}{2}\sum_{i=1}^{N} \frac{m_i d_i^2}{10}\Omega_i^2 \tag{78}$$

$$\mathcal{U} = -\mathcal{G}\sum_{i=1}^{N-1}\sum_{j=i+1}^{N} \frac{m_i\, m_j}{|\mathbf{r}_{ij}|} \tag{79}$$

$$\mathbf{H} = \frac{1}{M}\sum_{i=1}^{N-1}\sum_{j=i+1}^{N} m_i m_j \left(\mathbf{r}_{ij} \times \dot{\mathbf{r}}_{ij}\right) + \sum_{i=1}^{N} \frac{m_i d_i^2}{10}\omega_i \tag{80}$$

$$I_H = \frac{1}{M}\sum_{i=1}^{N-1}\sum_{j=i+1}^{N} m_i m_j \left(r_{ij}^2 - (\hat{\mathbf{H}}_r \cdot \mathbf{r}_{ij})^2\right) + \sum_{i=1}^{N} \frac{m_i d_i^2}{10} \tag{81}$$

Although we state several different quantities that are of interest, for our analysis of relative equilibrium and its stability we only need to know the two scalar quantities $I_H$ and $\mathcal{U}$. This discussion is restricted to bodies having equal sizes and densities. Thus, all particles have a common spherical diameter $d$ and mass $m$. Given this restriction the moment of inertia and potential energy take on simpler forms.

$$I_H = \frac{m}{N}\sum_{i=1}^{N-1}\sum_{j=i+1}^{N} r_{ij}^2 + \frac{N}{10}md^2 \tag{82}$$

$$\mathcal{U} = -\mathcal{G}m^2\sum_{i=1}^{N-1}\sum_{j=i+1}^{N} \frac{1}{r_{ij}} \tag{83}$$

where $m$ is the common mass of each body and $d$ the common diameter. For these spherical bodies the first term of $I_H$ denotes the moment of inertia of the point masses, with the assumption that they are all lying in the plane perpendicular to $\mathbf{H}$. The second terms of $I_H$ denotes the rotational moments of inertia of the spheres. We also note that the gravitational potential is unchanged from the $N$-body potential for these spherical bodies. Then $r_{ij} \geq d$ for all of the relative distances.

## 4.3 Normalization

Now introduce some convenient normalizations, scaling the moment of inertia by $md^2$ and scaling the potential energy by $\mathcal{G}m^2/d$. For the moment we use the overline symbol to denote normalization, however we will discard this notation below. Then the minimum energy function is

$$\overline{\mathcal{E}} = \frac{\overline{H}^2}{2\overline{I}_H} + \overline{\mathcal{U}} \tag{84}$$

where

$$\overline{\mathcal{E}} = \frac{\mathcal{E}d}{\mathcal{G}m^2} \tag{85}$$

$$\overline{H}^2 = \frac{H^2}{\mathcal{G}m^3 d} \tag{86}$$

$$\overline{I}_H = \frac{I_H}{md^2} \tag{87}$$

$$= \frac{1}{N}\sum_{i=1}^{N-1}\sum_{j=i+1}^{N} \overline{r}_{ij}^2 + \frac{N}{10} \tag{88}$$

$$\overline{\mathcal{U}} = \frac{\mathcal{U}d}{\mathcal{G}m^2} \tag{89}$$

$$= -\sum_{i=1}^{N-1}\sum_{j=i+1}^{N} \frac{1}{\overline{r}_{ij}} \tag{90}$$

and the constraint from the finite density assumption becomes $\overline{r}_{ij} \geq 1$.

In the following the $\overline{(-)}$ notation is dropped for $r_{ij}$ and $H$, as it will be assumed that all quantities are normalized.

# 5 $N = 2$

We first revisit a specialized version of the derivation for $N = 2$ from [9], only considering equal mass grains. The minimum energy function is explicitly written as

$$\mathscr{E} = \frac{H^2}{2\left(\frac{1}{2}r^2 + 2I_s\right)} - \frac{1}{r} \tag{91}$$

where we keep the term $I_s$ explicitly symbolic to aid in later interpretation. There is only a single degree of freedom in this system, $r$, which is subject to the constraint $r \geq 1$.

## 5.1 *Equilibria and Stability*

For the Full 2-body problem, there are two types of equilibria that occur—contact (or resting) equilibria and non-contact (or orbiting) equilibria. The addition of finite density changes the structure of equilibria in the 2-body problem drastically, and was thoroughly explored previously in [9]. In the following we first discuss the orbiting equilibria, noting that the inclusion of finite density creates up to two orbiting equilibria at the same level of angular momentum—one of which is energetically unstable. Then we discuss the resting equilibria, and note that conditions for them to exist are related to the unstable orbiting equilibria.

### 5.1.1 Non-contact Equilibria

First consider the non-contact equilibrium case. This is a single degree of freedom function so relative equilibria are found by taking a variation with respect to $r$

$$\delta\mathscr{E} = \left[ -\frac{H^2}{2\left(\frac{1}{2}r^2 + 2I_s\right)^2} + \frac{1}{r^3} \right] r\delta r \tag{92}$$

Setting the variation equal to zero yields an equation for $r$:

$$2\left[\frac{1}{2}r^2 + 2I_s\right]^2 - H^2 r^3 = 0 \tag{93}$$

$$r^4 - 2H^2 r^3 + 8I_s r^2 + 16I_s^2 = 0 \tag{94}$$

From the Descartes rule of signs we note that there are either 0 or 2 roots to this equation. We note that both conditions appear [9], thus to find the specific condition for when the roots bifurcate into existence we also consider the condition for a double root of the polynomial, when the derivative of this expression with respect to $r$ is also zero. Taking the derivative of the first expression and solving for $H^2$ yields

$$H^2 = \frac{4}{3r}\left[\frac{1}{2}r^2 + 2I_s\right] \tag{95}$$

Substituting this into the original expression yields an immediate factorization of the condition

$$\left[\frac{1}{2}r^2 + 2I_s\right]\left[2I_s - \frac{1}{6}r^2\right] = 0 \tag{96}$$

The only physical root is the second one, which yields the angular momentum and location where the roots come into existence

$$r^* = 2\sqrt{3I_s} \tag{97}$$

$$= 1.095\ldots \tag{98}$$

$$H^{*2} = \frac{16\sqrt{I_s}}{3\sqrt{3}} \tag{99}$$

$$= 0.9737\ldots \tag{100}$$

There are a few important items to note here. First, as $r^* > 1$ the two roots will bifurcate into existence at a finite separation and thus are both physical. Second, for $H^2 < H^{*2}$ there are no roots for $r > 1$, although this does not mean that there are no minimum energy configurations, just that they are not orbiting configurations.

To develop a more global view of the equilibria, solve for $H^2$ as a function of $r$, yielding

$$H^2 = \frac{2\left(\frac{1}{2}r^2 + 2I_s\right)^2}{r^3} \tag{101}$$

This has a minimum point, corresponding to the double root where the equilibria came into existence. For higher values of $H$ the system will then have two roots, one to the right of $r^*$ (the outer solution) and the other to the left (the inner solution). At the bifurcation point the system technically only has one solution.

The stability of each of these solutions is determined by inspecting the second variation of $\mathcal{E}$ with respect to $r$, evaluated at the relative equilibrium. Taking the variation and making the substitution for $H^2$ from Eq. 101 yields

$$\delta^2\mathcal{E} = \left[\frac{2}{\left(\frac{1}{2}r^2 + I_s\right)} - \frac{3}{r^2}\right]\frac{(\delta r)^2}{r} \tag{102}$$

Checking for when $\delta^2\mathcal{E} > 0$ yields the condition $r^2 > 12I_s$. Thus, the outer solution will always be stable while the inner solution will always be unstable, with the relative

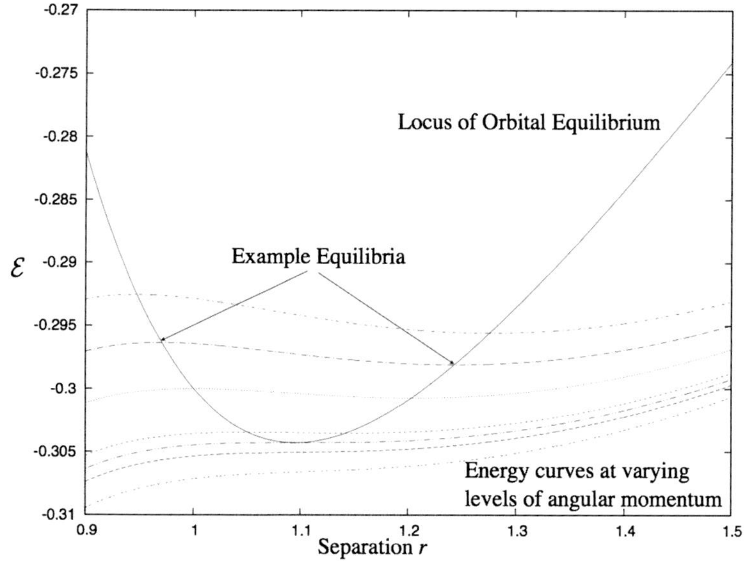


**Fig. 1** Locus of orbital equilibria across a range of angular momentum values. Plotted is the minimum energy function versus the system configuration, $r$, the distance between the two particles. This plot assumes equal masses and sizes of the two particles. For clarity, equilibrium solutions are shown below the physical limit $r \geq 1$

equilibria occurring at a minimum and maximum of the minimum energy function, respectively. Figure 1 shows characteristic energy curves and the locus of equilibria for different levels of angular momentum for the spherical full 2-body problem. This should be contrasted with the energy function for the point mass 2-body problem which only has one relative equilibrium.

For the inner, unstable solution the solution will only exist when $r \geq 1$, which corresponds to angular momentum values from $0.9737\ldots = H^{*2} \leq H^2 \leq 2\left(\frac{1}{2} + 2I_s\right)^2 = 0.98$. This is a somewhat narrow, but finite, interval.

### 5.1.2 Contact Equilibria

There is only one constraint that can be activated for the $N = 2$ case, $r \geq 1$. The condition is $\delta\mathscr{E} \geq 0$, so evaluating this variation at $r = 1$ given that $\delta r \geq 0$ yields

$$\left[-\frac{H^2}{2\left(\frac{1}{2} + 2I_s\right)^2} + 1\right] \geq 0 \tag{103}$$

$$H^2 \leq 2\left(\frac{1}{2} + 2I_s\right)^2 = 0.98 \tag{104}$$

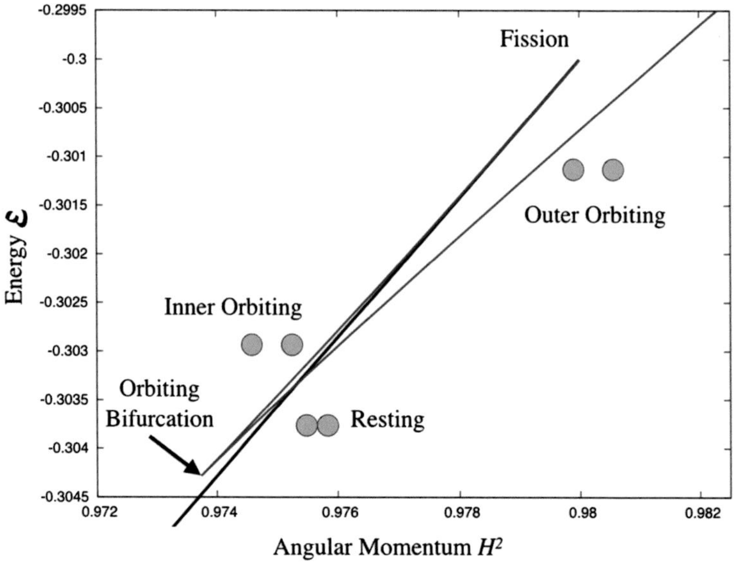


**Fig. 2** The energy-angular momentum diagram shows the relative equilibria in terms of the total angular momentum and corresponding energy

which is precisely the same angular momentum at which the inner orbital solution terminates. Thus the contact relative equilibria exists across $0 \leq H^2 \leq 0.98$, at which point it disappears by colliding with the unstable inner solution. For $H^2 > 0.98$ only the stable, outer solution exists. At the bifurcation point the energy of the system equals $\mathscr{E} = -0.3$. We denote this point as the "fission" point.

The relative equilibrium results are presented in an energy—angular momentum plot (Fig. 2) that plots the energy of the different equilibria as a function of the angular momentum. We note that the resting condition is just linear as a function of $H^2$ as $\mathscr{E} = H^2/1.4 - 1$. The Orbital energy curves are more complex and are plotted by generating the angular momentum and energy as a function of distance. The limiting energy as $r \to \infty$ is $\mathscr{E} \to -1/(2r)$, and thus steadily approaches zero.

## 5.2 *Hill Stability Conditions*

A key constraint can be placed on when the system is Hill stable, meaning that the two bodies cannot escape from each other. While a sufficient condition for escape is difficult to construct in general, the necessary conditions are simple to specify. To do so, assume that the two bodies escape with $r \to \infty$. The minimum energy function then takes on a value $\lim_{r \to \infty} \mathscr{E} = 0$, and the energy inequality remains $0 \leq E$, which is the necessary condition for mutual escape.

If the energy has a value $E < 0$, then the corresponding limit on the distance between the bodies can be developed by solving the implicit equation $\mathscr{E}(r) = E$. Rewriting this we see that it can be expressed as a cubic equation

$$r^3 - \frac{1}{|E|}r^2 + \left(\frac{H^2}{|E|} + 4I_s\right)r - \frac{4I_s}{|E|} = 0 \tag{105}$$

From the Descartes rule of signs we note that this will either have 1 or 3 positive roots. This can be linked to the system having the proper combination of energy and angular momentum. From observation of Fig. 1 it is clear that a line drawn at constant, negative energy will intersect the constant angular momentum lines either once or three times. The outermost intersection delineates the largest separation possible while the lower ones denote additional regions the body can be trapped in. We note that if the intersection occurs at $r < 1$ then the system is in a contact case.

# 6 $N = 3$

For the case where $N = 3$ the minimum energy function is explicitly written as

$$\mathscr{E} = \frac{H^2}{2\left[\frac{1}{3}\left(r_{12}^2 + r_{23}^2 + r_{31}^2\right) + 3I_s\right]} - \left[\frac{1}{r_{12}} + \frac{1}{r_{23}} + \frac{1}{r_{31}}\right] \tag{106}$$

where we again keep the term $I_s = 0.1$ explicitly symbolic to aid in later interpretation. In this problem the function has 3 degrees of freedom, which can be enumerated as $r_{12}$, $r_{23}$ and $r_{31}$ with the constraint that $r_{31} \leq r_{12} + r_{23}$. If this constraint is active, then it is sometimes more convenient to use the angle $\theta_{31}$, defined via the rule of cosines as $r_{31}^2 = r_{12}^2 + r_{23}^2 - 2r_{12}r_{23}\cos\theta_{31}$.

## 6.1 *Equilibria and Stability*

For this case there are 7 unique configurations that can result in a relative equilibrium, which are classified into 9 types of relative equilibrium in [9]. These are shown in Fig. 3 and described below. In [9] the existence and stability of these were proven. Here we provide a summary derivation of this result.

### 6.1.1 Static Resting Configurations

There are two static resting relative equilibrium configurations, the Euler Resting and Lagrange Resting configurations. The Lagrange Resting configuration is ener-

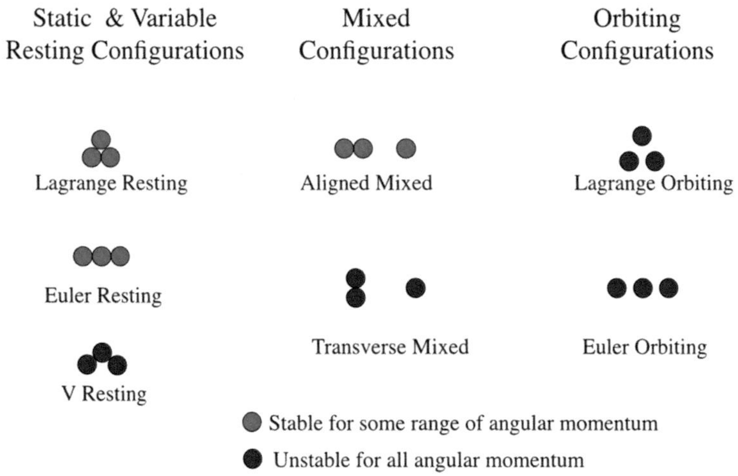


**Fig. 3** Diagram of all relative configurations for the Full 3-body problem. Those *shaded green* can also be stable, and each takes a turn as the global minimum energy configuration for a range of angular momentum

getically stable whenever it exists while the Euler Resting configuration transitions from unstable to stable as a function of angular momentum while it exists.

First consider the static resting configurations, defined as when all of the bodies are in contact and maintain a fixed shape over a range of angular momentum values. When in contact there is only one degree of relative freedom for the system, defined as the angle between the two outer particles as measured relative to the center of the middle particle and shown in Fig. 4. As defined the angle must always lie in the limit $60° \leq \theta \leq 300°$.

Given the geometric relationships in Fig. 4, the minimum energy function can be written as

$$\mathscr{E}_S = \frac{H^2}{2\left[\left(2 + 4\sin^2(\theta/2)\right)/3 + 0.3\right]} - 2 - \frac{1}{2\sin(\theta/2)} \tag{107}$$

where the $S$ subscript stands for "Static." The first variation is then

$$\delta\mathscr{E}_S = \sin\theta\left[-\frac{3H^2}{\left(2.9 + 4\sin^2(\theta/2)\right)^2} + \frac{1}{8\sin^3(\theta/2)}\right]\delta\theta \tag{108}$$

The second variations will be considered on a case-by-case basis. Since this system has a constraint on the angle $\theta$, both the free variations of $\theta$ and the constrained variations when at the limit must be considered.

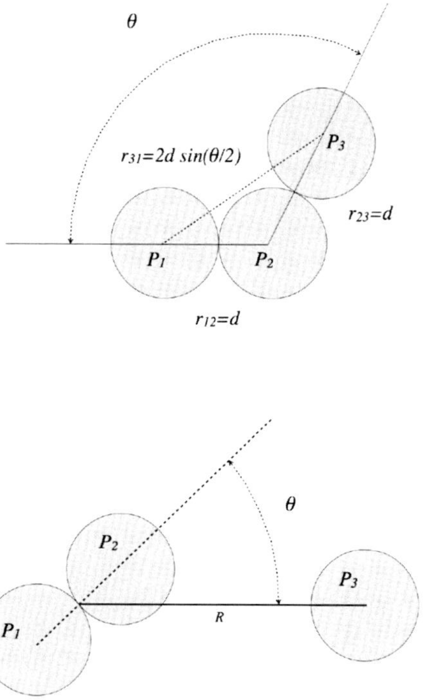


**Fig. 4** Generic description of the planar contact geometry between three equal sized particles (*left*) and the mixed configuration geometry (*right*)

### *Euler Rest Configuration*

If $\theta = \pi$ the minimum energy function will be stationary. Define this as the Euler Rest Configuration, which consists of all three particles lying in a single line. The stability of this configuration is evaluated by taking the second variation of the energy function and evaluating it at $\theta = \pi$, yielding

$$\delta^2 \mathscr{E}_S = -\frac{1}{8}\left[1 - \frac{24}{(6.9)^2}H^2\right]\delta\theta^2 \tag{109}$$

Recall that the stability condition is that the second variation of the energy be positive definite, yielding an explicit condition for stability as $H^2 > 1.98375$, with lower values of $H^2$ being definitely unstable. The Euler Rest Configuration energy can be specified as a function of angular momentum:

$$\mathscr{E}_{SE} = \frac{H^2}{2\,(2.3)} - \frac{5}{2} \tag{110}$$

*Lagrange Rest Configuration*

Now consider the constrained stationary point with $\theta = 60°$ (300°). Define this as the Lagrange Rest Configuration. Here it suffices to evaluate the first variation at the boundary condition, yielding

$$\delta \mathscr{E}_S|_{60^\circ} = \frac{\sqrt{3}}{2}\left[-\frac{3H^2}{(3.9)^2} + 1\right]\delta\theta \tag{111}$$

At the 60° constraint $\delta\theta \geq 0$ and the Lagrange Rest Configuration will exist and be stable for $H^2 < 5.07$, but beyond this limit an increase in $\theta$ will lead to a decrease in energy and the relative equilibrium will no longer exist. Note that if the $\theta = 300°$ limit is taken, the sign of the first variation switches but the constraint surface is now $\delta\theta \leq 0$, and the same results hold.

The Lagrange Rest Configuration energy can be specified as a function of angular momentum:

$$\mathscr{E}_{SL} = \frac{H^2}{2\,(1.3)} - 3 \tag{112}$$

Comparing the energy of these two rest configurations shows that the Lagrange configuration has lower energy for $H^2 < 2.99$ while the Euler configuration has a lower energy above this level of angular momentum.

### 6.1.2 Variable Contact Configurations

In addition to the static resting configurations it is also possible to have full contact configurations which change as the angular momentum varies. These are not fully static, as they depend on having a specific level of angular momentum, generating centripetal accelerations that balance the gravitational and contact forces. For these configurations, as the level of angular momentum varies the configuration itself shifts, adjusting to the new environment. For the $N = 3$ case there is only one such "variable contact" configuration when restricted to the plane. This particular configuration is always unstable, yet plays an important role in mediating the stability of the other configurations.

This equilibrium configuration yields the final way for a stationary value of the minimum energy function to exist, with the terms within the parenthesis of Eq. 108 equaling zero. Instead of solving the resulting quartic equation in $\sin(\theta/2)$ it is simpler to evaluate the angular momentum as a function of the system configuration to find

$$H^2 = \frac{\left(2.9 + 4\sin^2(\theta/2)\right)^2}{24\sin^3(\theta/2)} \tag{113}$$

The range of angular momenta that correspond to this configuration can be traced out by following the degree of freedom $\theta$ over its range of definition. Thus this equilibrium will exist for angular momentum values ranging from $H^2 = 1.98375$ at $\theta = 180°$ to $H^2 = 5.07$ at $\theta = 60°$. Note that the angles progress from $\theta = 180° \rightarrow 60°$ as the angular momentum increases, and that the limiting values occur when the Euler Rest Configuration stabilizes and the Lagrange Rest Configuration destabilizes. Note that a symmetric family moves from $\theta = 180°$ to $\theta = 300°$ at the same levels of angular momentum.

Taking the second variation and evaluating the sign of $\delta^2 \mathscr{E}_C$ along the V configuration for arbitrary variations shows that it is always negative definite over the allowable values of $\theta$ and thus that the V Rest Configuration is always unstable.

The energy of this configuration is found by setting $r_{12} = r_{23} = 1$ and $r_{31} = 2\sin(\theta/2)$. Then we find

$$\mathscr{E} = \frac{3H^2}{2\left(2.9 + 4\sin^2(\theta/2)\right)} - 2 - \frac{1}{2\sin(\theta/2)} \tag{114}$$

Substituting in the angular momentum then yields

$$\mathscr{E} = \frac{\left(2.9 - 4\sin^2(\theta/2)\right)}{16\sin^3(\theta/2)} - 2 \tag{115}$$

### 6.1.3 Mixed Configurations

Now consider mixed configurations where both resting and orbital states can co-exist. For $N = 3$ there is only one fundamental topology of this class allowed, two particles rest on each other and the third orbits. Further, from simple symmetry arguments two candidate states for relative equilibrium can be identified, a Transverse Configuration where the line joining the two resting particles is orthogonal to the third particle ($\theta = \pm 90°$), and an Aligned Configuration where a single line joins all of the mass centers ($\theta = 0°, 180°$). To enable a stability analysis a full configuration description of these systems is introduced which requires two coordinates: the distance from the center of the resting pair to the center of the third particle to be $R$, and the angle between the line $R$ and the line joining the resting pair as $\theta$ (see Fig. 4).

The distances between the different components can be worked out as

$$\begin{aligned} r_{12} &= 1 && \text{(11} \\ r_{23} &= \sqrt{R^2 - R\cos\theta + 0.25} && \text{(1} \\ r_{31} &= \sqrt{R^2 + R\cos\theta + 0.25} && \text{(} \end{aligned}$$

Thus the minimum energy function takes on the form

$$\mathscr{E}_M = \frac{3H^2}{4\left(1.2 + R^2\right)} - 1 - \frac{1}{\sqrt{R^2 - R\cos\theta + 0.25}} - \frac{1}{\sqrt{R^2 + R\cos\theta + 0.25}} \tag{119}$$

where the $M$ stands for "Mixed." Taking the variation with respect to $\theta$ yields

$$\delta_\theta \mathscr{E}_M = \frac{R\sin\theta}{2} \times \left[\frac{1}{\left(R^2 - R\cos\theta + 0.25\right)^{3/2}} - \frac{1}{\left(R^2 + R\cos\theta + 0.25\right)^{3/2}}\right]\delta\theta \tag{120}$$

As expected, the variation is stationary for the Aligned Configuration, $\theta = 0°, 180°$, and for the Transverse Configuration, $\theta = \pm 90°$. The variation in the distance yields

$$\delta_R \mathscr{E}_M = \left[-\frac{3H^2 R}{2(1.2 + R^2)^2} + \frac{2R - \cos\theta}{2\left(R^2 - R\cos\theta + 0.25\right)^{3/2}} + \frac{2R + \cos\theta}{2\left(R^2 + R\cos\theta + 0.25\right)^{3/2}}\right]\delta R \tag{121}$$

and is discussed in the following.

*Transverse Configurations*

First consider the Transverse Configurations with $\theta = \pm 90°$. Evaluating the variation of $\mathscr{E}_M$ with respect to $R$, setting this to zero, and substituting $\theta = 90°$ allows us to ...lve for the angular momentum explicitly as a function of the separation distance

$$H_{MT}^2 = \frac{4(1.2 + R^2)^2}{3\left(R^2 + 0.25\right)^{3/2}} \tag{122}$$

...n has a minimum value of angular momentum of $H_{MT}^2 \sim 4.002\ldots$ which ... $\sqrt{2.6}$. This is an allowable value of separation and thus this bifurcation ...ur. For higher values of angular momentum there are two relative ...th separation less than $\sqrt{2.6}$ and the other with separation larger ...olution touches the other two particles, forming a Lagrange-like ... $= \sqrt{3}/2$. Substituting this into the above equation for $H_{MT}^2$ ...t a value of 5.07, which is precisely the value at which the ...ion becomes unstable. Recall that this was also the value of ...h point the V Rest Configuration terminated by reaching ...h is also equal to the Lagrange Orbit Configuration ...ance, the inner Transverse Configuration family of solutions terminates. Conversely, the outer Transverse Configuration persists for all angular momentum values above the bifurcation level.

Now consider the energetic stability of this class of relative equilibria. First note that the cross partials, $\delta^2_{\theta R}\mathscr{E}_M$ are identically equal to zero for the Transverse Configuration. This can be easily seen by taking partials of Eq. 121 with respect to $\theta$ and inserting the nominal value $\theta = \pm 90°$. Next, taking the second partial of Eq. 121 with respect to $\theta$ and evaluating it at the nominal configuration yields

$$\delta^2_{\theta\theta}\mathscr{E}_{MT} = \frac{-3R^2}{2\left(R^2 + 0.25\right)^{5/2}}(\delta\theta)^2 \tag{123}$$

and $\delta^2_{\theta\theta}\mathscr{E}_{MT} < 0$. It is not necessary to check further as this tells us that none of the Transverse Configurations are energetically stable. The explicit energy of the Transverse Configurations is

$$\mathscr{E}_{MT} = \frac{(0.7 - R^2)}{\left(R^2 + 0.25\right)^{3/2}} - 1 \tag{124}$$

*Aligned Configurations*

Now consider the Aligned Configurations with $\theta = 0°, 180°$. Again solve for the angular momentum as a function of separation

$$H^2_{MA} = \frac{2(1.2 + R^2)^2}{3R}\left[\frac{1}{\left(R - \frac{1}{2}\right)^2} + \frac{1}{\left(R + \frac{1}{2}\right)^2}\right] \tag{125}$$

Finding the minimum point of this equation as a function of $R$ yields a cubic equation in $R^2$ without a simple factorization. Root finding shows that it bifurcates into existence at a distance of $R = 2.33696\ldots$ with a value of $H^2_{MA} = 5.32417\ldots$. Again, there is an inner and an outer solution. The inner solution continues down to a distance of $R = 3/2$, where the two groups touch and form an Euler configuration. The value of the angular momentum at this point equals 6.6125 and equals the value at which the Euler Rest Configuration terminates and the Euler Orbit Configuration is born. The outer solution continues its growth with increasing angular momentum.

Now consider the energetic stability of these solutions. Similar to the Transverse Configurations, the mixed partials of the minimum energy function are identically zero at these relative equilibria. The second partials of Eq. 121 with respect to $\theta$ yields

$$\delta^2_{\theta\theta}\mathscr{E}_{AM} = \frac{R}{2}\left[\frac{1}{(R - 0.5)^3} - \frac{1}{(R + 0.5)^3}\right](\delta\theta)^2 \tag{126}$$

which is always positive. The second partial of Eq. 121 with respect to $R$ is

$$\delta^2_{RR}\mathscr{E}_{AM} = 2\left[\frac{9H^2}{4(1.2+R^2)^3}\left(R^2-0.4\right) - \frac{1}{(R-0.5)^3} - \frac{1}{(R+0.5)^3}\right](\delta R)^2 \quad (127)$$

The resulting polynomial is of high order and is not analyzed. Alternately, inspecting the graph of this function shows that it crosses from negative to positive at the bifurcation point, as expected. Thus the outer Aligned Configurations are energetically stable while the inner Aligned Configurations are unstable, and remain so until they terminate at the Euler configuration. To make a final check, evaluate the asymptotic sign of the second energy variation. For $R \gg 1$, $H^2_{MA} \sim 4/3R$. Substituting this into the above and allowing $R \gg 1$ again yields $\delta^2_{RR}\mathscr{E}_{AM} \sim 1/R^3\delta R^2$, and thus the outer relative equilibria remain stable from their bifurcation onwards.

The explicit energy of the Aligned Configurations are

$$\mathscr{E}_{MA} = \frac{(1.2+R^2)}{2R}\left[\frac{1}{(R-0.5)^2} + \frac{1}{(R+0.5)^2}\right] - 1 - \frac{1}{(R-0.5)} - \frac{1}{(R+0.5)} \quad (128)$$

A direct comparison between $\mathscr{E}_{MA}$ and $\mathscr{E}_{MT}$ at the same levels of angular momentum shows that the Aligned Configurations always have a lower energy than the Transverse Configurations. This is wholly consistent with the energetic stability results found throughout.

### 6.1.4 Purely Orbital Configurations

There are 2 purely orbital relative equilibrium configurations, the Lagrange and Euler configurations. These are always energetically unstable.

Finally consider the purely orbital configurations for this case. As this is the sphere restricted problem, the orbital relative equilibria will be the same as exist for the point mass problem.

*Euler Solution*

For the Euler solution take the configuration where $r_{12} = r_{23} = R$ and $r_{31} = 2R$, $R \geq 1$, reducing the configuration to one degree of freedom. The minimum energy function then simplifies to

$$\mathscr{E}_{OE} = \frac{3H^2}{2\left(6R^2+0.9\right)} - \frac{5}{2R} \quad (129)$$

Taking the variation of the minimum energy function with respect to this configuration then yields

$$\delta_R \mathscr{E}_{OE} = -\frac{18H^2R}{\left(6R^2 + 0.9\right)^2} + \frac{5}{2R^2} \tag{130}$$

Set this equal to zero and solving for the corresponding angular momentum to find

$$H_{OE}^2 = \frac{5}{36}\frac{\left(6R^2 + 0.9\right)^2}{R^3} \tag{131}$$

It can be shown that there are two orbital Euler configurations for $H_{OE}^2 > 8\sqrt{5}/3$ and none for lower values. The non-existence of solutions at a given total angular momentum occurs due to the coupling of the rotational angular momentum of the different bodies. In this case, however, the lower solutions all exist at $R < 1$ and thus are not real for this system. In fact, given the constraint $R \geq 1$ there will be a single family of orbital Euler solutions at $H_{OE}^2 \geq 6.6125$ with corresponding radii ranging from $R = 1 \to \infty$ as $H_{OE}^2 = 6.6125 \to \infty$. The correspond energy of these Euler solutions as a function of $R$ is

$$\mathscr{E}_{OE} = -\frac{5}{24R^3}\left(6R^2 - 0.9\right) \tag{132}$$

Our simple derivation of the orbital Euler solutions only considers one-dimensional variations in the distance. However for a complete stability analysis it is necessary to consider variations of each component in turn. We provide a brief derivation here. First, at this configuration the constraint $r_{31} \leq r_{12} + r_{23}$ is active, meaning that it is better to use the degrees of freedom $r_{12}$, $r_{23}$ and $\theta_{31}$, assuming that body 2 is in the center. We know that the relative equilibrium occurs at $\theta_{31} = \pi$ and $r_{12} = r_{23} = R$ and $H^2$ as above, so we only need to consider the Hessian matrix of $\mathscr{E}$ evaluated at these conditions. The relevant partials are

$$\mathscr{E}_{r_{12}r_{12}} = \mathscr{E}_{r_{23}r_{23}} = \frac{5}{R(2R^2 + 0.3)} - \frac{37}{12R^3} \tag{133}$$

$$\mathscr{E}_{r_{12}r_{23}} = \frac{5}{R(2R^2 + 0.3)} - \frac{3}{2R^3} \tag{134}$$

$$\mathscr{E}_{r_{12}\theta_{31}} = \mathscr{E}_{r_{23}\theta_{31}} = 0 \tag{135}$$

$$\mathscr{E}_{\theta_{31}\theta_{31}} = \frac{7}{24R} \tag{136}$$

As the $\theta$ terms are decoupled and as $\mathscr{E}_{\theta_{31}\theta_{31}} > 0$ we only need to consider the 2-by-2 matrix for the radius variations. For the structure of this matrix, symmetric with equal diagonal components, the eigenvalues can be shown to equal the diagonal plus or minus the off-diagonal terms. Thus the eigenvalues of the Hessian matrix are

$$\mathscr{E}_{r_{12}r_{12}} + \mathscr{E}_{r_{12}r_{23}} = \frac{10}{R(2R^2 + 0.3)} - \frac{55}{12R^3} \tag{137}$$

$$\mathscr{E}_{r_{12}r_{12}} - \mathscr{E}_{r_{12}r_{23}} = -\frac{19}{12R^3} \tag{138}$$

The first eigenvalue can be shown to be positive when $R^2 > 1.65$ and negative for values less than this, while the second is obviously negative for all distances $R$. Thus the Euler orbiting solutions are always unstable.

*Lagrange Solution*

To find the conditions for the Lagrange solution take the configuration to be $r_{12} = r_{23} = r_{31} = R \geq 1$, again reducing the minimum energy function to a single degree of freedom.

$$\mathscr{E}_{OL} = \frac{3H^2}{2\left(3R^2 + 0.9\right)} - \frac{3}{R} \tag{139}$$

The variation now yields the condition

$$3R\left[\frac{1}{R^3} - \frac{3H^2}{\left(3R^2 + 0.9\right)^2}\right] = 0 \tag{140}$$

which can be solved for the angular momentum of the orbital Lagrange solutions as a function of orbit size

$$H_{OL}^2 = \frac{\left(3R^2 + 0.9\right)^2}{3R^3} \tag{141}$$

Again, two solutions exist for $H_{OL}^2 > 16/\sqrt{10}$, however the inner solution has radius $R < 1$ and is not allowed by this model. Thus, again for the constraint $R \geq 1$ there is a single family of Lagrange solution orbits that range from $R = 1 \rightarrow \infty$ as $H_{OL}^2 = 5.07 \rightarrow \infty$. The corresponding energy of these Lagrange solutions as a function of $R$ is

$$\mathscr{E}_{OL} = -\frac{1}{2R^3}\left(3R^2 - 0.9\right) \tag{142}$$

Our simple derivation of the orbital Lagrange solutions only considers one-dimensional variations in the distance, again. As before for a complete stability analysis it is be necessary to consider variations of each component in turn. First, at this configuration we have $r_{31} = r_{12} = r_{23}$ and the inequality constraint is not active, meaning that we can use the radii as the degrees of freedom, which simplifies the evaluation of the Hessian. We know that the relative equilibrium occurs at $r_{12} = r_{23} = r_{31} = R$ and $H^2$ as above, so we only need to consider the Hessian matrix of $\mathscr{E}$ evaluated at these conditions. Indeed, due to the symmetry we have $\mathscr{E}_{r_{12}r_{12}} = \mathscr{E}_{r_{23}r_{23}} = \mathscr{E}_{r_{31}r_{31}}$ and $\mathscr{E}_{r_{12}r_{23}} = \mathscr{E}_{r_{23}r_{31}} = \mathscr{E}_{r_{31}r_{12}}$. The relevant partials are

$$\mathscr{E}_{r_{12}r_{12}} = \frac{4}{3R(R^2+0.3)} - \frac{3}{R^3} \tag{143}$$

$$\mathscr{E}_{r_{12}r_{23}} = \frac{4}{3R(R^2+0.3)} \tag{144}$$

In this case the Hessian again is symmetric and has equal diagonal values, and has equal off-diagonal values. This matrix also has a simple eigenstructure and will have three eigenvalues, a repeated eigenvalue with value $\mathscr{E}_{r_{12}r_{12}} - \mathscr{E}_{r_{12}r_{23}}$ and a single eigenvalue with value $\mathscr{E}_{r_{12}r_{12}} + 2\mathscr{E}_{r_{12}r_{23}}$.

$$\mathscr{E}_{r_{12}r_{12}} - \mathscr{E}_{r_{12}r_{23}} = -\frac{3}{R^3} \tag{145}$$

$$\mathscr{E}_{r_{12}r_{12}} + 2\mathscr{E}_{r_{12}r_{23}} = \frac{4}{R(R^2+0.3)} - \frac{3}{R^3} \tag{146}$$

The first, double eigenvalue is negative making the Hessian negative definite and thus unstable. The second eigenvalue will be positive for $R^2 > 0.9$, making it positive for all possible configurations.

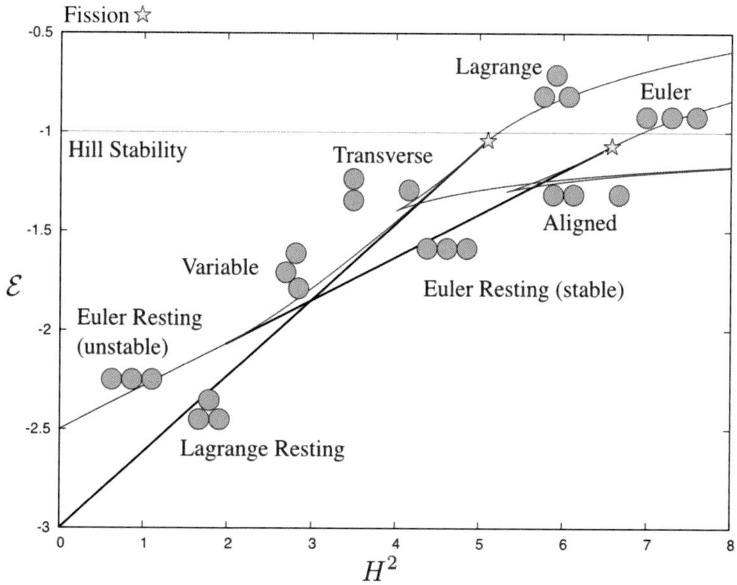


**Fig. 5** Bifurcation diagram showing the energy-angular momentum curves of all stable relative equilibria and their transition paths for increasing and decreasing angular momentum

As a final point, note that the energy of the Euler solutions is actually less than the energy of the Lagrange solutions when $R^2 < 63/60$, i.e., when $R$ is near unity. For larger values of $R$ the Lagrange solution is always lower energy.

The complete bifurcation chart of relative equilibria, minimum energy states, and global minimum energy states of the sphere restricted $N = 3$ full body problem as a function of angular momentum is graphically illustrated in Fig. 5.

## 6.2 Hill Stability Conditions

Now we consider conditions for the system to be bounded, or Hill stable, and unbounded. As with the 2 body case, we can easily derive the conditions for the complete dispersal of the system. If all components escape from each other we have $r_{12}, r_{23}, r_{31} \rightarrow \infty$ and the amended potential then goes to zero again, defining the necessary condition $E \geq 0$. Indeed, if $E < 0$ then there will necessarily be some components that are bound to each other, either resting or orbiting.

In [10] a sharper condition is derived for when a single body can escape from the system. This is summarized here. Assume that body 3 escapes, meaning that $r_{23}, r_{31} \rightarrow \infty$. Then the mutual potential becomes $\mathcal{E} = -\frac{1}{r_{12}} \leq E$. Here, however, we know that $r_{12} \geq 1$, which provides additional constraints, specifically that $-\frac{1}{r_{12}} \geq -1$. From this inequality we see that if $E \leq -1$ that the system is bound and Hill-stable, meaning that none of the bodies can escape. Note from Fig. 5 that all of the resting and mixed configurations exist at energies less than $-1$, meaning that these are Hill stable.

It is instructive to consider the energy of the aligned configuration for large separation distances. Equation 128 when $R \gg 1$ is approximated as

$$\mathcal{E}_{MA} \sim -1 - \frac{1}{R} + \cdots \tag{147}$$

using the fact that $H^2 \sim 4/3R + \cdots$. Thus we see that the aligned configurations are always Hill stable. Consider also the unstable transverse configurations. Again for $R \gg 1$ Eq. 124 can be expanded as

$$\mathcal{E}_{MT} = -1 - \frac{1}{R} + \cdots \tag{148}$$

and has the same asymptotic form. Thus these unstable configurations are also Hill stable asymptotically. Indeed, only the Orbital Lagrange and Euler solutions are Hill Unstable for large enough angular momentum.

# 7 $N = 4$

For the case where $N = 4$ the minimum energy function is explicitly written as

$$\mathscr{E} = \frac{H^2}{2\left(\frac{1}{4}\left(r_{12}^2 + r_{23}^2 + r_{31}^2 + r_{41}^2 + r_{42}^2 + r_{43}^2\right) + 4I_s\right)} - \left[\frac{1}{r_{12}} + \frac{1}{r_{23}} + \frac{1}{r_{31}} + \frac{1}{r_{41}} + \frac{1}{r_{42}} + \frac{1}{r_{43}}\right] \quad (149)$$

where we again keep the term $I_s = 0.1$ explicitly symbolic to aid in later interpretation. There are now a total of 6 degrees of freedom for this system, three more than the previous case. There are several different topologies in which these degrees of freedom can exists, depending on whether there is contact between the bodies or if they are unconstrained.

## 7.1 *Equilibria and Stability*

For the case of $N = 4$ the number of possible configurations grows significantly as compared to the 7 unique configurations identified above for the $N = 3$ body problem. First of all, for orbital configurations the full set of relative equilibria for all mass values is not known and only bounded [3]. However, none of these are expected to be energetically stable and thus can be left out of this analysis [5]. Based on this same premise, and as articulated in the Hypothesis in [9], it can also be surmised that the only energetically stable orbital configurations will have the system collected in

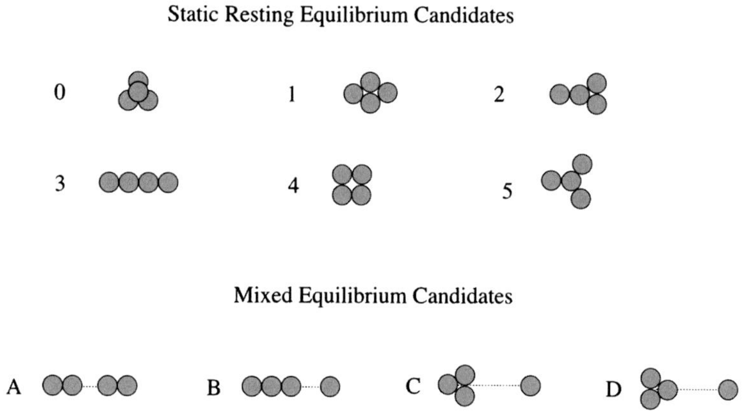


**Fig. 6** Candidate relative equilibria for the $N = 4$ Full-body problem. Many other configurations are expected to exist, however these appear to control the minimum energy landscape

two orbiting clusters, further restricting the space to be considered a priori. Beyond this, one can also rely on principles of symmetry to identify the potential relative equilibrium candidates. An album of possible relative equilibrium configurations for the equal mass and size case is shown in Fig. 6. These candidate configurations were identified by noting symmetries in the configurations but do not preclude the possibility of missed symmetric configurations or asymmetric configurations, which are sure to become more significant as the number of particles increases.

No assertion that all possible relative equilibria have been identified is being made, however the ones listed in Fig. 6 are hypothesized to control the minimum energy configurations. To carry out a detailed analysis would require the development of appropriate configuration variables for the different classes of motion and the formal evaluation of stationary conditions and second variations. This is tractable in general, as the different possible planar configurations of the contact structures can all be described by two degrees of freedom. Some specific examples are given later.

Instead of taking a first principles approach, as was done for the $N = 3$ case in [9] and reviewed above, a number of alternate and simpler approaches to determining the global minimum configurations as a function of angular momentum are developed in the paper [11], and are again summarized here.

### 7.2 *Static Rest Configurations*

Assuming that all of the relevant static rest relative equilibria have been identified, or at least those which may be a global minimum, the global minimum can be found by directly comparing the minimum energy functions of the various configurations as a function of angular momentum. By default the minimum energy state must be stable, independent of a detailed stability analysis. This approach cannot detect when a configuration is energetically stable but not the global minimum. For these static rest structures, the minimum energy function is affine in $H^2$ since the

**Table 1** Table of polar moments of inertia and gravitational energies for each static configuration

| Configuration | $I_i$ | $\mathcal{U}_i$ |
|---|---|---|
| 0 | 1.4 | $-6$ |
| 1 | 2.4 | $-\left[5 + \frac{1}{\sqrt{3}}\right]$ |
| 2 | $2.4 + \frac{\sqrt{3}}{2}$ | $-\left[4 + \frac{2}{\sqrt{2+\sqrt{3}}}\right]$ |
| 3 | 5.4 | $-\left[4 + \frac{1}{3}\right]$ |
| 4 | 2.4 | $-\left[4 + \sqrt{2}\right]$ |
| 5 | 3.4 | $-\left[3 + \sqrt{3}\right]$ |

polar moment of inertia and the potential take on constant values. Thus a graph of $(H^2, \min \sum_{i=1}^{M} H^2/(2I_i) + \mathcal{U}_i)$ will simply delineate the global minimum structures.

Alternately, one can directly determine the moments of inertia and the gravitational potentials of all of the different candidate configurations. Given two configurations, it is then simple to determine the angular momentum at which their minimum energies are equal, and thus where the transitions between these two configurations would occur independent of all other global results. Table 1 presents the computed polar moment of inertia and gravitational potential for each of the static configurations.

Finally, note that configuration "0" is a 3-dimensional configuration, and thus when its moment of inertia $I_H$ is computed a direction for evaluating the moment of inertia about its rotation axis must be defined, however the tetrahedron has a uniform moment of inertia which is equal about any arbitrary axis through its center of mass. All the other configurations lie in a plane with the rotation axis perpendicular to this plane (note that this always yields the maximum moment of inertia and thus minimizes the energy, all else being fixed).

Assume two candidate configurations, $i$ and $j$, then their minimum energy functions are equal for the same value of angular momentum if

$$\frac{H^2}{2I_i} + \mathcal{U}_i = \frac{H^2}{2I_j} + \mathcal{U}_j. \tag{150}$$

The angular momentum at which this occurs can be solved for as

$$H^2 = \frac{2\left(\mathcal{U}_j - \mathcal{U}_i\right) I_i I_j}{I_j - I_i} \tag{151}$$

and represents the angular momentum at which the minimum of the two switch. Inputing the values from Table 1 into this formula generates a transition table for the different static configurations. Table 2 shows the different transitions that occur between the static configurations. Figure 7 graphically shows the energy vs. angular momentum squared plot with the minimum energy configuration taking turns in number from 0 to 3.

With this approach it is not possible to identify the precise transition points or the excess energy when the different states switch, unlike the more detailed analysis that can be given for the $N = 3$ case [9]. Evaluation of these transitions requires that the variable resting configurations be identified, as they mediate the loss and gain of stability for the various resting configurations.

## 7.3 *Mixed Equilibrium Configurations*

To identify the global minimum energy configurations it is also necessary to consider the mixed equilibrium configurations. Each different candidate configuration can be

**Table 2** Transition values of $H^2$ between different static resting configurations, with transitions leading to or from global minima indicated in bold

| $\ddots$ | 0 | 1 | 2 | 3 | 4 | 5 |
|---|---|---|---|---|---|---|
| 0 | $\ddots$ | **2.8402** | 4.7278 | 6.300 | 3.9365 | 6.0354 |
| 1 | **2.8402** | $\ddots$ | **9.813** | 10.748 | × | 13.795 |
| 2 | 4.7278 | **9.813** | $\ddots$ | **11.603** | 6.860 | 50.265 |
| 3 | 6.300 | 10.748 | **11.603** | $\ddots$ | 9.339 | 7.320 |
| 4 | 3.9365 | × | 6.860 | 9.339 | $\ddots$ | 11.133 |
| 5 | 6.0354 | 13.795 | 50.265 | 7.320 | 11.133 | $\ddots$ |

Note that as configurations 1 and 4 have the same moment of inertia, they never cross

analyzed using a single degree of freedom. As an example, Fig. 8 shows configuration D with its single degree of freedom identified.

For any of the mixed configurations the polar moment of inertia and the gravitational potential as a function of the distance between the components can be defined as $d$ and represented as $I(d)$ and $\mathscr{U}(d)$. Then the minimum energy function is $\mathscr{E}(d) = H^2/(2I(d)) + \mathscr{U}(d)$. By definition, relative equilibrium will exist when

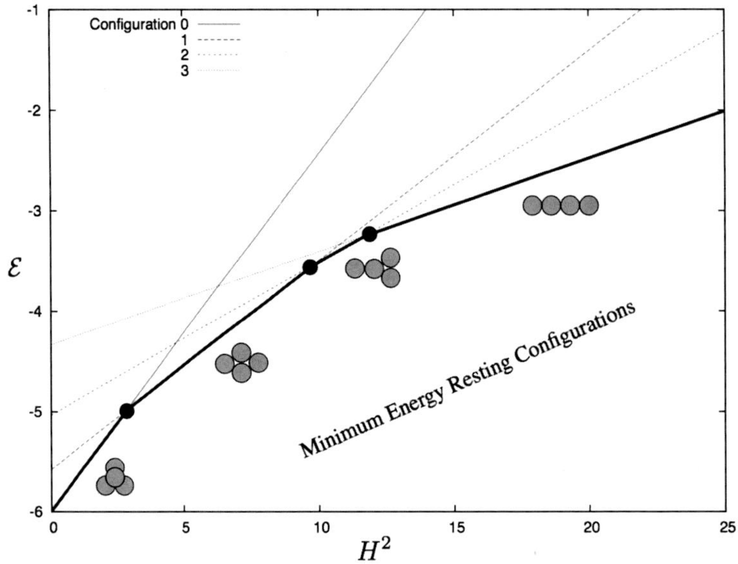


**Fig. 7** Energy-angular momentum graph showing the different transitions between minimum energy static resting states (*left*)

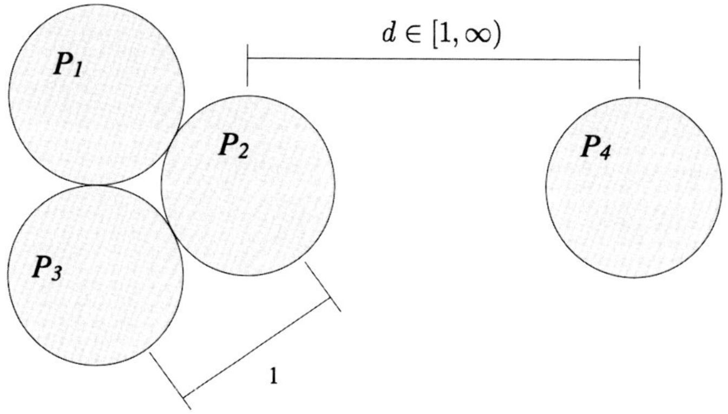


**Fig. 8** Mixed equilibrium configuration candidate D, showing its single degree of freedom, the distance $d$

$\partial \mathscr{E} / \partial d = \mathscr{E}_d = 0$. This is expressed as a function of $I$ and $\mathscr{U}$ as

$$\mathscr{E}_d = -\frac{H^2}{2I^2} I_d + \mathscr{U}_d \tag{152}$$

Equating this to zero solves for the angular momentum for a relative equilibrium as a function of the distance between the components, $d$. The functions $I(d)$ and $\mathscr{U}(d)$ and their partials are listed below for configurations A through D, which are the most relevant to the discussion. In addition to these values the value of $H^2$ is also given when the components are touching, which defines the angular momentum when a given static rest configuration (defined when $d = 1$) fissions into the given mixed equilibrium configuration, and the energy of the system when this occurs.

Configuration A:

$$I_A = \frac{1}{4}\left[6 + d^2 + (d+1)^2 + (d+2)^2\right] + 0.4 \tag{153}$$

$$I_{Ad} = \frac{3}{2}(d+1) \tag{154}$$

$$\mathscr{U}_A = -\left[2.5 + \frac{1}{d} + \frac{1}{d+1} + \frac{1}{d+2}\right] \tag{155}$$

$$\mathscr{U}_{Ad} = \frac{1}{d^2} + \frac{1}{(d+1)^2} + \frac{1}{(d+2)^2} \tag{156}$$

$$H_A^2\big|_{d=1} = 26.46 \tag{157}$$

$$\mathscr{E}_A\big|_{d=1} = -1.88333 \tag{158}$$

Configuration B:

$$I_B = \frac{1}{4}\left[2 + d^2 + 2(d+1)^2 + (d+2)^2\right] + 0.4 \tag{159}$$

$$I_{Bd} = 2(d+1) \tag{160}$$

$$\mathcal{U}_B = -\left[2 + \frac{1}{d} + \frac{2}{d+1} + \frac{1}{d+2}\right] \tag{161}$$

$$\mathcal{U}_{Bd} = \frac{1}{d^2} + \frac{2}{(d+1)^2} + \frac{1}{(d+2)^2} \tag{162}$$

$$H_B^2\Big|_{d=1} = 23.49 \tag{163}$$

$$\mathcal{E}_B|_{d=1} = -2.158333 \tag{164}$$

Configuration C:

$$I_C = \frac{1}{4}\left[3 + 2\left(d'^2 + \frac{1}{4}\right) + \left(d' + \frac{\sqrt{3}}{2}\right)^2\right] + 0.4 \tag{165}$$

$$I_{Cd} = \frac{1}{2}\left(3d' + \frac{\sqrt{3}}{2}\right) \tag{166}$$

$$\mathcal{U}_C = -\left[3 + \frac{2}{\sqrt{d' + \frac{1}{4}}} + \frac{1}{d' + \frac{\sqrt{3}}{2}}\right] \tag{167}$$

$$\mathcal{U}_{Cd} = \frac{2d'}{\left(d'^2 + \frac{1}{4}\right)^{3/2}} + \frac{1}{\left(d' + \frac{\sqrt{3}}{2}\right)^2} \tag{168}$$

$$H_C^2\Big|_{d'=\sqrt{3}/2} = 13.737 \tag{169}$$

$$\mathcal{E}_C|_{d'=\sqrt{3}/2} = -2.7155 \tag{170}$$

Configuration D:

$$I_D = \frac{1}{4}\left[3 + d^2 + 2\left(1 + \sqrt{3}d + d^2\right)\right] + 0.4 \tag{171}$$

$$I_{Dd} = \frac{3}{2}\left(d + \frac{1}{\sqrt{3}}\right) \tag{172}$$

$$\mathcal{U}_D = -\left[3 + \frac{1}{d} + \frac{2}{\sqrt{1 + \sqrt{3}d + d^2}}\right] \tag{173}$$

$$\mathcal{U}_{Dd} = \frac{1}{d^2} + \frac{2d + \sqrt{3}}{\left(1 + \sqrt{3}d + d^2\right)^{3/2}} \tag{174}$$

$$H_D^2\Big|_{d=1} = 13.684 \tag{175}$$

$$\mathcal{E}_D|_{d=1} = -2.9404 \tag{176}$$

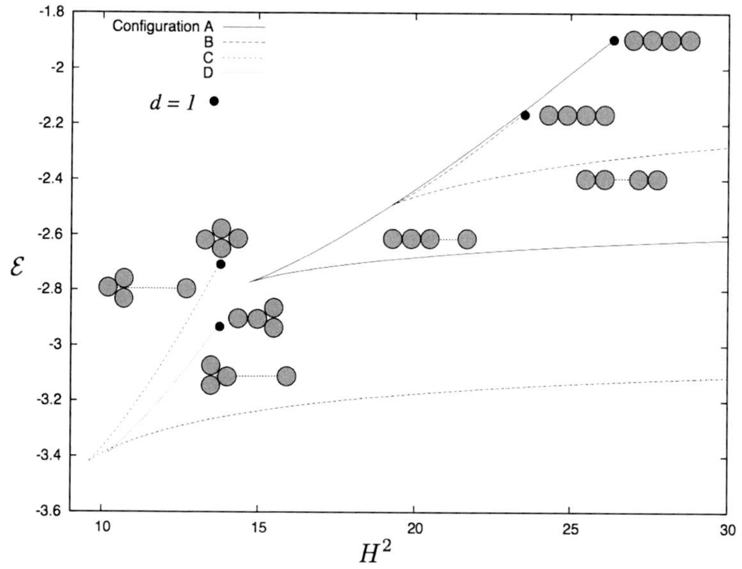


**Fig. 9** Energy versus angular momentum for the orbital relative equilibria $A$ through $D$. The end-points show where these families terminate by touching the static resting configurations. There are portions of these curves where there exist two of the orbital equilibria for a given angular momentum, although for large enough angular momentum there is only one member per family

Note that configuration C has a lower limit of $d = \sqrt{3}/2$. In the following plots the substitution $d' = d - 1 + \sqrt{3}/2$ is made for this configuration, allowing all of the energy functions and angular momenta to be compared across the same range of $d \in [1, \infty)$. When touching, each of these configurations is equivalent to one of the static resting configurations. Specifically, A and B are equivalent to 3, C is equivalent to 1, and D is equivalent to 2. The bifurcation values of $H^2$ and their associated energies and distances can be directly read off of the graphs. Also, their relative energy ordering is apparent (Fig. 9).

## 7.4 Transitions Between Resting and Mixed Configurations

To start to sketch out the more detailed picture of transitions as a function of angular momentum several specific transition states are investigated. Specifically the angular momentum values when static configurations 0, 1, 2 and 3 either become stable, lose stability, or both are found.

**Table 3** Static configurations and the orbital configurations they fission into

| Static configuration | Orbital configuration | Fission $H^2$ | Energy at fission |
|---|---|---|---|
| 1 | C | 13.737 | −2.7155 |
| 2 | D | 13.684 | −2.9404 |
| 3 | B | 23.49 | −2.15833 |

### 7.4.1 Fission Transitions

First, given the results on the mixed relative equilibria the angular momentum values at which the different static configurations no longer exist, i.e., when they fission, can be identified. Static configuration 1 terminates when the inner orbital configuration C collides with it. Similarly static configuration 2 terminates when the inner orbital configuration D collides with it. For static configuration 3, there are two possible configurations it could fission into, A or B. It is interesting to note that configuration A consistently has a lower energy than configuration B, however the static configuration 3 fissions into configuration B at a lower value of angular momentum. Thus, in terms of a sequence of local minimum energy configurations linked geometrically, A is isolated from the static configuration 3. This is discussed in more detail later. In Table 3 the fission conditions and the angular momentum and energy values at which these occur are listed.

### 7.4.2 Stability Transitions

Of additional interest are the stability transitions for the various static configurations, specifically, the range of $H^2$ where they are stable. Of specific interest are the values of angular momentum at which static configuration 0 becomes unstable and at which 1, 2, and 3 become stable. These are dealt with in turn in the following. Figure 10 shows the different degrees of freedom that are considered in the following analysis.

*Stability of Static Configurations 0 and 1*

The stability transitions of configurations 0 and 1 can be treated with the same model, with a single degree of freedom. Starting from the 1 configuration, if the outer two particles are brought up out of the plane the angle between them defines an allowable degree of freedom. In configuration 1 this angle equals 180°, while at the tetrahedron limit it defines an angle $\sin(\theta/2) = 1/\sqrt{3}$, or $\theta = 70.5288\ldots^\circ$. Take the angular momentum vector $\hat{\mathbf{H}}$ to be perpendicular to the plane of configuration 1 and assume that the two outermost particles symmetrically rise out of the plane, then a general description of the distance between these two particles is expressed as $\sqrt{3}\sin(\theta/2)$. At the lower limit of $\theta$ the distance between the particles is unity, forming the tetrahedron, while at 180° the total distance is $\sqrt{3}$. Then the moment of inertial about the normal to the planar direction and the gravitational potential as a function of $\theta$ is

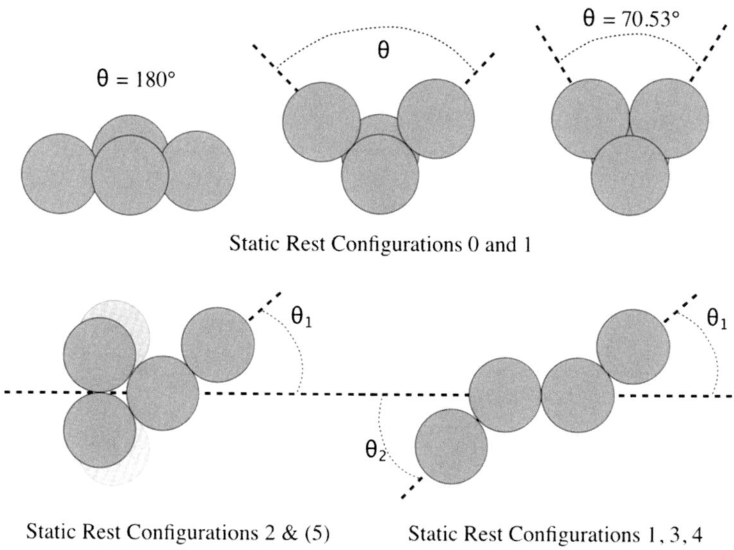


**Fig. 10** Degrees of freedom considered for stability analysis of the various static rest configurations

$$I_H = \frac{1}{2}\left[1 + 3\sin^2(\theta/2)\right] + 0.4 \tag{177}$$

$$\mathscr{U} = -\left[5 + \frac{1}{\sqrt{3}\sin(\theta/2)}\right] \tag{178}$$

The first variation of the minimum energy function with respect to this degree of freedom yields

$$\delta\mathscr{E} = \cos(\theta/2)\left[-\frac{3}{4}\frac{H^2}{I_H^2}\sin(\theta/2) + \frac{1}{2\sqrt{3}\sin^2(\theta/2)}\right]\delta\theta \tag{179}$$

First consider the stability of configuration 0, defined by setting $\sin(\theta/2) = 1/\sqrt{3}$. The condition for this static configuration to exist is that $\delta\mathscr{E} \geq 0$ for $\delta\theta \geq 0$. Then the explicit condition for stability is that the term within the brackets be positive, and for this to hold true the angular momentum $H^2$ must be limited. Setting the values at the configuration 0 values (where $I_H = 1.4$) the stability condition becomes

$$H^2 \leq 2(1.4)^2 = 3.92 \tag{180}$$

At values of angular momentum larger than this the 0 configuration does not exist.

Next setting $\theta = 180°$ yields the stationarity condition that corresponds to configuration 1. To evaluate the stability of this, take the second partial of $\mathscr{E}$ and set $\theta = 180°$ to find

$$\delta^2 \mathscr{E}_{\theta=\pi} = -\frac{1}{2}\left[-\frac{3}{4}\frac{H^2}{I_H^2} + \frac{1}{2\sqrt{3}}\right]\delta\theta^2 \tag{181}$$

where $I_H = 2.4$ now. Solving for when $\delta^2 \mathscr{E} \geq 0$ (note the minus sign in front) places the constraint on the angular momentum to be

$$H^2 \geq \frac{2(2.4)^2}{3\sqrt{3}} = 2.217\ldots \tag{182}$$

at a corresponding energy of $\mathscr{E} = -5.1155\ldots$. Note that configuration 1 becomes stable before configuration 0 goes unstable.

*Stability of Static Configuration 2*

For this configuration suppose that the degree of freedom of interest is the angle defined in the plane from the nominal single contact particle to its location as it rolls on the contact particle. For a positive angle the distance from this particle to the two particles at the far end are

$$d_a = \sqrt{2\left[1 - \cos(150 - \theta)\right]} \tag{183}$$

$$d_b = \sqrt{2\left[1 - \cos(150 + \theta)\right]} \tag{184}$$

Then the moment of inertia and the gravitational potential are found as

$$I_H = 2.4 + \frac{\sqrt{3}}{2}\cos\theta \tag{185}$$

$$\mathscr{U} = -\left[4 + \frac{1}{\sqrt{2(1-\cos(150-\theta))}} + \frac{1}{\sqrt{2(1-\cos(150+\theta))}}\right] \tag{186}$$

Evaluating $\delta\mathscr{E}$, this equals zero for $\theta = 0$, as expected. Evaluating the second variation and evaluating when it is positive yields the condition for stability of configuration 2:

$$H^2 \geq \frac{\left(8 + 6\sqrt{3}\right)\left(2.4 + \sqrt{3}/2\right)^2}{\left[2 + \sqrt{3}\right]^{5.2}} = 7.2913\ldots \tag{187}$$

and has a corresponding energy of $\mathscr{E} = -3.91904\ldots$.

*Stability of Static Configuration 3*

For configuration 3 the two general degrees of freedom allow the end particles to roll relative to the central pair. Define the outer-right body as 1, the outer left body as 2 and measure the angles $\theta_1$ and $\theta_2$ as defined in Fig. 10. The distances of the outer bodies to the furthest of the inner pair is

$$d_i = \sqrt{2(1 + \cos\theta_i)} \tag{188}$$

and the distance of these two bodies from each other is

$$d_{12} = \sqrt{3 + 2(\cos\theta_1 + \cos\theta_2) + 2\cos(\theta_1 + \theta_2)} \tag{189}$$

The moment of inertia and the potential energy are then

$$I_H = \frac{1}{2}\left[5 + 2(\cos\theta_1 + \cos\theta_2) + \cos(\theta_1 + \theta_2)\right] + 0.4 \tag{190}$$

$$\mathscr{U} = -\left[3 + \frac{1}{\sqrt{2(1 + \cos\theta_1)}} + \frac{1}{\sqrt{2(1 + \cos\theta_2)}} + \frac{1}{\sqrt{3 + 2(\cos\theta_1 + \cos\theta_2) + 2\cos(\theta_1 + \theta_2)}}\right] \tag{191}$$

Evaluating $\delta\mathscr{E}$, this equals zero for $\theta_1 = \theta_2 = 0$, again as expected. At this configuration the moment of inertia takes on a value of 5.4.

Evaluating the second variation and evaluating when it is positive yields the following entries that must be put into a matrix:

$$\delta^2_{\theta_i\theta_i}\mathscr{E} = \frac{3}{4}\frac{H^2}{I_H^2} - \frac{1}{8} - \frac{2}{27} \tag{192}$$

$$\delta^2_{\theta_i\theta_j}\mathscr{E} = \frac{1}{4}\frac{H^2}{I_H^2} - \frac{1}{27} \tag{193}$$

The two diagonal entries are equal as are the two off-diagonal entries. To be positive definite the following conditions must hold:

$$\delta^2_{\theta_i\theta_i}\mathscr{E} + \delta^2_{\theta_i\theta_j}\mathscr{E} > 0 \tag{194}$$

$$\delta^2_{\theta_i\theta_i}\mathscr{E} - \delta^2_{\theta_i\theta_j}\mathscr{E} > 0 \tag{195}$$

The controlling condition is the second one, and holds true for $H^2 > 9.45$ and has a corresponding energy of $\mathscr{E} = -3.458333\ldots$.

*Stability of Static Configuration 5*

Finally, it is also of interest to study the stability of static configuration 5. This is similar to configuration 2, except the masses are equally spaced about the central body. This modifies the distances to

$$d_a = \sqrt{2\left[1 - \cos(120 - \theta)\right]} \tag{196}$$

$$d_b = \sqrt{2\left[1 - \cos(120 + \theta)\right]} \tag{197}$$

Repeating the analysis we now find the controlling condition for this configuration to be stable relative to variations in $\theta$ to be

$$H^2 > \frac{10}{3^{3/2}} I_H^2 = 22.247 \tag{198}$$

However, if we analyze the variation $\delta_d \mathscr{E}$ with respect to the distance of the rightmost particle from the rest of the configuration we find that the condition for stability in this direction is

$$H^2 < \left(1 + \frac{1}{\sqrt{3}}\right) I_H^2 = 18.23 \tag{199}$$

Thus this configuration is never stable, as it would undergo fission prior to the angles becoming stabilized. It is important to note that this particular configuration was also found to never be a minimum energy configuration, which is now to be expected given this result.

## 7.5 *Compilation of Possible Relative Equilibria*

Figure 11 shows a global view of the identified stable static relative equilibria and stable and unstable orbital equilibria. Figure 12 shows a detail of this figure on the right indicating some of the transition points. At an angular momentum of zero Configuration 0 is the only stable relative equilibrium. As angular momentum is increased Configuration 1 becomes stable at $H^2 = 2.217\ldots$ and soon thereafter becomes the global minimum configuration. Configuration 0 remains stable until $H^2 = 3.92$ when it ceases to exist. At this point Configuration 1 is the only stable relative equilibrium and remains stable until fission into the unstable Orbital Configuration C occurs at $H^2 = 13.737$. During this evolution both Configurations 2 and 3 become stable. Further, Configuration 2, while the global minimum for an interval, fissions into Orbital Configuration D at $H^2 = 13.684$, before the fission of Configuration C occurs.

At $H^2 = 23.49$ Configuration 3 will fission into the inner, unstable Orbital Configuration B. At this angular momentum level there are four possible stable orbital configurations for the system. The energies of these relative equilibria are, in order, B, A, C, D. For configurations B, C or D the resulting systems will remain stable for arbitrarily large values of angular momentum, which yield low spin rates for the separate components. For Configuration A at high enough angular momentum the Euler rest configuration of the primary will become essentially decoupled from the distant grain and, once the total spin rate becomes slow enough will become unstable. We do not provide a precise calculation of this, but an order of magnitude analysis indicates that this will occur at a value of $H^2 > 30$. It is evident that stepping from $N = 3$ to $N = 4$ particles creates a much richer set of possible outcomes and removes the determinism that was present for the sphere-restricted $N \leq 3$ Full Body problem.

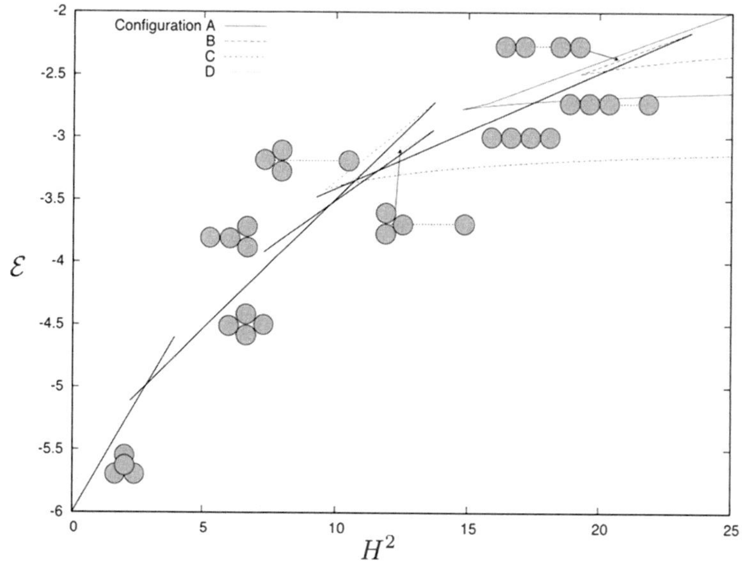


**Fig. 11** Energy versus angular momentum showing the relationship between the static resting configurations and the orbital configurations. The static configurations are only shown for when they exist and are stable. As angular momentum is increased or decreased there are transition points where multiple possible stable states exist

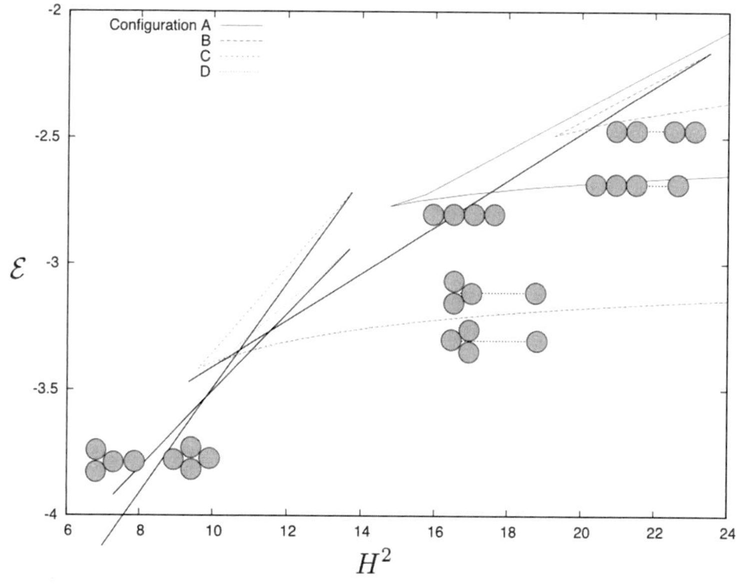


**Fig. 12** Detail of the energy-angular momentum curve

## 7.6 Hill Stability Conditions

Hill stability results can again be developed for this case. As was seen in the 3-body case, different criterion can be developed depending on how many of the bodies become dispersed. Starting with complete dispersal again, we see that a necessary condition for all $r_{ij} \rightarrow \infty$ is that $E \geq 0$. Thus, if the total energy is negative some of the bodies must remain bound to each other orbitally.

Now consider that two of the bodies, say 3 and 4, escape to infinity. Then we have $r_{23}, r_{31}, r_{41}, r_{42}, r_{43} \rightarrow \infty$ and the amended potential becomes $\mathscr{E} \rightarrow -\frac{1}{r_{12}}$. This leads to a necessary condition for this instability to occur as $E \geq -1$, and if $E < -1$ this cannot occur. We note that all of the non-orbital relative equilibrium configurations satisfy this inequality.

A slightly more restrictive case occurs when bodies 3 and 4 escape from bodies 1 and 2, but do not escape from each other. Then the distance $r_{43}$ remains bounded and the amended potential becomes $\mathscr{E} \rightarrow -\left[\frac{1}{r_{12}} + \frac{1}{r_{34}}\right] \geq -2$. Thus if $E < -2$ this form of escape cannot occur. Again we note that all non-orbital relative equilibria satisfy this inequality too.

The final case to consider is when only one body escapes, assume it is body 4. Then $r_{4i} \rightarrow \infty$ and the amended potential approaches $\mathscr{E} \rightarrow -\left[\frac{1}{r_{12}} + \frac{1}{r_{23}} + \frac{1}{r_{31}}\right] \geq -3$. Thus if $E < -3$ the system is Hill stable against any of the bodies escaping. Here the lower limit can be achieved if the three bodies are mutually touching in a resting Lagrange configuration, which is the minimum potential energy configuration of this system. It is relevant to note that in this case there are relative equilibria that exceed this level of energy and thus that they can lose one of their members through escape. All of the resting configurations shown in Fig. 12 that are brought to fission can thus lose one of their component bodies, but only one. This is the first instance where this can occur. We note that fission of configurations 1 and 2 will directly lose a single body, which then can escape from the system. We note that in doing so it must take angular momentum and energy from the spinning contact configuration, and thus will leave it in a slowly spinning configuration (as has been seen in the asteroid pair population [6]). In this aspect we note that configuration 2 fissions just barely above the energy level of $-3$, meaning that the primary would need to essentially stop spinning should escape occur. Another interesting aspect relates to the in-line configuration 3. This configuration fissions into two equal pieces, and is prevented from directly fissioning one body from its end. Thus, even though the body fissions, it is initially in a configuration which cannot escape (as the energy is less than $-2$). Thus, in order to lose the single body, one of the two components must itself split, which has been termed "secondary fission" and has been hypothesized as a geophysical process for rubble pile asteroids [4].

# 8 Summary

This chapter revisits the derivation of the Full $N$-body problem, introducing a new approach to deriving the amended potential and showing its direct utility for computing relative equilibria and evaluating their stability. The paper then reviews the equal mass and density spherical $N$-body problem and presents the detailed relative equilibria for $N = 2, 3, 4$. In addition it derives and discusses the Hill stability for each of these cases, providing sharp conditions for stability for when different numbers of bodies are ejected from the system. Taken together, this chapter provides a restatement of this problem that can be generalized for future studies and collects in one place the significant solutions which already exist to the question of relative equilibria, stability and minimum energy configurations in the Full $N$-body problem.

**Acknowledgments** The author acknowledges support from NASA grant NNX14AL16G from the Near Earth Objects Observation programs and from NASA's SSERVI program (Institute for the Science of Exploration Targets) through institute grant number NNA14AB03A.

# References

1. Cendra H, Marsden JE (2005) Geometric mechanics and the dynamics of asteroid pairs. Dyn. Syst. Int. J. 20:3–21
2. Greenwood, D.T.: Classical Dynamics. Dover (1997)
3. Hampton M, Moeckel R (2006) Finiteness of relative equilibria of the four-body problem. Inventiones Math 163(2):289–312
4. Jacobson SA, Scheeres DJ (2011) Dynamics of rotationally fissioned asteroids: source of observed small asteroid systems. Icarus 214:161–178
5. Moeckel R (1990) On central configurations. Math Z 205(1):499–517
6. Pravec P, Vokrouhlický D, Polishook D, Scheeres DJ, Harris AW, Galád A, Vaduvescu O, Pozo F, Barr A, Longa P et al (2010) Formation of asteroid pairs by rotational fission. Nature 466(7310):1085–1088
7. Scheeres DJ (2002) Stability in the full two-body problem. Celest Mech Dyn Astron 83(1):155–169
8. Scheeres DJ (2006) Relative equilibria for general gravity fields in the sphere-restricted full 2-body problem. Celest Mech Dyn Astron 94:317–349
9. Scheeres DJ (2012) Minimum energy configurations in the n-body problem and the celestial mechanics of granular systems. Celest Mech Dyn Astron 113(3):291–320
10. Scheeres DJ (2014) Hill stability in the full 3-body problem. Proc Int Astron Union 9(S310):134–137
11. Scheeres DJ (2015) Stable and minimum energy configurations in the spherical, equal mass full 4-body problem. In: International conference on computational and experimental engineering and sciences, special symposium on computational methods in celestial mechanics
12. Simo JC, Lewis D, Marsden JE (1991) Stability of relative equilibria. Part I: the reduced energy-momentum method. Arch Ration Mech Anal 115(1):15–59
13. Smale S (1970) Topology and mechanics. I. Inventiones Math 10(4): 305–331
14. Smale S (1970) Topology and mechanics. II. Inventiones Math 11(1): 45–64

# Station Keeping Strategies for a Solar Sail in the Solar System

**Ariadna Farrés and Àngel Jorba**

**Abstract** In this paper we focus on the station keeping around an equilibrium point for a solar sail in the Earth-Sun system. The strategies that we present use the information on the dynamics of the system to derive the required changes on the sail orientation to remain close to an equilibrium point for a long time. We start by describing the main ideas when we consider the RTBP with the effect of the SRP as a model. Then we will see how to extend these ideas when we consider a more complex dynamical model which includes the gravitational attraction of the main bodies in the solar system. One of the goals of the paper is to check the robustness of the algorithms in a more realistic setting and study the effect of errors both in the position determination of the probe and in the orientation of the sail.

**Keywords** Invariant manifolds · Control · Low-thrust · Libration Points

## 1 Introduction

Solar sails are a form of spacecraft propulsion that takes advantage of the Solar radiation pressure (SRP) to propel a probe. The idea is to provide a spacecraft with large ultra-thin mirrors such that the impact, and further reflection, of the photons of the Sun on the mirrors accelerate the probe in a continuous way. Solar sails offer the possibility of low-cost operations, combined with long operating lifetimes. This capability is extremely interesting for long interplanetary transfers, but also offers advantages in Lagrange Point Orbit (LPO) missions, as we can artificially displace equilibria and periodic orbits with an appropriate sail orientation.

A. Farrés (✉) · À. Jorba
Departament de Matemàtica Aplicada i Anàlisi, Universitat de Barcelona,
Gran Via de Les Corts Catalanes 585, 08007 Barcelona, Spain
e-mail: ariadna.farres@maia.ub.es

À. Jorba
e-mail: angel@maia.ub.es



B. Bonnard and M. Chyba (eds.), *Recent Advances in Celestial and Space Mechanics*, Mathematics for Industry 23, DOI 10.1007/978-3-319-27464-5_3

The concept of Solar sailing has already been tested successfully by JAXA in 2010 with their probe IKAROS,[1] NASA with NanoSail-D2[2] in 2011, and recently June 2015 with LightSail[3] by the Planetary Society. These have been test missions to validate the solar sail technology, we still need a complete operational mission to consider solar sailing a reality. One of the main advantages of solar sails is that they open a new range of challenging mission applications that cannot be achieved by a traditional spacecraft [26, 29]. For instance, Robert L. Forward in 1990 proposed to use a solar sail to hover one of the Earth's poles [15]. He proposed to place a solar sail high above the ecliptic plane in such a way that the SRP would counteract the Earth's gravitational attraction. He called it "Statite": the spacecraft that does not move. Nowadays, these ideas are being reconsider in the Pole-Sitter and/or the Polar Observer missions [2, 25]. This mission concept would enable to have constant monitoring of the Polar regions of the Earth for climatology studies.

Another interesting proposal is the so called Geostorm mission [25, 33] now being considered by NASA as the Sunjammer.[4] The goal is to place a solar sail at an equilibrium point closer to the Sun than the Lagrangian point $L_1$ and displaced about 5° from the Earth-Sun line, enabling observations of the Sun's magnetic field having a constant communication with the Earth. This would enable to alert of Geo-magnetic storms, doubling the actual alert time from ACE (the Advanced Composition Explorer[5]) spacecraft, that is now orbiting on a Halo orbit around $L_1$.

Both of these missions require to maintain a solar sail in a fixed location. Nevertheless, all of these equilibria are unstable, hence a station keeping strategy is required to maintain a solar sail close to equilibria for a long time. In previous works [6, 7, 12, 13] we used dynamical systems tools to develop a station keeping strategy for this purpose using as models the Circular and the Elliptical Restricted Three Body Problem. Here we want to see how to extend these ideas when we consider a more complex model for the motion of a solar sail in the Solar system.

The main ideas behind these strategies are: to know the relative position between the sail and the stable and unstable manifolds for a fixed sail orientation, and understand how the manifolds vary when the sail orientation is changed. This information can be used to derive a sequence of changes on the sail orientation that keep the trajectory close to equilibria. We have already tested these algorithms with the GeoStorm and Polar Observer missions [6, 7, 12]. In our simulations we considered the RTBP as a model, including the effect of the solar radiation pressure. We also included random errors on the position and velocity determination as well as on the sail orientation to test the robustness of these algorithms. We found that the most relevant sources of errors (the ones with more impact on the dynamics) are the errors on the sail orientation.

---

[1] http://www.isas.jaxa.jp/e/enterp/missions/ikaros/index.shtml.

[2] http://www.nasa.gov/mission_pages/smallsats/nanosaild.html.

[3] http://sail.planetary.org/.

[4] http://www.sunjammermission.com/AboutSunjammer.

[5] http://www.srl.caltech.edu/ACE/.

Here we want to test the robustness of these strategies when other perturbations are included into the system. To have a more realistic model for the dynamics, one should include the gravitational attraction of the main bodies in the solar system. Another improvement can be introduced by considering a more realistic approximation to the sail performance, taking into account its shape and intrinsic properties.

We have organised this paper as follows, in Sect. 2 we introduce the different dynamical models that we use and how to model the acceleration given by the solar sail. In Sect. 3 we do a review on some of the most relevant dynamical properties of the RTBP when the effect of the solar sail is included. In Sect. 4 we describe the station keeping strategies that we have developed. First, in Sect. 4.1 we explain the main ideas on the station keeping strategy considering the RTBPS as a model, and in Sect. 4.2 how to extend these ideas when we consider a more complete dynamical model. Finally, in Sect. 5 we study the robustness of these strategies for the Sunjammer mission and end up with some conclusions in Sect. 6.

## 2 Dynamical Models

To describe the motion of a solar sail we must include in our model the gravitational attraction of the Sun and the other planets plus the effect of the solar radiation pressure (SRP) on the sail. For the gravitational part we consider two models, the Restricted Three Body Problem (to account for the effect of Earth and Sun) and the $N-$Body problem (to include the effect of the full Solar system not only on the probe, but also on the motion of Earth and Sun). For the effect of the SRP we will assume the sail to be flat and perfectly reflecting (i.e. we include only the reflection of the photons on the surface of the sail).

### *2.1 Restricted Three Body Problem for a Solar Sail*

When we consider the motion of a spacecraft in the Earth's vicinity, one of the classical models in astrodynamics is the Restricted Three Body Problem (RTBP) [32], where we consider the spacecraft as a mass-less particle which is only affected by the gravitational attraction of two major bodies, in our case Earth and Sun. We assume that these two bodies are point masses that move around their mutual centre of mass in a circular way. We must also include the effect of the SRP due to the fact that the spacecraft is propelled by a solar sail. The acceleration given by the solar sail will depend on its performance and orientation, details on how to model this acceleration are given in Sect. 2.3.

We use normalised units of mass, distance and time, so that the total mass of the system is 1, the Earth-Sun distance is 1 and their orbital period is $2\pi$. In these units the gravitational constant is equal to 1, the mass of the Earth is given by $\mu = 3.00348060100486 \times 10^{-6}$, and $1-\mu$ is the mass of the Sun. We take a rotating reference system where the origin is the centre of mass of the Earth-Sun system and such that the Earth and Sun are fixed on the $x$-axis, the $z$-axis is perpendicular to the ecliptic plane and $y$-axis defines an orthogonal positive oriented reference system. In this reference frame the Sun is fixed at $(\mu, 0, 0)$ and the Earth at $(1 - \mu, 0, 0)$.

With these assumptions, the equations of motion in the rotating reference system are:

$$\begin{aligned}
\ddot{x} - 2\dot{y} &= x - (1-\mu)\frac{x-\mu}{r_{ps}^3} - \mu\frac{x-\mu+1}{r_{pe}^3} + a_x, \\
\ddot{y} + 2\dot{x} &= y - \left(\frac{1-\mu}{r_{ps}^3} + \frac{\mu}{r_{pe}^3}\right) + a_y, \\
\ddot{z} &= -\left(\frac{1-\mu}{r_{ps}^3} - \frac{\mu}{r_{pe}^3}\right) z + a_z,
\end{aligned} \tag{1}$$

where $\mathbf{r} = (x, y, z)$ is the position of the solar sail, $r_{ps} = \sqrt{(x-\mu)^2 + y^2 + z^2}$ is the Sun-sail distance, $r_{pe} = \sqrt{(x-\mu+1)^2 + y^2 + z^2}$ is the Earth-sail distance, and $\mathbf{a} = (a_x, a_y, a_z)$ is the acceleration due to the solar sail.

## 2.2 N-Body Problem for a Solar Sail

In the scenario of a real mission we use a more realistic model which includes the gravitational effect of all the planets in the solar system and the Moon. Again the spacecraft is assumed to be a mass-less particle which is affected by the gravitational attraction of all these bodies but does not affect them.

The equations of motion for the solar sail are:

$$\ddot{x} = \sum_{i=0}^{n} Gm_i \frac{x_i - x}{r_{is}^3} + a_x, \quad \ddot{y} = \sum_{i=0}^{n} Gm_i \frac{y_i - y}{r_{is}^3} + a_y, \quad \ddot{z} = \sum_{i=0}^{n} Gm_i \frac{z_i - z}{r_{is}^3} + a_z, \tag{2}$$

where $\mathbf{r} = (x, y, z)$ is the position of the solar sail, $\mathbf{r_i} = (x_i, y_i, z_i)$ are the positions for each of the bodies that we consider and $m_i$ are their masses, $r_{is} = \sqrt{(x_i - x)^2 + (y_i - y)^2 + (z_i - z)^2}$ are the body-sail distances, $G = 6.67428 \times 10^{-11} \mathrm{m^3 kg^{-1} s^{-2}}$ stands for the universal gravitational constant and $\mathbf{a} = (a_x, a_y, a_z)$ is the acceleration given by the solar sail.

To fix notation we consider that the planets are ordered by their distance to the Sun, where $i = 0$ corresponds to the Sun and $i = 1, \ldots, 9$ to the planets from Mercury to Neptune and the Moon. Hence, *0 = Sun, 1 = Mercury, 2 = Venus, 3 = Earth, 4 = Mars, 5 = Jupiter, 6 = Saturn, 7 = Uranus, 8 = Neptune, 9 = Moon*. The position and velocities of the planets and Moon along time will be taken from the DE405

**Table 1** Table with the mass parameters of the different bodies included in the NBP model

| | |
|---|---|
| $Gm_0$ = 2.959122082855911E-04 | $Gm_5$ = 2.825345909524226E-07 |
| $Gm_1$ = 4.912547451450812E-11 | $Gm_6$ = 8.459715185680659E-08 |
| $Gm_2$ = 7.243452486162703E-10 | $Gm_7$ = 1.292024916781969E-08 |
| $Gm_{em}$ = 8.997011346712499E-10 | $Gm_8$ = 1.524358900784276E-08 |
| $Gm_4$ = 9.549535105779258E-11 | $Gm_9$ = 2.188699765425970E-10 |
| $Gm_9$ = $Gm_{em}$/(1+EMRAT)* | $Gm_3$ = $Gm_{em}$EMRAT/(1+EMRAT)* |

Here: 0 = Sun, 1 = Mercury, 2 = Venus, 3 = Earth, 4 = Mars, 5 = Jupiter, 6 = Saturn, 7 = Uranus, 8 = Neptune, 9 = Moon, and *em* stands for the Earth-Moon couple
*EMRAT=0.813005600000000E+02

JPL ephemerides.[6] We will use the same reference frame used in the DE405 JPL ephemerides, which is, equatorial coordinates (J2000) centred at the Solar System barycentre. In Table 1 we give the values of $Gm_i$ used, that have also been taken from the JPL ephemerides.

## 2.3 The Solar Sail

The acceleration given by the sail depends on both, its orientation and efficiency. As a first approach, one can consider only the force due to the reflection of the photons emitted by the Sun on the surface of the sail [27]. For a more realistic model, one should also include the force produced by the absorption of photons by the sail material [1, 5]. The force produced due to reflection, $\mathbf{F_r}$, is directed along the normal direction to the surface of the sail, while the absorption, $\mathbf{F_a}$, is in the direction of the SRP:

$$\mathbf{F_r} = 2PA\langle \mathbf{r_s}, \mathbf{n}\rangle^2 \mathbf{n}, \qquad \mathbf{F_a} = PA\langle \mathbf{r_s}, \mathbf{n}\rangle \mathbf{r_s}.$$

where, $P = P_0(R_0/R)^2$ is the SRP magnitude at a distance $R$ from the Sun ($P_0 = 4.563\,\text{N/m}^2$, the SRP magnitude at $R_0 = 1\text{AU}$), $A$ is the area of the solar sail, $\mathbf{r_s}$ is the direction of SRP and $\mathbf{n}$ is the normal direction to the surface of the sail.

If we denote by $a$ the absorption coefficient and by $\rho$ the reflectivity coefficient, we have $a + \rho = 1$. Hence, the solar sail acceleration in this simplified non-perfectly reflecting model SNPR [5] is given by:

$$\mathbf{a} = \frac{2PA}{m}\langle \mathbf{r_s}, \mathbf{n}\rangle \left(\rho\langle \mathbf{r_s}, \mathbf{n}\rangle \mathbf{n} + 0.5(1-\rho)\mathbf{r_s}\right). \tag{3}$$

Notice that, $\rho = 1$ corresponds to a perfectly reflecting solar sail, and $\rho = 0$ to a perfect solar panel where the absorption of the panels is the only effect. According

[6]DE405 JPL ephemerides: http://ssd.jpl.nasa.gov/?ephemerides#planets.

**Table 2** Values of the sail lightness number $\beta$ for different sail missions according to data from https://directory.eoportal.org/web/eoportal/satellite-missions/

| Mission | $m$ (kg) | $A$ (m$^2$) | $\sigma = m/A$ | $\beta$ |
|---|---|---|---|---|
| IKAROS | 307 | $14 \times 14$ | 1530.61 | $\sim 0.001$ |
| NanoSail-D2 | 4 | 10 | 400 | $\sim 0.00385$ |
| LightSail | 31 | 32 | 968.75 | $\sim 0.00158$ |
| Sunjammer | 32 | $38 \times 38$ | 22.16 | $\sim 0.069$ |

to [5] a solar sail with a highly reflective aluminium-coated side has an estimated value of $\rho \approx 0.88$.

As the SRP is proportional to the inverse square of the distance to the Sun, it is common to write its effect as a correction of the Sun's gravitational attraction:

$$\mathbf{a} = \beta \frac{Gm_s}{r_{ps}^2} \langle \mathbf{r_s}, \mathbf{n} \rangle \left( \rho \langle \mathbf{r_s}, \mathbf{n} \rangle \mathbf{n} + 0.5(1-\rho)\mathbf{r_s} \right), \tag{4}$$

where $G$ is the universal gravitational constant, $m_s$ is the mass of the Sun and $r_{ps}$ is the Sun-sail distance, and $\beta$ is a constant, defined as the *sail lightness number* which accounts for the effectiveness of the solar sail.

Here,

$$\beta = \sigma^*/\sigma, \qquad \sigma = m/A \qquad \text{and} \qquad \sigma^* = \frac{2P_0R_0^2}{Gm_s} = 1.53\text{g/m}^2,$$

where $\sigma$ is the area-to-mass ratio of the solar sail. The values for the sail lightness number that are being considered for the Sunjammer mission are between 0.0388 – 0.0455 [19]. For comparison in Table 2 we show the values of IKAROS,[7] NanoSail-D,[8] and LightSail.[9]

The sail orientation is given by the normal vector to the solar sail, $\mathbf{n}$, which can be parametrised by two angles, $\alpha$ and $\delta$, that can be defined in many ways [23, 27, 30]. We have chosen to relate the angles $\alpha$ and $\delta$ to the horizontal and vertical displacement of the normal direction, $\mathbf{n}$, with respect to the Sun-sail line, $\mathbf{r_s}$. In other words, $\alpha$ is the angle between the projection of $\mathbf{r_s}$, and $\mathbf{n}$, on the ecliptic plane; and $\delta$ is the angle between the projection $\mathbf{r_s}$ and $\mathbf{n}$, on the $y = 0$ plane (see Fig. 1).

If we consider $(x, y, z)$ to be the position of the solar sail and $(x_0, y_0, z_0)$ the position of the Sun, then it is clear that $\mathbf{r_s} = (x - x_0, y - y_0, z - z_0)/||\mathbf{r_s}||$. In spherical

[7] http://www.jspec.jaxa.jp/e/activity/ikaros.html.

[8] http://www.nasa.gov/mission_pages/smallsats/nanosaild.html.

[9] http://sail.planetary.org.

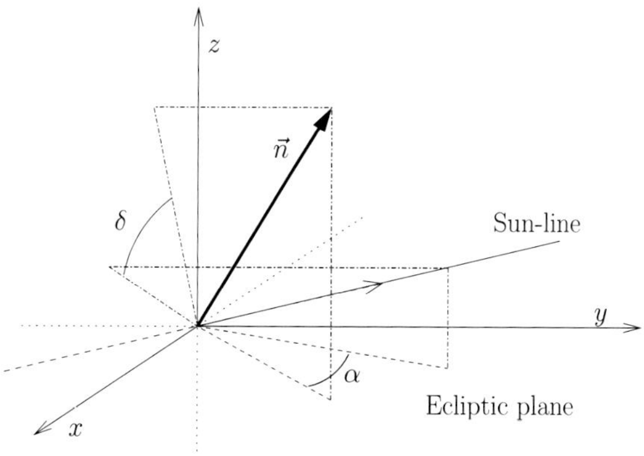


**Fig. 1** Schematic representation of the two angles that define the sail orientation: $\alpha$ is the angle between the projection of $\mathbf{r}_s$ and $\mathbf{n}$ on the ecliptic plane, and $\delta$ the angle between them on the $y = 0$ plane

coordinates we have that $\mathbf{r_s} = (\cos\phi(x,y)\cos\psi(x,y,z),\ \sin\phi(x,y)\cos\psi(x,y,z),\ \sin\psi(x,y,z))$, where

$$\phi(x,y) = \arctan\left(\frac{y-y_0}{x-x_0}\right), \qquad \psi(x,y,z) = \arctan\left(\frac{z-z_0}{\sqrt{(x-x_0)^2+(y-y_0)^2}}\right).$$

Following the definitions given above for $\alpha$ and $\delta$ we have that $\mathbf{n} = (n_x, n_y, n_z)$ is:

$$\begin{aligned} n_x &= \cos(\phi(x,y)+\alpha)\cos(\psi(x,y,z)+\delta),\\ n_y &= \sin(\phi(x,y)+\alpha)\cos(\psi(x,y,z)+\delta),\\ n_z &= \sin(\psi(x,y,z)+\delta), \end{aligned}$$

which can be rewritten as:

$$\begin{aligned} n_x =& \frac{x-x_0}{r_{ps}}\cos\alpha\cos\delta - \frac{(x-x_0)(z-z_0)}{r_2 r_{ps}}\cos\alpha\sin\delta - \frac{y-y_0}{r_{ps}}\sin\alpha\cos\delta\\ &+ \frac{(y-y_0)(z-z_0)}{r_2 r_{ps}}\sin\alpha\sin\delta,\\ n_y =& \frac{y-y_0}{r_{ps}}\cos\alpha\cos\delta - \frac{(y-y_0)(z-z_0)}{r_2 r_{ps}}\cos\alpha\sin\delta + \frac{x-x_0}{r_{ps}}\sin\alpha\cos\delta\\ &- \frac{(x-x_0)(z-z_0)}{r_2 r_{ps}}\sin\alpha\sin\delta,\\ n_z =& \frac{z-z_0}{r_{ps}}\cos\delta + \frac{r_2}{r_{ps}}\sin\delta, \end{aligned}$$

where $r_2 = \sqrt{(x-x_0)^2+(y-y_0)^2}$ and $r_{ps} = \sqrt{(x-x_0)^2+(y-y_0)^2+(z-z_0)^2}$.

# 3 Background on the RTBPS

In this section we want to give a quick overview on some of the phase space properties of the RTBPS. We will describe some of the interesting invariant objects that appear in the system, such as equilibrium points and periodic orbits. These objects are of interest for mission applications and will be our targets to test the station keeping strategies.

## *3.1 Equilibrium Points*

It is well-known that, when the radiation pressure is discarded, the RTBP has five equilibrium points: three of them ($L_{1,2,3}$) are on the axis joining the two primaries and their linear dynamics is of the type centre $\times$ centre $\times$ saddle; the other two ($L_{4,5}$) lie on the ecliptic plane forming an equilateral triangle with the two primaries and their linear dynamics totally elliptic (centre $\times$ centre $\times$ centre) if $\mu$ is below the critical Routh value $\mu_R = \frac{1}{2}\left(1 - \frac{\sqrt{69}}{9}\right) \approx 0.03852$ [32].

If we consider the sail to be perpendicular to the Sun-sail line ($\alpha = \delta = 0$), we have a similar phase portrait as in the RTBP. Notice that we are essentially changing the attracting force of the Sun on the sail (but not on the Earth). This system is still Hamiltonian and has 5 equilibrium points, $SL_{1,\dots,5}$, which are closer to the Sun than the classical $L_{1,\dots,5}$. The dynamics around these displaced equilibria ($SL_{1,\dots,5}$) is qualitatively the same as the one around their "brothers" $L_{1,\dots,5}$ (i.e. $SL_{1,2,3}$ are centre $\times$ centre $\times$ saddle while $SL_{4,5}$ are centre $\times$ centre $\times$ centre).

For a fixed sail lightness number, $\beta$, we can artificially displace these equilibria by changing the sail orientation, having a 2D family of equilibrium points parameterised by the two angles, $\alpha$, $\delta$, that define the sail orientation [27, 28, 30]. In Fig. 2 we show two slices of these families for $\beta = 0.01, 0.02, 0.03$ and $0.04$.

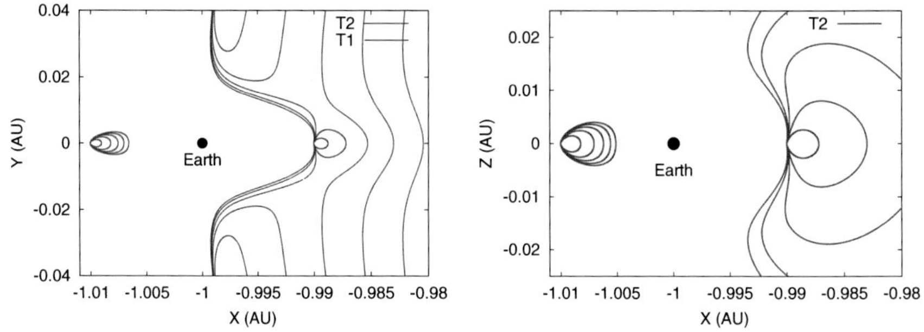


**Fig. 2** Position of the family of "artificial" equilibria close to $L_1$ and $L_2$ for $\beta = 0.01, 0.02, 0.03$ and $0.04$. The *blue points* correspond to class $T_1$ instability, while *red points* to class $T_2$ instability. *Right* Fixed points for $Z = 0$. *Left* Fixed points for $Y = 0$

Most of these “artificial” equilibria are linearly unstable [28]. We note that if the sail is not perpendicular to the Sun-sail line, i.e. $\alpha, \delta \neq 0$, the RTBPS is no-longer Hamiltonian. Hence, the eigenvalues of the differential of the flow at equilibria will not have the Hamiltonian restrictions. We can distinguish two kind of linear behaviours around the equilibria. Class $T_1$, where there are 3 pair of complex eigenvalues $\nu_{1,2,3} \pm i\omega_{1,2,3}$; and class $T_2$ where there are 2 pair of complex eigenvalues $\nu_{2,3} \pm i\omega_{2,3}$ and a pair of real eigenvalues $\lambda_1 > 0, \lambda_2 < 0$. In general $|\nu_i|$ is small, hence we can say that the points of class $T_1$ are practically stable as trajectories will take long time to escape from a close vicinity of equilibria [6, 12], and the instability of the class $T_2$ equilibria is given by the saddle.

These “artificial” equilibria, due to their interesting location, open a wide new range of possible mission applications that cannot be achieved by a traditional spacecraft. Two examples are the *Geostorm Warning Mission* [24, 33] (now renamed as Sunjammer [19]), and the *Polar-Sitter Mission* [2, 25]. The Geostorm mission places a sail around an equilibrium point between the Sun and the Earth, closer to the Sun than $L_1$ and shifted 5° from the Earth-Sun line, making observations of the Sun geomagnetic activity while keeping a constant communication with the Earth (Fig. 3 top). On the other hand, the Pole Sitter mission aims to place a sail at a fixed point high above the ecliptic plane, being able to constantly observe one of the Earth Poles (Fig. 3 bottom).

The suitable equilibrium points for these two missions are unstable and of class $T_2$, so station keeping manoeuvres must be done to remain close to them. The station keeping strategies that we describe in Sect. 4 are specific for class $T_2$ equilibria.

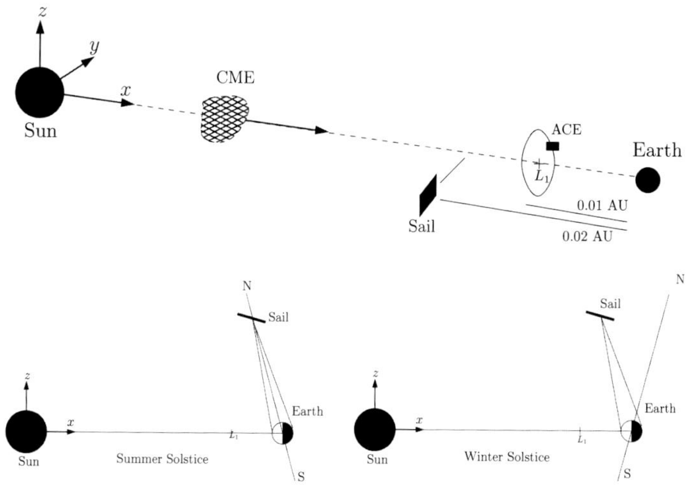


**Fig. 3** Schematic representation of the solar sail relative position for the Geostorm/Sunjammer mission (*top*) and the Polar Observer mission (*bottom*)

## 3.2 Periodic Orbits

To find periodic and quasi-periodic motion around the equilibrium points in the RTBPS we must restrict to the case $\alpha = 0$ and $\delta \in [-\pi/2, \pi/2]$ (i.e. the sail orientation only varies vertically w.r.t. the Sun-sail line direction). Now the system is time-reversible by the symmetry $R : (t, X, Y, Z, \dot{X}, \dot{Y}, \dot{Z}) \rightarrow (-t, X, -Y, Z, -\dot{X}, \dot{Y}, -\dot{Z})$, which means that under certain constraints the flow will behave locally as a Hamiltonian system [22, 31]. This is not the case for $\alpha \neq 0$, where further studies on the non-linear dynamics around the "artificial equilibria" should be done to see if some periodic and quasi-periodic motions persist.

When $\alpha = 0$, we have five 1D family of equilibria parametrised by $\delta$. Three of these families lie on the $Y = 0$ plane and are related to the classical $L_{1,2,3}$ Lagrange points, the linear dynamics around these equilibria is centre $\times$ centre $\times$ saddle. The reversible character of the system ensures the existence of periodic and quasi-periodic motion around them. More concretely, around each equilibrium point there exist two continuous families of Lyapunov periodic orbits, each one related to one of the oscillations of the linear part. The coupling between these two oscillations gives rise to a Cantor family of invariant tori [10].

For $\delta = 0$, one of the families of periodic orbits emanating from the equilibrium point, $p_0$, are totally contained in the $Z = 0$ plane, and are centre $\times$ saddle. At a certain point, a pitchfork bifurcation takes place and two new families periodic orbits are born. These orbits are the Halo orbits for a solar sail when the sail is perpendicular to the Sun-sail line (Fig. 4 left). The Halo orbits inherit the centre $\times$ saddle behaviour and, the rest of the orbits on the $Z = 0$ plane become saddle $\times$ saddle. The other family of periodic orbits are similar to the vertical Lyapunov orbits having a bow tie shape (Fig. 5 left). These orbits are centre $\times$ saddle and do not suffer any bifurcation for energies close to $p_0$.

For $\delta \neq 0$, the family of periodic orbits emanating from the equilibrium point $p_1$ are no longer contained in the $Z = 0$ plane. But one of the two families is almost planar for $\delta$ small, and the orbits are also centre $\times$ saddle. Due to the symmetry breaking of the system for $\delta \neq 0$ [4], there is no longer a pitchfork bifurcation, and

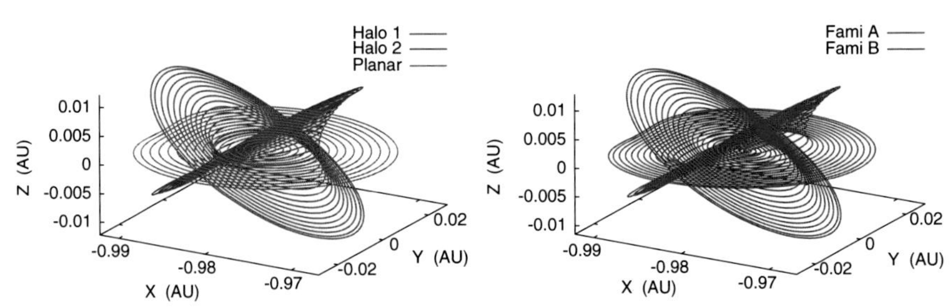


**Fig. 4** Projections of the $X, Y, Z$ plane of the P-Lyapunov family of periodic orbits and related Halo-type orbits close to $SL_1$ for $\beta = 0.05$ and $\delta = 0$ rad (*left*), $\delta = 0.01$ rad (*right*)

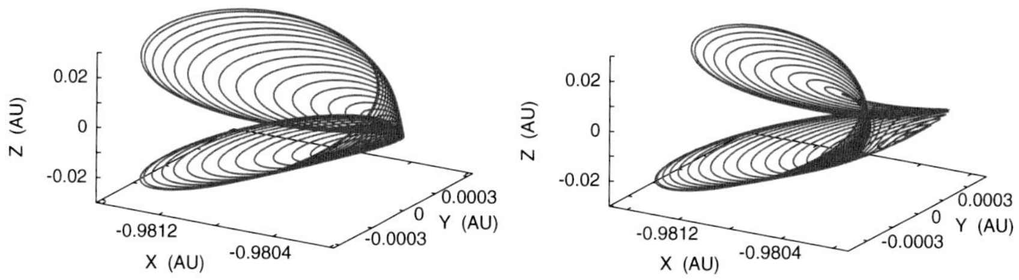


**Fig. 5** Projections on the $X, Y, Z$ plane of the V-Lyapunov family of periodic orbits close to $SL_1$ for $\delta = 0$ red (*left*) and $\delta = 0.01$ (*right*)

the two branches defining the Halo orbits split. We can still find families of Halo-type orbits which are centre $\times$ saddle and almost planar orbits that are saddle $\times$ saddle (Fig. 4 right). The vertical family of periodic orbits also suffers some changes, the orbits still have a bow tie shape but the loops are no longer symmetric (Fig. 5 right).

We note that for $\delta$ small, one of the two complex eigendirections has a much wider vertical oscillation than the other. To fix a criteria we call the P-*Lyapunov family* to the family of periodic orbits emanating from an equilibrium point $p_0$ whose planar oscillation is wider that the vertical one and V-*Lyapunov family* to the other family of periodic orbits. In Fig. 4 we show the $X, Y, Z$ projections of the *P-Lyapunov family* and the associated Halo-type orbits for $\delta = 0$ (left) and $\delta = 0.01$ (right). As we can see that the qualitative behaviour of the phase space does not vary much for $\alpha = 0$ and $\delta$ small. In Fig. 5 we have the $X, Y, Z$ projections of the V-*Lyapunov family* for $\delta = 0$ (left) and $\delta = 0.01$ (right).

Halo orbits around $SL_1$ are of interest for missions within the philosophy of the Geostorm mission [19]. On the other hand, the vertical Lyapunov orbits around $SL_2$ have been proposed for the Pole Sitter mission by Ceriotti and McInnes [3], as for certain values of $\beta$ these orbits spend some time above and below the Earths' poles.

# 4 Station Keeping Strategy

In previous papers [7–9] we discussed how to derive station keeping strategies around unstable equilibria and periodic orbits in the circular RTBP using dynamical system tools. We also tested them and discussed their robustness when different sources of errors were included in the simulations (both on position and velocity determination and on the sail orientation).

In this paper we want to check the robustness of these strategies when other perturbations are added, such as the fact that the two primaries (Sun and Earth) actually orbit around their centre of mass in an elliptic way and that the solar sail is also affected by the gravitational attraction of the Moon and the other planets in the Solar system. These perturbations are small, but could compromise a long-term mission if they are not taken into account. In this paper we want to explain how to adapt our strategies to a more realistic model.

We will mainly focus on the station keeping around equilibrium points, but they can easily be extended to deal with unstable periodic orbits as in [13, 14]. In Sect. 4.1 we will start by reviewing the ideas behind our station keeping strategies in the RTBPS. Next, in Sect. 4.2, we will show how to extend these strategies for a more complex dynamical model, and test their robustness for the Sunjammer mission.

## 4.1 Station Keeping in the RTBPS

The station keeping strategies we propose in previous works [7, 12–14] takes advantage of the dynamical properties of the system to control a solar sail. We do not use optimal control theory algorithms, but rather dynamical system tools for our purpose. As the propellant of a solar sail is, in principal, unlimited and our goal is to remain close to an equilibrium point or a periodic orbit, there is apparently, no cost function to minimise. In other words, what we do is understand the geometry of the phase space and how this one is affected by variations on the sail orientation. Then use this information to derive a strategy that allows us to remain close to an equilibrium point or a periodic orbit for a long time. Most of the ideas behind our approach are based on the previous works by Gómez et al. [16, 18] on the station keeping around Halo orbits using a classical chemical thruster.

### 4.1.1 Ideas Behind the Station Keeping Strategies

Let us start by focusing on the dynamics close to an equilibrium point. In Sect. 3.1 we saw that the potentially interesting equilibria are unstable and the linear dynamics is a cross product between one saddle and either two sources, two sinks or one of each (i.e. the eigenvalues of the differential of the flow are: $\lambda_1 > 0$, $\lambda_2 < 0$ and real, $\lambda_3 = \nu_1 + i\omega_1$, $\lambda_4 = \bar{\lambda}_3$ and $\lambda_5 = \nu_2 + i\omega_2$, $\lambda_6 = \bar{\lambda}_5$). As the instability given by the sources and the dissipation due to the sinks are very small compared to the saddle ($|\lambda_1| \gg |\nu_1|, |\nu_2|$), to describe the dynamics we will assume that the linear dynamics is given by the cross product of one saddle and two centres (see Fig. 6).

This means that for an equilibrium point $p_0$ with a fixed sail orientation $(\alpha_0, \delta_0)$, a trajectory starting close to $p_0$ escapes along the unstable manifold, $\mathscr{W}_u(p_0)$, while rotating around the centre projections. If we change slightly the sail orientation $\alpha_1 = \alpha_0 + \Delta\alpha$, $\delta_1 = \delta_0 + \Delta\delta$, the qualitative phase space behaviour is the same, but the relative position of the new equilibria $p_1$ (and its stable and unstable manifolds which dominate the dynamics) is shifted. Hence, the trajectory escapes along the new unstable invariant manifold $\mathscr{W}_u(p_1)$ [6, 7].

In order to control the sail's trajectory we need to find a new sail orientation, such that $\mathscr{W}_u(p_1)$ brings the trajectory close to the stable manifold of $p_0$, $\mathscr{W}_s(p_0)$. Once the trajectory is close to $\mathscr{W}_s(p_0)$ we restore the sail orientation to $(\alpha_0, \delta_0)$ and let

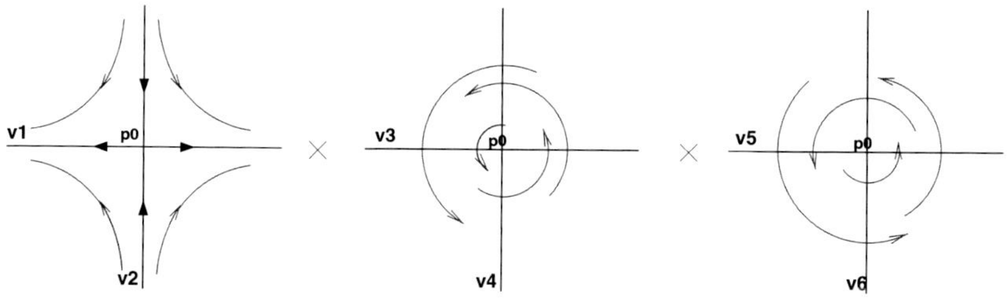


**Fig. 6** Schematic representation of the linear dynamics and the trajectory of the sail around an equilibrium point of type saddle × centre × centre

the natural dynamics act. We will repeat this process over and over to control the instability due to the saddle during the missions life-time. However, we must also take into account the centre projection of the sail's trajectory. Where a sequence of changes in the sail orientation derive in a sequence of rotations around different equilibrium points, which can become unbounded (see Fig. 7 for a schematic representation of these phenomena).

In the case of a periodic orbit things work in a very similar way. In Sect. 3.2 we have seen that for certain sail orientations ($\alpha_0 = 0, \delta_0$) there are planar and vertical Lyapunov orbits, and Halo orbits. Most of these orbits are unstable, and the linear dynamics is the cross product of a saddle, a centre and a neutral direction corresponding to the fact that the orbits come in a 1-parametric family. If $P_0(t)$ is a periodic orbit for a fixed sail orientation, when we are close to the orbit the trajectory escapes along the unstable manifold $\mathscr{W}_u(P_0(t))$. If we change slightly the sail orientation $\alpha_1 = \alpha_0 + \Delta\alpha$, $\delta_1 = \delta_0 + \Delta\delta$, although there might not be a new periodic orbit for this set of parameters, the instability of the system remains, and the trajectory escapes along a new "unstable manifold" $\mathscr{W}_u^1$. We want to find a new sail orientation that brings the sail close to the stable manifold of the target periodic

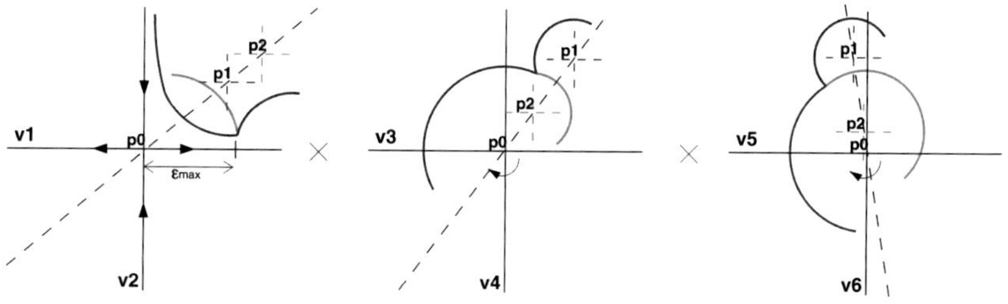


**Fig. 7** Schematic representation of the trajectories of a solar sail for small changes on the sail orientation in the saddle × centre × centre planes. The trajectory in *blue* is the one the sail follows close to $p_0$ for $\alpha = \alpha_0$, $\delta = \delta_0$. If we change the sail orientation and to $\alpha_1$, $\delta_1$ and the new equilibrium point is $p_1$ the sail will follow the *red* trajectory, while if the sail orientation is $\alpha_2$, $\delta_2$ with $p_2$ as equilibrium point, the sail will follow the *green* trajectory. In order to remain close to $p_0$ we are interested in sail orientations that produce the effects like $p_2$

orbit $\mathscr{W}_s(P_0(t))$. Once we are close to $\mathscr{W}_s(P_0(t))$ we restore the sail orientation to $\alpha_0, \delta_0$. We repeat this process until the end of the mission. As before, we need to choose the new sail orientation in order to come close to $\mathscr{W}_s(P_0(t))$ and keep the centre projection bounded.

The key point of this approach relies on finding the appropriate new sail orientation, which we will discuss in Sect. 4.1.3. First we will focus on how to derive a reference frame which lets us track the relative position between the sail's trajectory and the saddle and centre projections when we are close to an equilibrium point.

### 4.1.2 Reference Frame

We use a particular reference system to track the trajectory and make decisions on when and how we have to change the sail orientation. To fix notation, if $\varphi(t_0)$ is the position and velocity of the solar sail at time $t_0$, then

$$\varphi(t_0) = p_0 + \sum_{i=1}^{6} s_i(t_0)\mathbf{v_i},$$

where $p_0$ is the position and velocity of the equilibrium point, and $\{\mathbf{v_1}, \ldots, \mathbf{v_6}\}$ are a basis that gives the projection of the trajectory in the saddle and centre components.

It is well known that the local behaviour around an equilibrium points is given by the linearised equations at the point. To fix notation, if $\dot{x} = f(x, \alpha, \delta)$ are the equations of motion for the solar sail (i.e. Eq. (1) in compact form), and $p_0$ is an equilibrium point for $\alpha = \alpha_0$, $\delta = \delta_0$ (i.e. $f(p_0, \alpha_0, \delta_0) = 0$), the linearised system is given by:

$$\dot{x} = D_x f(p_0, \alpha_0, \delta_0)x.$$

In Sect. 3.1 we mentioned that the eigenvalues $(\lambda_{1,\ldots,6})$ of $D_x f(p_0, \alpha_0, \delta_0)$ satisfy: $\lambda_1 > 0$, $\lambda_2 < 0$ are real, $\lambda_3 = \nu_1 + i\omega_1$, $\lambda_4 = \bar{\lambda}_3$ and $\lambda_5 = \nu_2 + i\omega_2$, $\lambda_6 = \bar{\lambda}_5$ are complex. These three pairs of eigenvalues and their associated eigenvectors $(\mathbf{e_1}, \ldots, \mathbf{e_6})$ have the following geometrical meaning:

- The first pair $(\lambda_1, \lambda_2) \in \mathbb{R}$ are associated to the hyperbolic character of the equilibrium point. The eigenvector $\mathbf{e_1}$ (corresponding to the eigenvalue $\lambda_1 > 0$) gives the most expanding direction: at the equilibrium point, $\mathbf{e_1}$ is tangent to the unstable manifold $\mathscr{W}_u(p_0)$. In the same way, the eigenvector $\mathbf{e_2}$ (corresponding to $\lambda_2 < 0$) is associated to the stable manifold of the equilibrium point $\mathscr{W}_s(p_0)$.
- The second and third pairs $(\lambda_3, \lambda_4), (\lambda_5, \lambda_6) \in \mathbb{C}$ are complex conjugate ($\lambda_3 = \bar{\lambda}_4$, $\lambda_5 = \bar{\lambda}_6$). Due to the non-Hamiltonian structure of the system, they might not be purely imaginary. The linearised dynamics restricted to the invariant plane generated by the real vectors $\{\mathrm{Re}(\mathbf{e_3}), \mathrm{Im}(\mathbf{e_4})\}$ are spirals with a rotation rate given by $\Gamma_1 = \arctan(\mathrm{Im}(\lambda_3)/\mathrm{Re}(\lambda_3))$ and an increase or decrease rate given by $\mathrm{Re}(\lambda_3)$. The same happens in the plane given by $\{\mathrm{Re}(\mathbf{e_5}), \mathrm{Im}(\mathbf{e_5})\}$, having a rotating rate $\Gamma_2 = \arctan(\mathrm{Im}(\lambda_5)/\mathrm{Re}(\lambda_5))$ and an expansion rate given by $\mathrm{Re}(\lambda_5)$. We always

choose the second pair such that the vertical oscillation of $\mathbf{e_5}$ is larger that the one of $\mathbf{e_3}$ (i.e. $|\mathbf{e_5}|_{\dot{z}} \gg |\mathbf{e_3}|_{\dot{z}}$).

If we consider the reference frame (with origin at the equilibrium point) given by $\{\mathbf{v_1} = \mathbf{e_1}/|\mathbf{e_1}|, \mathbf{v_2} = \mathbf{e_2}/|\mathbf{e_2}|, \mathbf{v_3} = \mathrm{Re}(\mathbf{e_3})/|\mathbf{e_3}|, \mathbf{v_4} = \mathrm{Im}(\mathbf{e_3})/|\mathbf{e_3}|, \mathbf{v_5} = \mathrm{Re}(\mathbf{e_5})/|\mathbf{e_5}|, \mathbf{v_6} = \mathrm{Im}(\mathbf{e_5})/|\mathbf{e_5}|\}$, the linearised system $\dot{x} = D_x f(p_0)x$ takes the form $\dot{y} = J_1 y$, where $x = J_1 y$ and

$$J_1 = \begin{pmatrix} \lambda_1 & & & & & \\ & \lambda_2 & & & 0 & \\ & & \nu_1 & -\omega_1 & & \\ & & \omega_1 & \nu_1 & & \\ & 0 & & & \nu_2 & -\omega_2 \\ & & & & \omega_2 & \nu_2 \end{pmatrix}.$$

In this new set of coordinates, the dynamics around an equilibrium point can be easily described: on the plane generated by $\mathbf{v_1}, \mathbf{v_2}$ the trajectory escapes along the unstable direction $\mathbf{v_1}$, on the plane generated by $\mathbf{v_3}, \mathbf{v_4}$ the trajectory rotates (in a spiral form) around the equilibrium point, and the same behaviour happens on the plane generated by $\mathbf{v_5}, \mathbf{v_6}$ as describe in Fig. 6.

In the case of periodic orbits the linear behaviour is given by the monodromy matrix of the system, and we can use the Floquet Modes to derive the appropriate reference frame. This reference frame will be $T$-periodic, being $T$ the period of the orbit. For further details [13, 14].

### 4.1.3 Finding the New Sail Orientation

As we have described in Sect. 4.1.1 when the trajectory of the solar sail is far from the target point, $p_0$, we want to find a new sail orientation $\alpha_1, \delta_1$ that brings the trajectory close to $p_0$. The idea is to shift the phase space in such a way that the new unstable manifold brings us close to the stable manifold of $p_0$ without letting the centre projections grow. Hence, we need to know how a small change on the sail orientation will affect the sails trajectory.

The first order variational flow gives information on how small variations on the initial conditions affect the final trajectory. In the same way the first order variational equations with respect to the two angles defining the sail orientation describes how small variations on the sail orientation affect the final trajectory. We will use this to decide which is the appropriate sail orientation that brings the trajectory close to the target point.

If $\phi_h(t_0, x_0, \alpha_0, \delta_0)$ is the flow of our vector field at time $t = t_0 + h$ of our vector field starting at time $t_0$ for $(x_0, \alpha_0, \delta_0)$, then

$$F(\Delta\alpha, \Delta\delta, h) = \phi_h(t_0, x_0, \alpha_0, \delta_0) + \frac{\partial \phi_h}{\partial \alpha}(t_0, x_0, \alpha_0, \delta_0) \cdot \Delta\alpha + \frac{\partial \phi_h}{\partial \delta}(t_0, x_0, \alpha_0, \delta_0) \cdot \Delta\delta, \tag{5}$$

is the first order approximation of the final state if a change $\Delta\alpha$, $\Delta\delta$ is made at time $t = t_0$. This is an explicit expression for the final states of the trajectory as a function of the two angles and time.

Let us assume that at time $t = t_1$ we have $|s_1(t_1)|$ (the component along the unstable direction) large and we want to change the sail orientation to correct it. $F(\Delta\alpha, \Delta\delta, \Delta t)$ gives a map of how a small change in the sail orientation $\Delta\alpha$, $\Delta\delta$ at time $t = t_1$ will affect the sail trajectory at time $t = t_1 + \Delta t$. We want to find $\Delta\alpha_1$, $\Delta\delta_1$ and $\Delta t_1$ so that the flow at time $t = t_1 + \Delta t_1$ is close to the stable manifold (i.e. $|s_1(t_1 + \Delta t_1)|$ small) and the centre projections do not grow (i.e. $||(s_3(t_1 + \Delta t_1), s_4(t_1 + \Delta t_1))||_2 \leq ||(s_3(t_1), s_4(t_1))||_2, ||(s_5(t_1 + \Delta t_1), s_6(t_1 + \Delta t_1))||_2 \leq ||(s_5(t_1), s_6(t_1))||_2$). There are different ways to solve this problem, we proceed as follows:

1. We take a set of equally spaced times, $\{t_i\}$, in the time interval $[t_1 + \Delta t_{min}, t_1 + \Delta t_{max}]$. For each $t_i$ we compute the variational map $F(\Delta\alpha, \Delta\delta, \Delta t_i)$, $\Delta t_i = t_i - t_1$. Where $\Delta t_{min}$ is the minimum time allowed between manoeuvres, which depends on the solar sail restrictions, and $\Delta t_{max}$ is the maximum time between manoeuvres, which depends on the accuracy of $F(\Delta\alpha, \Delta\delta, \Delta t_{max})$.
2. For each $t_i$ we find $\Delta\alpha_i$, $\Delta\delta_i$ such that, $s_1(t_i) = s_5(t_i) = s_6(t_i) = 0$. Notice that this reduces to solve a linear system with 2 unknowns and 3 equations. We use the least squares method to solve this linear system and find the minimum norm solution. At the end we have a set of $\{t_i, \Delta\alpha_i, \Delta\delta_i\}$ such that, $||(s_1(t_i), s_5(t_i), s_6(t_i))||_2$ is small.
3. From the set of $\{t_i, \Delta\alpha_i, \Delta\delta_i\}_{i=1,\dots,n}$ found in step 2 we choose $j$ such that $||(s_3(t_j), s_4(t_j))||_2 \leq ||(s_3(t_i), s_4(t_i))||_2 \;\; \forall\, i \neq j$.

The desired set of parameters that bring the sail back to the equilibrium point are:

$$\alpha_1 = \alpha_0 + \Delta\alpha_j, \quad \delta_1 = \delta_0 + \Delta\delta_j, \quad \Delta t_1 = t_j - t_1. \tag{6}$$

*Remark 1* It is not evident that we can always find $\Delta t_1, \alpha_1, \delta_1$ which bring back the trajectory to $\mathscr{W}_s$, as we are in a 6D phase space and we only have three parameters to play with.

If we look at Fig. 7 we can see that the new unstable manifold will bring the trajectory back if the new equilibria is placed at the correct spot on the phase space. Using $\frac{\partial\phi_t}{\partial\alpha}, \frac{\partial\phi_t}{\partial\delta}$ one can check if variations on the two sail orientations allows to place a new equilibrium point on the desired saddle component. This allows us to check the controllability of the point.

*Remark 2* Notice that instead of steps 2 and 3 one could solve the linear system $s_1(t_i) = 0, s_3(t_i) = 0, s_4(t_i) = 0, s_5(t_i) = 0, s_6(t_i) = 0$. Where we would have a linear system with 5 equations and 2 unknowns that we can solve using the least square methods. This would find the minimum solution but it does not guarantee that $||(s_3(t_1), s_4(t_1))||_2$ and $||(s_5(t_1), s_6(t_1))||_2$ are small. We prioritise to control the size of $(s_3, s_4)$ over $(s_5, s_6)$, by solving the system using steps 2 and 3. In this way we can guarantee that $||(s_3(t_1), s_4(t_1))||_2$ will be as small as possible.

*Remark 3* When we find $\Delta t_1, \alpha_1, \delta_1$ following steps 2 and 3 we solve a linear system to minimise $||(s_1(t_i), s_5(t_i), s_6(t_i))||_2$ and then take $(\Delta\alpha_i, \Delta\delta_i, t_i)$ with the smallest $||(s_3(t_i), s_4(t_i))||_2$. Notice that we could switch the role of the two centres, i.e. solve the system to minimise $||(s_1(t_i), s_3(t_i), s_4(t_i))||_2$ and take $(\Delta\alpha_i, \Delta\delta_i, t_i)$ with the smallest $||(s_5(t_i), s_6(t_i))||_2$.

The approach presented here gives better results because $(s_5(t_i), s_6(t_i))$ are related to vertical oscillations around equilibria, which are compensated by moving $\delta$, while $(s_3(t_i), s_4(t_i))$ are related to the planar oscillations which are compensated with variations of $\alpha$ which also affects the saddle $s_1(t_i), s_2(t_i)$.

For points close to the $Z = 0$ plane variations on $\delta$ do not affect the saddle behaviour.

*Remark 4* The value of $\Delta t_{max}$ is strongly related to the validity of $F(\Delta\alpha, \Delta\delta, h)$, i.e. how good it approximates the behaviour of trajectories close to the reference orbit. If we consider a larger Taylor expansion in terms of $\alpha$ and $\delta$ we could be able to get larger times and probably better choices for $\Delta\alpha$, $\Delta\delta$.

### 4.1.4 Station Keeping Algorithm

For each mission we will define 3 parameters which depend on the mission requirements and the dynamics of the system around the equilibrium point. These are: $\varepsilon_{max}$, the maximum distance to the stable direction allowed, which we use to decide when to change the sail orientation; $\Delta t_{min}$ and $\Delta t_{max}$ the minimum and maximum time between manoeuvres allowed, which depends on the mission requirements and on the validity of the variational flow.

We recall that we use the reference frame described in Sect. 4.1.2 to look at the trajectory of the solar sail:

$$\phi(t) = p_0 + \sum_{i=1}^{6} s_i(t)\mathbf{v_i},$$

where $p_0$ is the equilibrium point we want to remain close to, and $\{\mathbf{v_i}\}_{i=0,\dots,6}$ are the basis defining the reference frames described in Sect. 4.1.2.

We always start the mission close to the target point $p_0$ with a fixed sail orientation $\alpha = \alpha_0$, $\delta = \delta_0$, and let the natural dynamics act. When we are far from $p_0$, and by this we mean $|s_1(t)| > \varepsilon_{max}$, we choose (as described in Sect. 4.1.3) a new sail orientation, $\alpha_1$ and $\delta_1$, which after some time, $\Delta t_1$, will bring the trajectory close to the stable manifold of $p_0$, i.e. $|s_1(t)|$ small. Once the trajectory is back to $p_0$ we restore the sail orientation. A sketch of the code is:

```
if(who == 0){ /* alfa0 and delta0 are acting */
  if(|s1(ti)| < epsmax){
    find_new_sail(&dalf,&ddel,&dt); /*(section\,4.1.3)*/
    alfa = alfa0 + dalf; delta = delta0 + ddel;
    tend = ti + dt;
    who = 1;
  }
}else if(who == 1){ /* alfa1 and delta1 are acting */
  if(ti > tend){
    alfa = alfa0; delta = delta0;
    who = 0;
      }
}
```

We must mention that all the strategies described here use information of the linear dynamics of the system to make decisions on the changes of the sail orientation, but the complete set of equations is taken into account during the simulations.

### 4.1.5 Example Mission

To illustrate the performance of these strategies we consider the **Sunjammer mission** where the solar sail needs to remain close to a an equilibrium point. The Sunjammer mission aims to make observations of the Sun, hence the equilibrium point must be placed on the ecliptic plane and displaced 5° from the Sun-Earth line (Fig. 3 left). The sail efficiency we take is $\beta = 0.0388$ which is considered as realistic for the Sunjammer mission [19]. And the target equilibria corresponds to: $p_0 = (-0.98334680272, -0.00146862443, 0.00000000000)$ (AU) for $\alpha_0 = 0.023954985$, $\delta_0 = 0.000000$ (rad).

In this example mission we consider $\varepsilon_{max} = 5 \times 10^{-5} \approx 7479.89$ km, $\Delta t_{min} = 0.02$ UT $\approx 1.1626$ days and $\Delta t_{max} = 2$ UT $\approx 116.26$ days. We have taken random initial conditions and performed the station keeping strategy to remain close to equilibria for 10 years

In Fig. 8 we have the controlled trajectory of the solar sail in the $XY$-plane (left), $YZ$-plane (middle) and the $XYZ$ projection (right). As we can see the trajectory remains close to the equilibrium point for all time.

In Fig. 9 we show the projection of the controlled trajectory of the solar sail in the saddle × centre × centre reference frame. Notice how the projection on the saddle plane (left) every time the trajectory reaches $|s_1(t)| > \varepsilon_{max}$ the trajectory is corrected to return to the stable direction, i.e. $|s_1(t)| \approx 0$. On the other hand, the trajectory on the first centre component (middle) is a succession of rotations which remain bounded, as desired. While the vertical oscillation (right) is completely cancelled out, i.e. $(s_5(t), s_6(t)) \to (0, 0)$

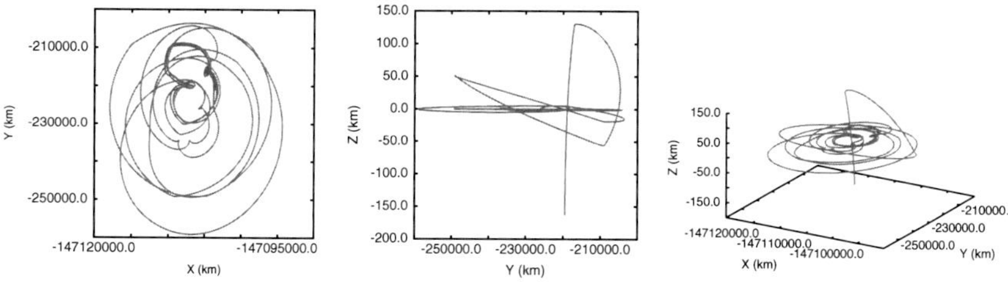


**Fig. 8** For the Sunjammer mission, trajectory of the controlled solar sail for 10 years on the: XY-plane (*left*), YZ-plane (*middle*) and XYZ-plane (*right*)

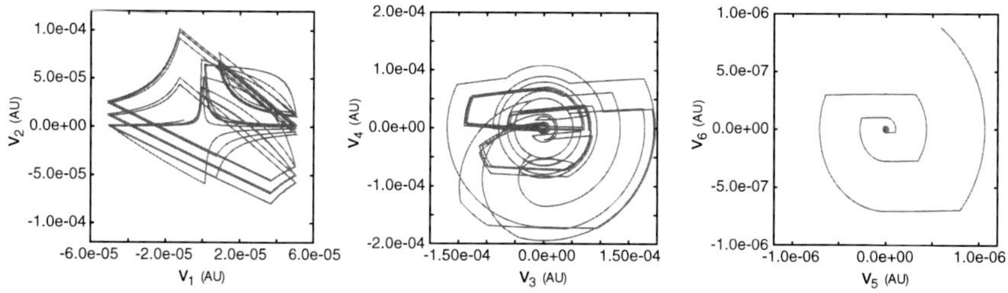


**Fig. 9** For the Sunjammer mission, trajectory of the controlled solar sail for 10 years on the: saddle plane (*left*), centre plane generated by $(\mathbf{v_3}, \mathbf{v_4})$ (*middle*) and centre plane generated by $(\mathbf{v_5}, \mathbf{v_6})$ (*right*)

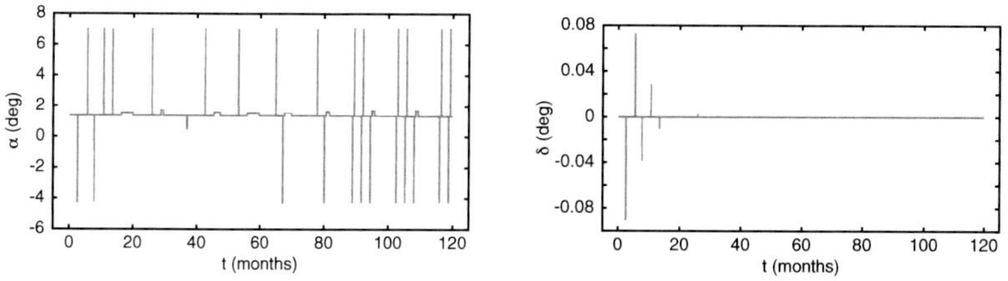


**Fig. 10** For the Sunjammer mission, variation of the sail orientation along time: $\alpha$ variation (*left*); $\delta$ variation (*right*)

Finally in Fig. 10 we see the variation of the two angles defining the sail orientation $\alpha$ (left) and $\delta$ (right) along time. As we can see, $\alpha$ has variations of order $\sim 5^\circ$ each time we need to correct the trajectory, while $\delta$ is almost zero.

## 4.2 *Station Keeping in the NBPS*

In this paper we look at the NBPS as a perturbation of the RTBPS. There exists several works considering other perturbations of the RTBPS. For instance, as discussed in [12], in the Elliptic RTBPS, we no longer have artificial equilibria, these are replaced by $2\pi$-periodic orbits. In the same way periodic orbits in the Circular RTBP are replaced by 2D invariant tori. This is because the Elliptic RTBPS can be seen as a $2\pi$-periodic perturbation of the Circular RTBP. Nevertheless, if the perturbation is small these orbits remain close to the equilibrium point or periodic orbits in the Circular RTBP and share the same qualitative behaviour.

When we include other kind of perturbations, such as the gravitational effect of the other planets, the system is no longer a periodic perturbation of the Circular RTBP, hence these periodic orbits no longer exist. Nevertheless, there still exist natural trajectories of the system that remain close to the periodic orbits of the unperturbed system [20, 21]. Moreover, the qualitative behaviour around these trajectories is similar to the behaviour around an equilibrium point in the CRTBP. We will use these natural trajectories, also called "dynamical substitute", as target orbits for our station keeping.

As we will discuss in Sect. 4.2.2 the dynamics around these "dynamical substitute" is equivalent to the one of the equilibrium points. We also have one expanding and one contracting direction along the orbit, and two almost centres, i.e. the trajectories close to these orbits will slightly spiral inwards or outwards on the centre projections.

In order to remain close to these orbits we will use the same ideas behind the RTBPS approach we have already explained in Sect. 4.1. We know that the trajectories will escape along the unstable manifold. When we are far from the target orbit we choose a new sail orientation that brings the trajectory close to the stable manifold and keeping the other two centre projections bounded. In order to find the appropriate new sail orientation we will use the algorithm explained in Sect. 4.1.3. The only technical detail is how to define an appropriate reference frame that allows us to know at all time what is the relative position between the solar sail and the target orbit and its stable and unstable manifolds. We will explain this in Sect. 4.2.3.

### 4.2.1 Dynamical Substitute

First of all we need to compute a good target orbit, i.e. the "dynamical substitute" of the equilibrium point in the RTBPS. For this we have simply implemented a parallel shooting method to get a solution in NBP model (Eq. 2) that stays close to the equilibrium point for all the considered time span [17]. The parallel shooting method is a classical numerical method used to compute periodic orbits that are very unstable or with a long period. Let us summarise how it works.

As in this case we want to compute an orbit for a long time, we split it in several pieces, and we ask that all these pieces match in a single orbit. In other words, we want to find $n+1$ sets of initial data $(t_i, x_i)$, $i = 0, \ldots, n$ such that the orbit starting at $(t_i, x_i)$ reaches $(t_{i+1}, x_{i+1})$ at time $t_{i+1} = t_i + \Delta t$:

$$\begin{aligned} \phi_{\Delta t}(t_0, x_0, \alpha_0, \delta_0) \; - \; x_1 \quad &= 0, \\ &\vdots \\ \phi_{\Delta t}(t_{n-1}, x_{n-1}, \alpha_0, \delta_0) \; - \; x_n &= 0, \end{aligned} \tag{7}$$

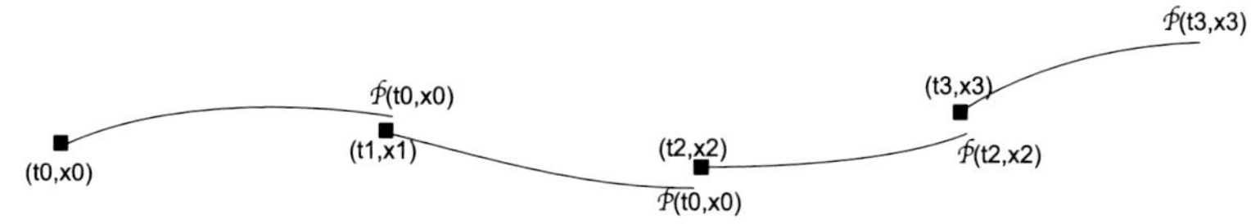


**Fig. 11** Schematic representation of the Parallel Shooting idea

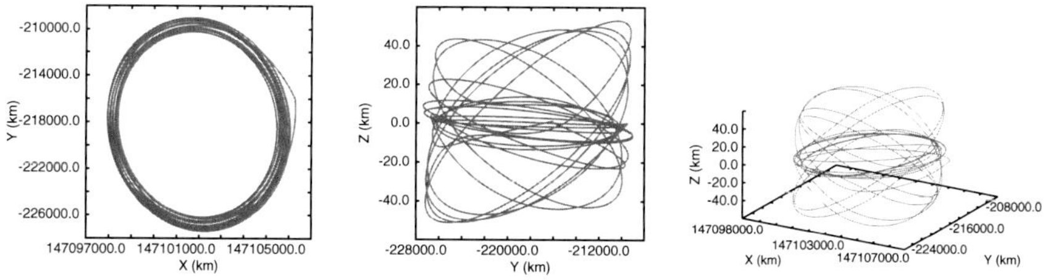


**Fig. 12** Dynamical substitute for $\beta = 0.0388$, $p_0 = (-0.98334680272, -0.00146862443, 0.00000000000)$ (AU) for $\alpha_0 = 0.023954985$, $\delta_0 = 0.000000$ (rad)

where $\phi_{\Delta t}(t_i, x_i, \alpha_0, \delta_0)$ denotes the solution of the system at time $t_{i+1} = t_i + \Delta t$ that starts at $t = t_i$ with $(x_i, \alpha_0, \delta_0)$ (see Fig. 11).

We split the mission time span $[0, T_{end}]$ (i.e. where we want to find the target orbit) into several equally spaced intervals $[t_i, t_{i+1}]$, $i = 0, \ldots, n-1$, verifying $t_0 = 0$, $t_n = T_{end}$ and $\Delta t = t_{i+1} - t_i = T_{end}/n$. We will say that the couples $(t_i, x_i)$ belong to the dynamical substitute if Eq. 7 is satisfied. This means that we need to solve a non-linear equation with $6n$ equations and $6n+6$ unknowns $(x_0, \ldots, x_n)$. Notice that we have more unknowns than equations, hence we add six more conditions to ensure the uniqueness of the solutions. In our case we choose to fix the initial positions (the first three components of $x_0$) and the final position (the first three components of $x_n$), but other options are possible [17]. In order to solve Eq. 7 we will use a standard Newton method taking as initial condition $x_i = p_0$ for $i = 1, \ldots, n$, where $p_0$ is the equilibria in the RTBPS for $\alpha_0, \delta_0$. As the orbit we are looking for is close to the equilibrium it is reasonable to use the point as initial guess for the Newton method.

In Fig. 12 we show different projections of the dynamical substitute of the equilibrium point for the Sunjammer mission used in Sect. 4.1.5. Here we have considered $T_{end} = 10$ years (the maximum duration of our mission) and $\Delta t = 1/2$ years, hence $n = 20$.

### 4.2.2 Linear Dynamics Around the Dynamical Substitute

It is well know that one should look at the first order variational equations in order to understand the behaviour close to a given trajectory. If $\dot{x} = F(t, x, \alpha, \delta)$ represents Eq. 2 in its compact form, then the first order variational equations are given by,

$$\dot{A} = D_x F(t, x, \alpha, \delta) A, \quad A \in \mathscr{L}(\mathbb{R}^6, \mathbb{R}^6), \tag{8}$$

taking as initial condition $A(0) = Id$.

Let $\phi$ denote the flow associated to Eq. 2 and $\phi_t(t_0, x_0, \alpha_0, \delta_0)$ the image of the point $x_0 \in \mathbb{R}^6$ after $t$ units of time. The solution of Eq. 8, $A(t) = D_x\phi_t(t_0, x_0, \alpha_0, \delta_0)$, is the differential flow of $\phi_t(t_0, x_0, \alpha_0, \delta_0)$ with respect to the initial condition $x_0$. For $h \in \mathbb{R}^6$ we have,

$$\phi_t(t_0, x_0 + h, \alpha_0, \delta_0) = \phi_t(t_0, x_0, \alpha_0, \delta_0) + D_x\phi_t(t_0, x_0, \alpha_0, \delta_0) \cdot h + O(|h|^2).$$

Therefore, $\phi_t(t_0, x_0, \alpha_0, \delta_0) + A(t) \cdot h$, gives a good approximation of $\phi_t(t_0, x_0 + h, \alpha_0, \delta_0)$ provided that $\|h\|$ is small enough. Hence, the linear dynamics around the target orbit computed in the previous section will be determined by the matrix $M = A(T_{end})$, in the sense that $M$ is the differential of the final point of the orbit with respect to the initial point $(t_0, x_0)$, so that their eigenvalues give information about how fast nearby orbits approach/escape from the base orbit, and their eigenvectors give the corresponding arriving/escaping directions. These eigenvectors can be used as initial data for the variational flow to obtain the linear approximation to the stable/unstable manifold. Finally, note that this analysis is only valid on the finite time span for which the orbit has been computed.

To avoid problems in the integration of the variational flow due to the instability of the system, we have split the reference orbit into $N$ pieces (i.e. we call reference orbit to the dynamical substitute computed following the scheme described in Sect. 4.2.1, which is the orbit be want to stay close) Each piece corresponds to one revolution of the Earth around the Sun, hence from now on we will refer to each piece of orbit as 1 revolution of the target orbit. Associated to each revolution we have the variational matrix $A_k$ in normalised coordinates. It is easy to check that $M = A_N \times A_{N-1} \times \cdots \times A_1$.

Due to the large value of the unstable eigenvalue of each one of the matrices $A_k$ (roughly 396) it is not possible to perform a direct computation of the eigenvalues of $M$ because of the possible overflow during the computation of $M$. We must take into account that the dominant eigenvalue of $M$ could be of the order of $396^N$. There exist procedures [16] that can be done to deal with this problem and find all the eigenvalues and eigenvectors of $M$.

Instead, we have decided to compute for each of the individual matrices $A_k$ their eigenvalues and eigenvectors, and use them to describe the linear dynamics for each revolution of the target orbit. The qualitative behaviour will be the same for each revolution, although there might be some small quantitative differences, i.e. the size of the eigenvalues and the directions of the eigenvector.

For each revolution, the eigenvalues $(\lambda_{1,\dots,6})$ of the $A_k$ are very similar, and satisfy: $\lambda_1 > 1$, $\lambda_2 < 1$ are real, $\lambda_3 = \nu_1 + i\omega_1$, $\lambda_4 = \bar{\lambda}_3$ and $\lambda_5 = \nu_2 + i\omega_2$, $\lambda_6 = \bar{\lambda}_5$ are complex. Each of these three pairs of eigenvalues have the following geometrical meaning:

- The first pair $(\lambda_1, \lambda_2)$ are related to the (strong) hyperbolic character of the orbit. The value $\lambda_1$ is the largest in absolute value, and is related to the eigenvector $\mathbf{e_1}(0)$, which gives the most expanding direction. Using $D_x\phi_t$ we can get the image of this vector under the variational flow: $\mathbf{e_1}(t) = D_x\phi_t\mathbf{e_1}(0)$. At each point of the orbit, the vector $\mathbf{e_1}(t)$ together with the tangent vector to the orbit, span a plane that is tangent to the local unstable manifold ($W^u_{loc}$). In the same way $\lambda_2$ and its related eigenvector $\mathbf{e_2}(0)$ are related to the stable manifold and $\mathbf{e_2}(t) = D_x\phi_t\mathbf{e_2}(0)$.

- The other two couples ($\lambda_3, \lambda_4 = \bar{\lambda}_3$) and ($\lambda_5, \lambda_6 = \bar{\lambda}_5$) are complex conjugate and their modulus is close to 1. The matrix $M$, restricted to the plane spanned by the real and imaginary parts of the eigenvectors associated to $\lambda_3, \lambda_4$ (and $\lambda_5, \lambda_6$) is a rotation with a small dissipation or expansion, so that the trajectories on these planes spiral inwards or outwards. $A_k$ restricted to these planes has the form,

$$\begin{pmatrix} \Delta_i \cos\Gamma_i & -\Delta_i \sin\Gamma_i \\ \Delta_i \sin\Gamma_i & \Delta_i \cos\Gamma_i \end{pmatrix},$$

where $\Delta_{1,2}$ denotes the modulus of $\lambda_3$ and $\lambda_5$ respectively, and are related to the rates of expansion and contraction, and $\Gamma_{1,2}$ denotes the argument of $\lambda_3$ and $\lambda_5$ respectively, which account for the rotation rate around the orbit.

As we did with the equilibrium points in Sect. 4.1.2 we always choose the second pair of complex eigenvalues such that the vertical oscillation of $\mathbf{e_5}$ is larger that the one of $\mathbf{e_3}$ (i.e. $|\mathbf{e_5}|_z \gg |\mathbf{e_3}|_z$). We also recall that $|\lambda_{3,4,5,6}| \ll |\lambda_1|$, hence the most expanding direction (by far) is given by $e_1(t)$.

To sum up, in a suitable basis the variational flow, $A_k$, associated to the $k$th revolution of the target orbit can be written as,

$$B_k = \left(\begin{array}{cc|cc|cc} \lambda_{k,1} & & & & & \\ & \lambda_{k,2} & & & 0 & \\ \hline & & \Delta_{k,1}\cos\Gamma_{k,1} & -\Delta_{k,1}\sin\Gamma_{k,1} & & \\ & & \Delta_{k,1}\sin\Gamma_{k,1} & \Delta_{k,1}\cos\Gamma_{k,1} & & \\ \hline & 0 & & & \Delta_{k,2}\cos\Gamma_{k,2} & -\Delta_{k,2}\sin\Gamma_{k,2} \\ & & & & \Delta_{k,2}\sin\Gamma_{k,2} & \Delta_{k,2}\cos\Gamma_{k,2} \end{array}\right), \tag{9}$$

and the functions $\mathbf{e_i}(t) = D_x\phi_t \cdot \mathbf{e_i}(0)$, $i = 1, \ldots, 6$, give an idea of the variation of the phase space properties in a small neighbourhood of the target orbit. We will use a modification of them, the so called Floquet modes [11, 12, 18] $\bar{\mathbf{e}}_\mathbf{i}(t)$ to track a trajectory close to the reference orbit and give a simple description of its dynamics.

### 4.2.3 Reference Frame

Here the Floquet modes are six (1 year)-periodic time-dependent vectors, $\bar{\mathbf{e}}_\mathbf{i}(t)$, $i = 1, \ldots, 6$, such that, if we call $P_k(t)$ the matrix that has the vectors $\bar{\mathbf{e}}_\mathbf{i}(t)$ as columns, then the change of variables $x = P_k(t)z$, turns the linearised equation around the $k$th revolution of the target orbit, $\dot{x} = A_k(t)x$, into an equation with constant coefficients $\dot{z} = B_k z$ (where $B_k$ is the matrix in Eq. (9)).

One of the main advantages of using the Floquet basis, is the fact that they are periodic, and they can be easily stored using a Fourier series. We can compute these Floquet basis for each of the revolutions, and use the as reference system, as we did for periodic orbits in [13, 14]. We need to keep in mind that after each revolution, when we change from one piece of the orbit to another, there will be a small discontinuity in our reference systems, this translates into having a small jump in the phase space made by the sail trajectory.

We define the first and second Floquet mode taking into account that the escape and contraction rate after one revolution along the unstable and stable manifolds is exponential:

$$\bar{\mathbf{e}}_1(t) = \mathbf{e}_1(t^k)\exp\left(-\frac{t^k}{\Delta t}\ln\lambda_1\right),$$
$$\bar{\mathbf{e}}_2(t) = \mathbf{e}_2(t^k)\exp\left(-\frac{t^k}{\Delta t}\ln\lambda_2\right).$$

Notice that using this definition after one revolution $\bar{\mathbf{e}}_1(t)$ and $\bar{\mathbf{e}}_2(t)$ are unitary.

The other two pairs are computed taking into account that after one revolution the plane generated by the real and imaginary parts of the eigenvectors associated to $(\lambda_3, \lambda_4)$ and $(\lambda_5, \lambda_6)$ is a rotation of angle $\Gamma_{1,2}$ and a dissipation/expansion by a factor of $\Delta_{1,2}$:

$$\bar{\mathbf{e}}_3(t) = \left[\cos\left(-\Gamma_1\frac{t^k}{\Delta t}\right)\mathbf{e}_3(t^k) - \sin\left(-\Gamma_1\frac{t^k}{\Delta t}\right)\mathbf{e}_4(t^k)\right]\exp\left(-\frac{t^k}{\Delta t}\ln\Delta_1\right),$$
$$\bar{\mathbf{e}}_4(t) = \left[\sin\left(-\Gamma_1\frac{t^k}{\Delta t}\right)\mathbf{e}_3(t^k) + \cos\left(-\Gamma_1\frac{t^k}{\Delta t}\right)\mathbf{e}_4(t^k)\right]\exp\left(-\frac{t^k}{\Delta t}\ln\Delta_1\right),$$
$$\bar{\mathbf{e}}_5(t) = \left[\cos\left(-\Gamma_2\frac{t^k}{\Delta t}\right)\mathbf{e}_5(t^k) - \sin\left(-\Gamma_2\frac{t^k}{\Delta t}\right)\mathbf{e}_6(t^k)\right]\exp\left(-\frac{t^k}{\Delta t}\ln\Delta_2\right),$$
$$\bar{\mathbf{e}}_6(t) = \left[\sin\left(-\Gamma_2\frac{t^k}{\Delta t}\right)\mathbf{e}_5(t^k) + \cos\left(-\Gamma_2\frac{t^k}{\Delta t}\right)\mathbf{e}_6(t^k)\right]\exp\left(-\frac{t^k}{\Delta t}\ln\Delta_2\right).$$

where $t^k = t - k\cdot\Delta t$ is a re-normalised time as the Floquet modes are for $t \in [0, \Delta t]$, and $k$ stands for the orbital revolution that we are considering.

To build our reference frame, again we split the time interval of the mission duration $[0, T_{end}]$ into $N$ revolutions, having $N$ time intervals $I_i = [t_i, t_{i+1}]$, $i = 0, \ldots, N-1$, where $t_0 = 0$, $t_i = t_{i-1} + \Delta t$ and $\Delta t = T_{end}/N$. In all of our examples we have considered $T_{end} = 10$ years (the maximum duration of our mission) and $N = 10$ (i.e. $\Delta t = 1$ year $= 1$ revolution).

For each time interval $I_k$ we compute the Floquet modes associated to the variational flow $A_k$ and store them via their Fourier series so they can be easily recomputed, and we define the reference system as:

$$\{\, N_0(t);\ \mathbf{v}_1(t), \mathbf{v}_2(t), \mathbf{v}_3(t), \mathbf{v}_4(t), \mathbf{v}_5(t), \mathbf{v}_6(t)\,\}, \tag{10}$$

where $N_0(t)$ are the positions and velocities of the target orbit at time $t$, and $\mathbf{v}_{\mathbf{1},\ldots,\mathbf{6}}(t) = \bar{\mathbf{e}}^{\mathbf{k}}_{\mathbf{1},\ldots,\mathbf{6}}(t)$ corresponds to the Floquet modes of $A_k$ for $t \in I_k$. This can be formally defined as,

$$\mathbf{v}_{\mathbf{i}}(t) = \sum_{k=0}^{N} \chi(I_k)\bar{\mathbf{e}}^{\mathbf{k}}_{\mathbf{i}}(t),$$

where $\chi(t) = \{1 \text{ if } t \in I_k,\ 0 \text{ if } t \notin I_k\}$ and $\bar{\mathbf{e}}^{\mathbf{k}}_{\mathbf{i}}(t)$ are the Floquet mode associated to the $k$th orbital revolution.

Notice that the directions in this reference frame are discontinuous at each revolution. This means that at each revolution there will be a small jump of the trajectory in the phase space. Nevertheless the difference between the different eigenvectors of $A_k$ is very small and these jumps will be negligible.

### 4.2.4 Station Keeping Algorithm

Using the reference system described in the previous section the dynamics around the target orbit is simple. $N_0(t_0)$ denotes the point on the target orbit at time $t_0$ closer to the solar sail position, and $\mathbf{v}_{\mathbf{1}}(t_0)$, $\mathbf{v}_{\mathbf{2}}(t_0)$ are the unstable and stable directions. When the base point $N_0(t)$ moves along the target orbit, the vectors $\mathbf{v}_{\mathbf{1}}(t)$, $\mathbf{v}_{\mathbf{2}}(t)$ moves along the orbit following the (two-dimensional) unstable and stable manifolds. In the same way, these two directions generate a plane that moves along the orbit, on which the dynamics is a saddle.

For each point on the target orbit, the couple $\mathbf{v}_{\mathbf{3}}(t_0)$, $\mathbf{v}_{\mathbf{4}}(t_0)$ span a plane that is tangent to another invariant manifold of the orbit. This plane spans a three-dimensional manifold when the base point moves along the orbit. The dynamics on this manifold can be visualised as a spiral motion (towards the target orbit) on the plane $\{\mathbf{v}_{\mathbf{3}}(t_0),\ \mathbf{v}_{\mathbf{4}}(t_0)\}$ at the same time that the plane moves along the orbit. In a similar way, the couple $\mathbf{v}_{\mathbf{5}}(t_0)$, $\mathbf{v}_{\mathbf{6}}(t_0)$ spans another three-dimensional manifold, on which the dynamics is again a spiral motion (but now escaping from the reference orbit) composed with the motion along the orbit.

The growing (or compression) of these spiral motions is due to the real part of $\lambda_{3,4}$ and $\lambda_{5,6}$, which are nonzero but very small. For this reason the spiralling motion is very small (almost circular) and as we did in the RTBPS to decide on the manoeuvres we will assume that this motion is not an spiral but a simple rotation.

Notice that with this reference frame at each instant of time $t_1$ we have 3 planes where the dynamics is the same as the one in the RTBPS. We will use the same station keeping strategy described in Sect. 4.1.4 but looking at the solar sail trajectory in this time-dependent reference frame. Hence, we will set a fixed sail orientation $\alpha_0, \delta_0$ until the trajectory is about to escape along the unstable direction $|s_1(t)| > \varepsilon_{max}$. Then we choose a suitable new sail orientation $\alpha_1, \delta_1$ which make the trajectory of the solar sail get close to the stable manifold keeping the centre projections bounded. We will use the same ideas described in Sect. 4.1.3 to find this appropriate new sail orientation.

The main difference between the station keeping strategy used in the NBPS and the RTBPS is the reference frame that we use. In the case of the NBPS we have a time-dependent reference frame along the target orbit (or dynamical substitute), while in the RTBPS we have a fixed reference frame. But the projections of the trajectories in the two reference frames are very similar. Moreover, given the fact that the perturbations from the other planets are small the dynamics of the system is very similar and the performance of the control will be very similar. Presenting similar results for the controllability of the solar sail and robustness towards different sources of errors during the station keeping.

## 5 Mission Application

In this section we want to test the robustness of the station keeping strategy when we include the perturbing effect of the whole Solar system. For this purpose we have taken the Sunjammer mission and done several Monte Carlo simulations including different sources of error.

As in Sect. 4.1.5 we consider a solar sail with a sail lightness number $\beta = 0.0388$, which is considered to be a realistic value according to Sunjammer mission [19]. We compute the dynamical substitute of the target equilibria $p_0 = (-0.98334680272, -0.00146862443, 0.00000000000)$ (AU) for $\alpha_0 = 0.023954985$, $\delta_0 = 0.000000$ (rad), which can be seen in Fig. 12, and its associated Floquet reference frame described in Sect. 4.2.3. The mission goal is to remain close to the dynamical substitute for 10 years.

As mission parameters we have considered $\varepsilon_{max} = 5 \times 10^{-5} \approx 7479.89$ km, $\Delta t_{min} = 8$ days and $\Delta t_{max} = 100$ days. We have taken random initial conditions and performed the station keeping strategy during the lifetime of the mission (10 years).

In Fig. 13 we have in red the controlled trajectory of the solar sail and in green the dynamical substitute, in the XY-plane (left), YZ-plane (middle) and the XYZ-projection (right). As we can see the solar sail trajectory remains close to the target orbit.

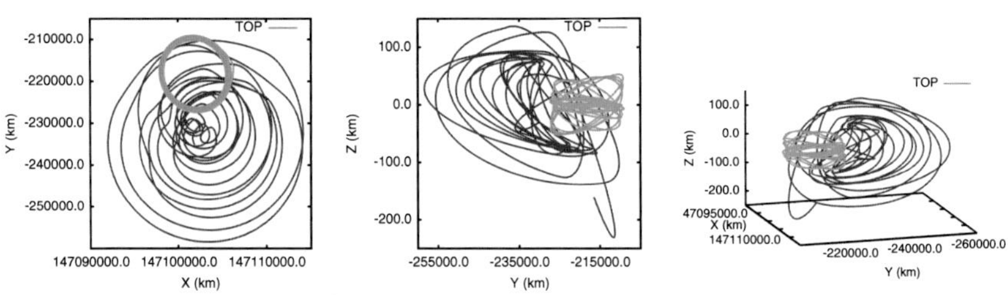


**Fig. 13** For the Sunjammer mission, (*red*) trajectory of the controlled solar sail for 10 years, (*green*) trajectory of the dynamical substitute: XY-plane (*left*), YZ-plane (*middle*) and XYZ-plane (*right*)

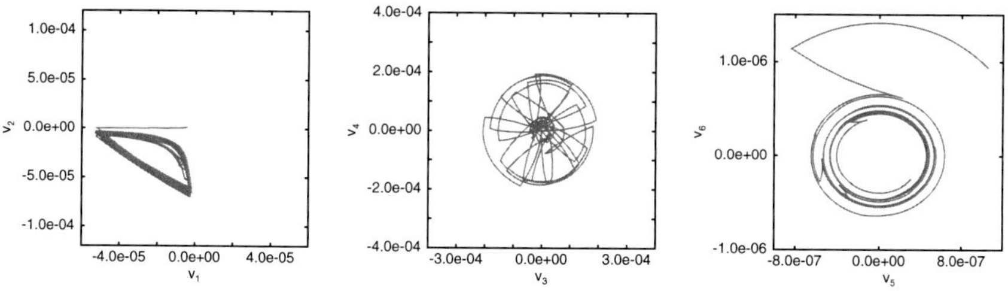


**Fig. 14** For the Sunjammer mission, trajectory of the controlled solar sail for 10 years on the: saddle plane (*left*), centre plane generated by $(\mathbf{v_3}, \mathbf{v_4})$ (*middle*) and centre plane generated by $(\mathbf{v_5}, \mathbf{v_6})$ (*right*)

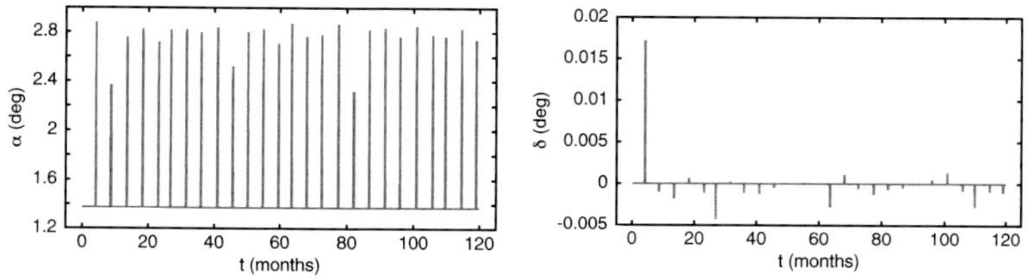


**Fig. 15** For the Sunjammer mission, variation of the sail orientation along time: $\alpha$ variation (*left*); $\delta$ variation (*right*)

In Fig. 14 we have the projection of the controlled trajectory in the saddle $\times$ centre $\times$ centre reference frame along the orbit. Where we can see how the projection in the saddle plane (left) is a connection of saddle motions that remain always bounded. On the other hand, the trajectory on the two centre components (middle and right) is a connection of rotations around different centres and remains bounded through time.

Finally in Fig. 15 we see the variation of the two angles defining the sail orientation $\alpha$ (left) and $\delta$ (right). Notice that this variations is very similar to the one observed for this mission using as dynamical model the RTBPS (Fig. 10).

### 5.1 *Mission Results*

In order to test the robustness of the strategy we have done some Monte Carlo simulations. We also want to test its sensitivity to different sources of error. It is a known fact that during a mission the position and velocity of the probe are not determined exactly, this will have an effect on the decisions taken by the control algorithm. Errors on the sail orientation can also be made, which have an impact on the sails trajectory. We will include these two sources of error in our simulations and discuss their effects.

We have taken 1000 random initial conditions close to the nominal orbit, following a normal distribution with zero mean and variance one whit a maximum size of the dispersion of $10^{-5} \approx 1495.99$ km. For each initial condition we have done a simulation considering no sources of errors, one only considering errors on the position and velocities and another considering also errors on the sail orientation. For each simulation we will check if the station keeping is able to keep the solar sail trajectory close to the target orbit. We will also measure the maximum and minimum time between manoeuvres, and the maximum and minimum size of the variations on the sail orientation $\Delta\alpha$ and $\Delta\delta$. The results are summarised in Table 3. The first column shows the success percentage, the second and third columns the maximum and minimum time between manoeuvres and the forth and fifth columns the range of variations for the two sail angles, $\alpha$ and $\delta$ respectively.

If we look at the first row in Table 3, results when no errors are considered, we see that the mean maximum and minimum time between manoeuvres are 34.05 months and 1.27 months respectively. If we look at the variation of the sail orientation we have that the average maximum variation is around 1.49° for $\alpha$, while the variation in $\delta$ is almost zero.

We have used standard values for the errors in position determination: they are assumed to follow a normal distribution with zero mean, with a precision on the position of the probe of ≈1 m in the space slant and ≈2–3 milli-arc-seconds in the angle determination of the probe. The precision in speed is around 20–30 $\mu$/s. These errors are introduced every time the control algorithm asks for the position of the probe to decide if a manoeuvre must be done, hence errors made on the measurement of the probes position will make, the algorithm change the sail orientation when not desired and the new fixed points position will also be modified. If these errors are not very big the difference between changing the sail orientation a little before or after in time will not affect the control of the probe. As we can see in Table 3 (second row) the effect of these errors turns out to be almost negligible.

Let us focus on the errors due to the sail orientation, we will see that these errors have an important effect on the sail trajectory and the controllability of the probe. Each time we change the sail orientation an error is introduced ($\alpha = \alpha_1 + \varepsilon_\alpha$, $\delta = \delta_1 + \varepsilon_\delta$),

**Table 3** Statistics on the Monte Carlo simulations for 1000 random initial conditions

| Succ (100 %) | $\Delta t_{max}$ (months) | $\Delta t_{min}$ (months) | $\Delta\alpha$ (deg) | $\Delta\delta$ (deg) |
|---|---|---|---|---|
| No Err | 34.05 | 1.27 | 1.49–0.41 | 0.06–0.00 |
| PV Err | 33.61 | 1.27 | 1.49–0.41 | 0.06–0.00 |
| SS Err* | 35.78 | 0.34 | 3.79–0.17 | 1.18–0.00 |
| SS Err† | 25.91 | 0.31 | 9.62–0.01 | 4.58–0.00 |
| SS Err‡ | 22.16 | 0.31 | 12.2–0.01 | 8.02–0.01 |

Results considering: No errors during the manoeuvres (first line), Errors only on the position and velocity determination (second line), Errors on the position and velocity determination and on the sail orientation (third to fifth line). The maximum error on the sail orientation is considered: 0.1° in Err*, 0.5° in Err† and 1.0° in Err‡ then the new fixed point $p_1$ is shifted $p(\alpha, \delta) = p(\alpha_1, \delta_1) + \varepsilon_p$ and so do the stable and unstable directions $\mathbf{v}_{1,2}(\alpha, \delta) = \mathbf{v}_{1,2}(\alpha_1, \delta_1) + \varepsilon_v$. Due to the sensitivity of the position of the equilibria to changes on the sail orientation, these errors can make the probes trajectory escape as the new equilibria can be placed on the incorrect side of the saddle or the central behaviour can blow up.

As solar sailing is a relatively new technology and there have been few demonstration missions, there is no information on estimates for the errors in the sail orientation. This is why we have considered different magnitudes for this error, in order to see which is the maximum error we can afford. We have considered $\varepsilon_\alpha = \varepsilon_\delta$ to follow a normal distribution with zero mean and $\varepsilon_{max} = 0.1°, 0.5°$ and $1°$. As we can see in Table 3 errors of order $0.1°$ are easily absorbed, but now we have that the average minimum time between manoeuvres is 0.34 months, more than half the size of the minimum time when no errors are considered. This means that the algorithm is obliged to do faster changes on the sail orientation to compensate the errors made. We also see that the maximum variation in $\alpha$ and $\delta$ are $3.79°$ and $1.18°$ larger than for the no error simulations.

If we look at the results for $\varepsilon_{max} = 0.5°$ and $1°$ we see that the maximum variation in $\alpha$ is $9.62°$ and $12.2°$ respectively. These variations are very large and despite we have a 100 % of success in the simulations we can say that we are at the verge of the controllability. As we will see in the following figures the station keeping strategy has to do very drastic changes on the sail orientation to control the trajectory and these changes might not be feasible.

In Fig. 16 we see the variation on the sail orientation for $\varepsilon_{max} = 0.1$ (top) and $\varepsilon_{max} = 0.5$ (bottom), in both cases we can see the effect of the errors on the sail

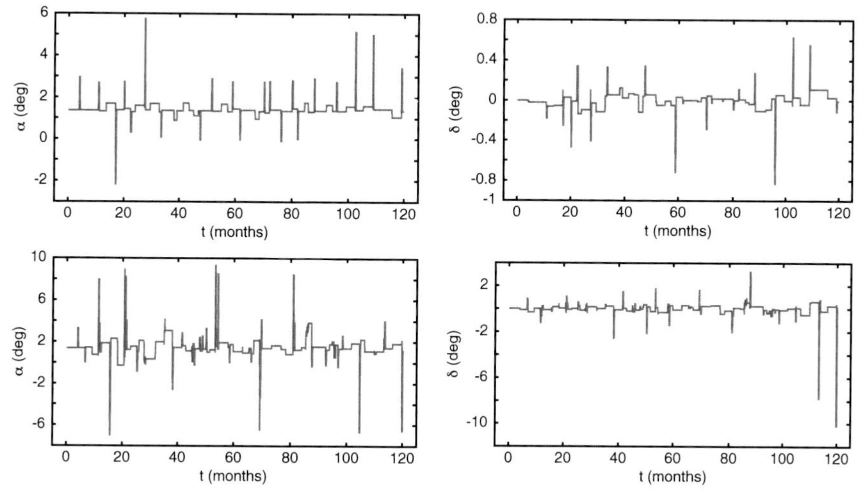


**Fig. 16** For the Sunjammer mission, variation of the sail orientation along time: $\alpha$ variation (*left*); $\delta$ variation (*right*). Simulations with errors on the sail orientation: $\varepsilon_{max} = 0.1$ (*top*) and $\varepsilon_{max} = 0.5$ (*bottom*)

orientation. Notice that for simulations considering $\varepsilon_{max} = 0.5$ there are times when we have a succession of quick changes on the sail orientation.

In Fig. 17 we have the trajectory the sail follows for different projection in the XYZ-plane. Notice how the trajectory still remains close to the target orbit (green). In Fig. 18 we have the projection of the trajectory on the saddle and centre planes.

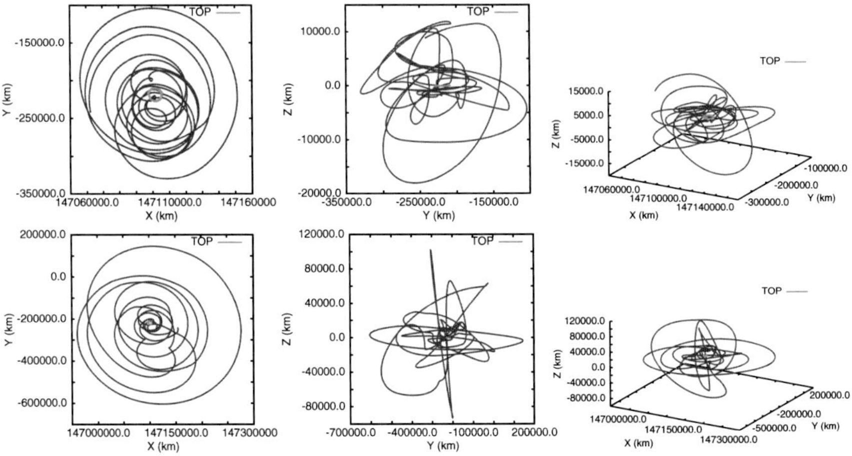


**Fig. 17** For the Sunjammer mission, (*red*) trajectory of the controlled solar sail for 10 years, (*green*) trajectory of the dynamical substitute: XY-plane (*left*), YZ-plane (*middle*) and XYZ-plane (*right*). Simulations with errors on the sail orientation: $\varepsilon_{max} = 0.1$ (*top*) and $\varepsilon_{max} = 0.5$ (*bottom*)

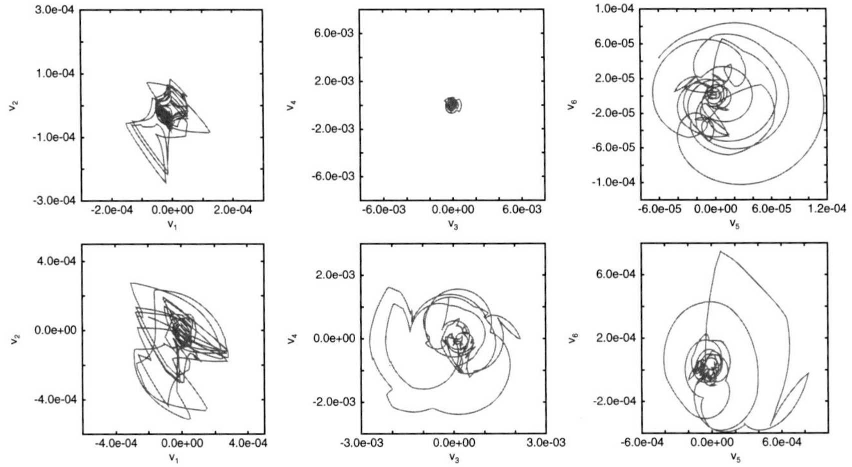


**Fig. 18** For the Sunjammer mission, trajectory of the controlled solar sail for 10 years on the: saddle plane (*left*), centre plane generated by $(\mathbf{v_3}, \mathbf{v_4})$ (*middle*) and centre plane generated by $(\mathbf{v_5}, \mathbf{v_6})$ (*right*). Simulations with errors on the sail orientation: $\varepsilon_{max} = 0.1$ (*top*) and $\varepsilon_{max} = 0.5$ (*bottom*)

Here we clearly see the effect of the errors on the sail orientation. In some cases, when the trajectory should return to the stable manifolds, the error on the orientation will lead to a different behaviour. Nevertheless the station keeping strategy is able to compensate these errors in both cases. If we look at the results for $\varepsilon_{max} = 0.5^\circ$ (Fig. 18 bottom) we can see that in some cases the trajectory does not really follow a saddle motion but rather a fussy one. This is due to the quick changes on the sail orientation trying to compensate the divergence of the sail trajectory.

# 6 Conclusions

In this paper we present a detailed description on how to use the information on the natural dynamics of the RTBP to derive station keeping strategies for a solar sail around an equilibrium point [10, 11]. These strategies are general enough and can be extended to deal with the station keeping of a periodic orbit [13].

Moreover, we have shown how to extend these ideas when we deal with a real mission scenario, i.e. when we include the effect of all the main bodies in the Solar system. For this purpose we need to compute the dynamical substitute of the equilibrium points and reference frame along the orbit to know the relative position between the solar sail trajectory and the stable and unstable invariant manifolds.

We have tested the robustness of these strategies for the Sunjammer mission, where we have performed several Monte Carlo simulations including different sources of error. Errors in the position and velocity determination and errors on the solar sail orientation. We have seen that the most relevant errors are those regarding the sail orientation, as small changes on the sail orientation can derive on big changes on the phase portrait. The station keeping strategy is able to deal with errors up to 1°.

In order to improve these results we propose to use higher order variationals to define the $\mathscr{F}(\Delta\alpha, \Delta\delta, \Delta t)$ map (Sect. 4.1.3) in order to represent more accurately larger variations in $\alpha$ and $\delta$ as we believe might be the main limiting factor.

**Acknowledgments** This work has been supported by the MEC grant MTM2012-32541, the AGAUR grant 2014 SGR 1145 and the AGAUR postdoctoral fellowship Beatriu de Pinós (BP-B 00142-2011).

# References

1. Aliasi G, Mengali G, Quarta A (2011) Artificial equilibrium points for a generalized sail in the circular restricted three-body problem. Celest Mech Dyn Astron 110:343–368
2. Ceriotti M, McInnes C (2010) A near term pole-sitter using hybrid solar sail propulsion. In: Kezerashvili R (ed) Proceedings of the second international symposium on solar sailing, pp 163–169
3. Ceriotti M, McInnes C (2012) Natural and sail-displaced doubly-symmetric lagrange point orbits for polar coverage. Celest Mech Dyn Astron 114(1–2):151–180
4. Crawford J (1991) Introduction to bifurcation theory. Rev Modern Phys **64** (1991)

5. Dachwald B, Seboldt W, Macdonald M, Mengali G, Quarta A, McInnes C, Rios-Reyes L, Scheeres D, Wie B, Görlich M, et al (2005) Potential solar sail degradation effects on trajectory and attitude control. In: AIAA guidance, navigation, and control conference and exhibit, vol 6172
6. Farrés A (2009) Contribution to the dynamics of a solar sail in the earth-sun system. PhD thesis, Universitat de Barcelona
7. Farrés A, Jorba À (2008) Dynamical system approach for the station keeping of a solar sail. J Astronaut Sci 58(2):199–230
8. Farrés A, Jorba À (2008) Solar sail surfing along families of equilibrium points. Acta Astronaut 63:249–257
9. Farrés A. Jorba À (2010) On the high order approximation of the centre manifold for ODEs. Discrete Contin Dyn Syst Ser B (DCDS-B), **14**:977–1000
10. Farrés A, Jorba À (2010) Periodic and quasi-periodic motions of a solar sail around the family $SL_1$ on the sun-earth system. Celest Mech Dyn Astron 107:233–253
11. Farrés A, Jorba À (2010) Sailing between the earth and sun. In: Kezerashvili R (ed) Proceedings of the second international symposium on solar sailing, pp 177–182
12. Farrés A, Jorba À (2011) On the station keeping of a solar sail in the elliptic sun-earth system. Adv Space Res 48:1785–1796. doi:10.1016/j.asr.2011.02.004
13. Farrés A, Jorba À (2014) Station keeping of a solar sail around a halo orbit. Acta Astronaut 94(1):527–539
14. Farrés A, Matteo C (2012) Solar sail station keeping of high-amplitude vertical lyapunov orbits in the sun-earth system. In: Proceedings of the 63rd international astronautical congress, naples, campania, Italy
15. Forward RL (1990) Statite: a spacecraft that does not orbit. J Spacecraft 28(5):606–611
16. Gómez G, Jorba À, Masdemont J, Simó C (2001) Dynamics and mission design near libration points—volume iii: advanced methods for collinear points, volume 4 of world scientific monograph series in mathematics. world scientific
17. Gómez G, Jorba À, Masdemont J, Simó C (2001) Dynamics and mission design near libration points—volume iv: advanced methods for triangular points, volume 5 of world scientific monograph series in mathematics. world scientific
18. Gómez G, Llibre J, Martínez R, Simó C (2001) Dynamics and mission design near libration points—volume i: fundamentals: the case of collinear libration points, volume 2 of world scientific monograph series in mathematics. world scientific
19. Heiligers J, Diedrich B, Derbes B, McInnes C (2008) Sunjammer : preliminary end-to-end mission design. In: Proceedings of the AIAA/AAS astrodynamics specialist conference, vol 2014
20. Jorba À, Simó C (1996) On quasiperiodic perturbations of elliptic equilibrium points. SIAM J Math Anal 27(6):1704–1737
21. Jorba A, Villanueva J (1997) On the persistence of lower dimensional invariant tori under quasi-periodic perturbations. J Nonlinear Sci 7:427–473. doi:10.1007/s003329900036
22. Lamb J, Roberts J (1998) Time-reversal symmetry in dynamical systems: a survey. Phys D 112:1–39
23. Lawrence D, Piggott S (2004) Solar sailing trajectory control for Sub-L1 stationkeeping. AIAA **2004–5014**
24. Lisano M (2005) Solar sail transfer trajectory design and station keeping control for missions to Sub-L1 equilibrium region. In: 15th AAS/AIAA space flight mechanics conferece, colorado (2005). AAS paper 05–219
25. Macdonald M, McInnes C (2004) A near-term road map for solar sailing. In: 55th international astronautical congress, Vancouver, Canada
26. Macdonald M, McInnes C (2011) Solar sail science mission applications and advancement. Adv Space Res 48:1702–1716. doi:10.1016/j.asr.2011.03.018
27. McInnes C (1999) Solar sailing: technology. Dynamics and Mission Applications, Springer-Praxis

28. McInnes C, McDonald A, Simmons J, MacDonald E (1994) Solar sail parking in restricted three-body system. J Guidance Control Dyn 17(2):399–406
29. McKay R, Macdonald M, Biggs J, McInnes C (2011) Survey of highly non-keplerian orbits with low-thrust propulsion. J Guidance Control Dyn 34(3):645–666. doi:10.2514/1.52133
30. Rios-Reyes L, Scheeres D (2005) Robust solar sail trajectory control for large pre-launch modelling errors. In: 2005 AIAA guidance, navigation and control conference
31. Sevryuki M (1986) Reversible systems. Springer, Berlin
32. Szebehely V (1967) Theory of orbits. Academic Press, The restricted problem of three bodies
33. Yen C-WL (2004) Solar sail geostorm warning mission design. In: 14th AAS/AIAA space flight mechanics conference, Hawaii

# Minimum Fuel Round Trip from a $L_2$ Earth-Moon Halo Orbit to Asteroid 2006 $RH_{120}$

**Monique Chyba, Thomas Haberkorn and Robert Jedicke**

**Abstract** The goal of this paper is to design a spacecraft round trip transfer from a parking orbit to asteroid 2006 $RH_{120}$ during its geocentric capture while maximizing the final spacecraft mass or, equivalently, minimizing the delta-v. The spacecraft begins in a halo "parking" orbit around the Earth-Moon $L_2$ libration point. The round-trip transfer is composed of three portions: the approach transfer from the parking orbit to 2006 $RH_{120}$, the rendezvous "lock-in" portion with the spacecraft in proximity to and following the asteroid orbit, and finally the return transfer to $L_2$. An indirect method based on the maximum principle is used for our numerical calculations. To partially address the issue of local minima we restrict the control strategy to reflect an actuation corresponding to up to three thrust arcs during each portion of the transfer. Our model is formulated in the circular restricted four-body problem (CR4BP) with the Sun considered as a perturbation of the Earth-Moon circular restricted three body problem. A shooting method is applied to numerically optimize the round trip transfer, and the 2006 $RH_{120}$ rendezvous and departure points are optimized using a time discretization of the 2006 $RH_{120}$ trajectory.

**Keywords** Asteroid 2006 $RH_{120}$ · Sun perturbed Earth-Moon bicircular restricted four body problem · Geometric optimal control · Shooting method

M. Chyba
Department of Mathematics, University of Hawaii, Honolulu, HI 96822, USA
e-mail: chyba@math.hawaii.edu

T. Haberkorn (✉)
MAPMO-Fédération Denis Poisson, University of Orléans, 45067 Orléans, France
e-mail: thomas.haberkorn@univ-orleans.fr

R. Jedicke
Institute for Astronomy, University of Hawaii, Honolulu, HI 96822, USA
e-mail: jedicke@hawaii.edu



B. Bonnard and M. Chyba (eds.), *Recent Advances in Celestial and Space Mechanics*, Mathematics for Industry 23, DOI 10.1007/978-3-319-27464-5_4

# 1 Introduction

A population of geocentric near Earth asteroids, Temporarily Captured Orbiters (TCOs), may be the lowest delta-v targets for spacecraft missions. The TCOs are defined by the following simultaneous conditions [1]:

1. negative geocentric Keplerian energy $E_{\text{planet}}$;
2. geocentric distance less than three Earth Hill radii ($3R_{H,\oplus} \sim 0.03$ AU);
3. make at least one full revolution around the Earth in the Sun-Earth rotating frame.

They are often referred to as *minimoons* but we will use TCO here. In [1], the authors generated, pruned and integrated a very large random sample of "test-particles" from the near Earth object (NEO) population to determine the steady-state orbit distribution of the TCO population. Of the 10 million integrated test-particles over 16,000 became TCOs. These capture statistics imply that at any moment there is a 1 m diameter TCO orbiting Earth. The advantages presented by the TCOs for space missions have been discussed in several papers [1–4] and we will not repeat the arguments here. In particular, their location within the Earth-Moon system is very attractive and their geocentric energy will reduce the thrust required to reach them relative to otherwise equivalent heliocentric objects passing through cis-lunar space.

The orbits of the TCOs presented in [1] exhibit a wide range of behaviors with capture duration ranging from a few weeks to a few months. In this paper we focus on designing a round trip minimum fuel transfer to the only known TCO, 2006 $RH_{120}$. It is a few meters diameter asteroid that was discovered by the Catalina Sky Survey in September 2006. Its orbit from June 1st 2006 to July 31st 2007 is represented on Figs. 1 and 2. The period June 2006 to July 2007 was chosen to include the time during which the asteroid was energetically bound in the Earth-Moon system. 2006 $RH_{120}$ approaches as close as 0.72 Earth-Moon distance from the Earth-Moon barycenter.

Motivated by the successful Artemis mission and prior numerical simulations on the approach transfer [5], while awaiting the discovery of a suitable TCO we assume the spacecraft is hibernating on a halo orbit around the Earth-Moon $L_2$ libration point with a $z$-excursion of 5000 km (Fig. 3). This orbit is similar to those successfully used for the Artemis mission [6, 7]. The highest $z$-coordinate of the halo orbit is $q_{HaloL_2} \approx (1.119, 0, 0.013, 0, 0.180, 0)$ and its period is $t_{HaloL_2} \approx 3.413$ normalized time units or 14.84 days.

The round trip is composed of an approach transfer to bring the spacecraft to 2006 $RH_{120}$, followed by a lock-in phase where the spacecraft travels with the asteroid, and finally a return transfer to the hibernating orbit. Clearly, this optimization problem presents a very large set of variables including the departure time, the target rendezvous point, the lock-in duration on the asteroid and the return transfer duration. To simplify our approach we first decompose the round trip into an approach transfer and a return transfer that we address separately. We consider the global mission once these two optimization sub-problems are solved.

The spacecraft's departure time from the hibernating orbit must occur after the TCO's detection time. We will vary the departure time or/and the detection time to

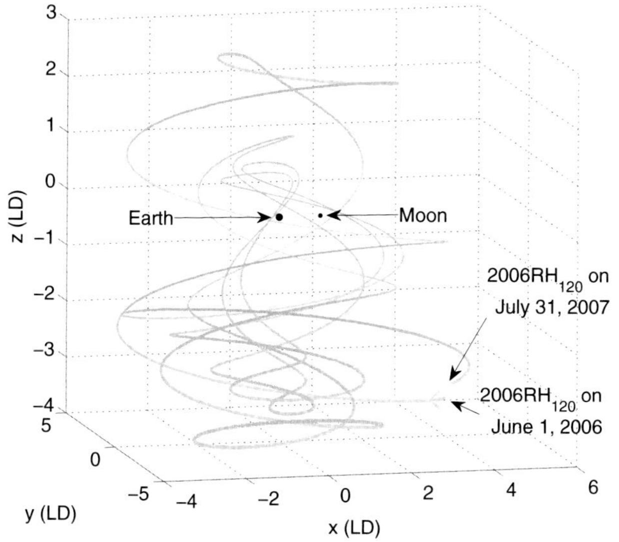


**Fig. 1** 2006 $RH_{120}$ trajectory between June 1st 2006 and July 31st 2007 in the Earth-Moon rotating frame (Ephemerides generated using the jet propulsion laboratory's HORIZONS database.)

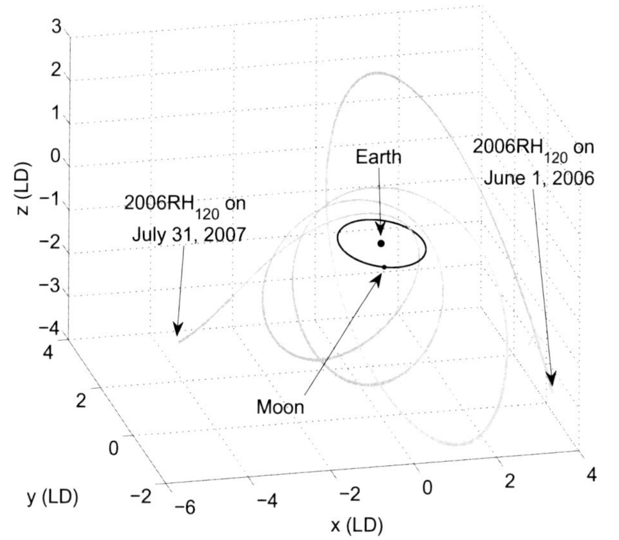


**Fig. 2** 2006 $RH_{120}$ trajectory between June 1st 2006 and July 31st 2007 in an Earth-centered inertial frame (Ephemerides generated using the jet propulsion laboratory's HORIZONS database.)

analyze its impact on fuel consumption even though it is actually known for 2006 $RH_{120}$. In addition to the transfer duration discretization, we also discretize the TCO orbit to optimize the rendezvous point. The combination of discretizations results in more than 5000 optimization problems.

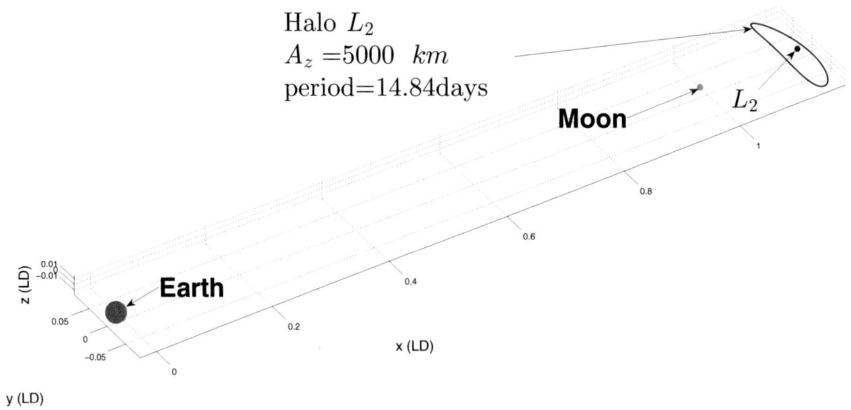


**Fig. 3** Parking halo orbit (in the EM rotating frame) around $L_2$ with a $z$-excursion of 5000 km

The return transfer optimizes the trajectory from 2006 $RH_{120}$ to the Earth-Moon $L_2$ libration point rather than the halo orbit to reduce the number of calculations (it can be expanded easily if necessary). This problem will be solved with fixed transfer durations and we will study the influence of the departure point on the 2006 $RH_{120}$ trajectory and of the transfer duration on the fuel consumption. As for the approach transfer, there are more than 5000 return transfers to be considered.

The global round trip will be analyzed based on the results of the rendezvous and return transfers. The best transfers in each case can be connected together to minimize the fuel consumption with the additional, obvious constraint that the return trip must begin after the rendezvous trip ends. The result provides us the optimal duration of the lock-in phase for mission planning purposes.

We use indirect shooting methods [3–5, 8] to solve the optimization problems associated with our mission. The main difficulty with these methods is the initialization of the algorithm and the existence of numerous local minima. To partially reduce the number of local minima, we fix the control structure to be composed of at most three constant norm thrust arcs separated in time by 2 ballistic arcs. We have two motivations for imposing this control structure. First, preliminary calculations on a set of random TCOs with a free control structure for a similar control problem in [5] provided results mimicking transfers with at most three impulsive thrust maneuvers: more thrusting periods reduced the cost only rarely. Second, since the parking orbit is not a periodic orbit of the CR4BP we have to impose an initial impulse to leave the halo orbit. Indeed, starting the transfer with a ballistic arc is extremely unlikely to be efficient or even possible since the time duration of the transfer is fixed. This strongly suggests a strategy with one impulse to leave the hibernating orbit, a second one to redirect the spacecraft toward the rendezvous point, and a final one to match the position and the velocity of the asteroid at rendezvous. Based on the GRAIL spacecraft's characteristics we consider a chemical propulsion spacecraft with maximum thrust $T_{\max} = 22\,\mathrm{N}$, specific impulse $I_{sp} = 230\,\mathrm{s}$, and initial mass $m_0 = 350\,\mathrm{kg}$. Information about monopropellant engine types can be found in [9].

The novelty of this work is at least threefold. First, the target object is a TCO with a "crazy straw" geocentric trajectory quite different from the periodic elliptical orbits considered in the literature of minimum fuel transfers. Second, we consider synchronized transfers to produce a global round trip mission and add a practical detection constraint. The existence of efficient round trip transfers enables multiple rendezvous scenario with successive TCOs to maximize the spacecraft utility. Third, the calculation of all the possible approach transfers with respect to departure time and rendezvous point, and all possible return transfers with respect to the departure point on the TCO trajectory, makes this study comprehensive. The trade-off is our restricting the mission to a three-thrust strategy.

The techniques presented here can be applied to any TCO. Asteroid 2006 $RH_{120}$ was chosen as a test-bed to illustrate the algorithm since it is currently the only known TCO. Future discoveries of TCOs should be common as [2] predict that the Large Synoptic Survey Telescope (LSST) could detect about 1.5 TCOs/lunation—more than a dozen per year. This would provide ample population candidates for a real asteroid space mission. Furthermore [10] have shown that LSST will efficiently detect the largest TCOs that are arguably the most interesting spacecraft targets.

The outline of the paper is as follows: Sect. 2 presents the equations of motion used as the dynamics of the optimal control problems. Section 3 gives the exact formulation of the two optimal control problems, the necessary conditions satisfied by their solutions, and the numerical method used for the calculations. Section 4 provides the numerical results for the two optimal transfers and discusses the complete round trip problem. We conclude with a discussion of future research opportunities.

## 2 Equations of Motion

We expected that the spacecraft would remain within or near Earth's Hill sphere during its entire mission in which case the Earth-Moon circular restricted three body problem (CR3BP) would provide a good first-order approximation for the equations of motion (see [11]). In the CR3BP setting the spacecraft has negligible mass and responds to the gravitational fields of 2 primaries, $P_1$ and $P_2$, of respective masses $M_1$ and $M_2$ with $M_1 > M_2$. In addition, the two primaries follow circular orbits around their barycenter. A normalization to obtain a dimensionless system is introduced by setting the mass unit to $M_1 + M_2$, the unit of length as the constant distance between the two primaries, and the unit of time so that the period of the primaries around their barycenter is $2\pi$. The only parameter in the model is then $\mu = M_1/(M_1 + M_2)$. Table 1 provides numerical values for some of the parameters of our CR3BP model.

Finally, we introduce a rotating reference frame with origin at the center of mass with the $x$-axis oriented from $P_1$ to $P_2$. The $z$-axis is orthogonal to the orbital plane of the 2 primaries and the $y$-axis is orthogonal to both in a conventional right-hand coordinate system. In this reference frame the potential energy of the spacecraft with position and velocity $q = (x, y, z, \dot{x}, \dot{y}, \dot{z})$ is

**Table 1** Numerical values for the CR3BP and sun perturbed CR3BP

| CR3BP parameters | | Sun perturbed parameters | |
|---|---|---|---|
| $\mu$ | 0.01215361914 | $\mu_S$ | $3.289 \cdot 10^5$ |
| 1 norm. dist. (LD) | 384400 km | $r_S$ | $3.892 \cdot 10^2$ (LD) |
| 1 norm. time | 104.379 h | $\omega_S$ | $-0.925$ rad/norm. time |

$$\Omega_3(x, y, z) = \frac{x^2 + y^2}{2} + \frac{1-\mu}{\rho_1} + \frac{\mu}{\rho_2} + \frac{\mu(1-\mu)}{2},$$

with $\rho_1$ (resp. $\rho_2$) the distance from the spacecraft to the first (resp. second) primary, that is

$$\rho_1 = \sqrt{(x-\mu)^2 + y^2 + z^3}, \quad \rho_2 = \sqrt{(x - 1 + \mu)^2 + y^2 + z^2}.$$

The ballistic motion of the spacecraft in our 3-body CR3BP is then given by

$$\ddot{x} - 2\dot{y} = \frac{\partial \Omega_3}{\partial x}, \qquad \ddot{y} + 2\dot{x} = \frac{\partial \Omega_3}{\partial y}, \qquad \ddot{z} = \frac{\partial \Omega_3}{\partial z}. \tag{1}$$

It is well known that there exists 5 equilibrium points in this system, the so called Lagrange points $L_1$, $L_2$, $L_3$, $L_4$ and $L_5$. The points $L_{1,2,3}$ are distributed along the $x$-axis of the frame while $L_4$ and $L_5$ form an equilateral triangle with the two primaries in the $xy$-plane. We will focus on $L_2$, motivated by the existence of periodic orbits around this point that can be used as hibernating orbits for a spacecraft awaiting detection of a TCO. We could also choose $L_1$ or $L_3$ but preliminary computation suggested that $L_2$ is a better choice. Halo orbits around $L_2$ are periodic orbits that are isomorphic to circles (see [11] for a proof of their existence and how to compute them).

Even though the TCO's orbit is in a vicinity of the Earth-Moon system during its capture it can be as far as 5 normalized distance units from the CR3BP origin. Our preliminary calculations showed that efficient transfers might require that the spacecraft range even further from the CR3BP origin to maximize the thrust impact on the spacecraft's motion. To increase the accuracy of our model in these cases we use an extended CR3BP as in [12] and employ a Sun-perturbed Earth-Moon CR3BP in which the Sun follows a circular orbit around the Earth-Moon center of mass without modifying their circular orbits. In this 4-body case the potential energy of the spacecraft is $\Omega_4 = \Omega_3 + \Omega_S$ with

$$\Omega_S(x, y, z, \theta) = \frac{\mu_S}{r_S} - \frac{\mu_S}{\rho_S^2}(x\cos\theta + y\sin\theta), \tag{2}$$

where $\theta$ is the time dependent angular position of the Sun in the rotating frame, $r_S$ is the constant distance from the Sun to the center of the reference frame, $\rho_S$ is the distance from the spacecraft to the Sun ($\rho_S = \sqrt{(x - r_S\cos\theta)^2 + (y - r_S\sin\theta)^2 + z^2}$) and $\mu_S$ is the Sun's normalized mass ($\mu_S = M_{\text{Sun}}/(M_1 + M_2)$). As the Sun is assumed to follow a circular orbit in the rotating frame its angular position is $\theta(t) = \theta_0 + t\omega_S$ with $\omega_S$ the angular speed of the circular orbit and $\theta_0$ the angular position of the Sun at time 0. The equations of motion take the same form as in (1) but with the perturbed potential $\Omega_4$ replacing $\Omega_3$. The values of the new parameters are given in Table 1.

We assume the spacecraft thrusters can produce a thrust of at most $T_{\max}$ Newton in any direction of $\mathbb{R}^3$. With the thrust direction represented by $u = (u_1, u_2, u_3) \in \bar{B}(0,1) \subset \mathbb{R}^3$, and when $m(\cdot)$ is the mass of the spacecraft, the controlled equations of motion are an affine control system

$$\dot{q}(t) = F_0(t, q(t)) + \frac{\tilde{T}_{\max}}{m(t)} \sum_{i=1}^{3} F_i u_i(t) \tag{3}$$

where the drift is given by:

$$F_0(t,q) = \begin{pmatrix} \dot{x} \\ \dot{y} \\ \dot{z} \\ 2\dot{y} + x - \frac{(1-\mu)(x+\mu)}{\rho_1^3} - \frac{\mu(x-1+\mu)}{\rho_2^3} - \frac{(x-r_S\cos\theta(t))\mu_S}{\rho_S^3(t)} - \frac{\mu_S\cos\theta(t)}{r_S^2} \\ -2\dot{x} + y - \frac{(1-\mu)y}{\rho_1^3} - \frac{\mu y}{\rho_2^3} - \frac{(y-r_S\sin\theta(t))\mu_S}{\rho_S^3(t)} - \frac{\mu_S\sin\theta(t)}{r_S^2} \\ -\frac{(1-\mu)z}{\rho_1^3} - \frac{\mu z}{\rho_2^3} - \frac{z\mu_S}{\rho_S^3(t)} \end{pmatrix}. \tag{4}$$

The vector field $F_1$ (resp. $F_2$ and $F_3$) is the vector of the canonical base $e_4$ (resp. $e_5$ and $e_6$) of $\mathbb{R}^6$ and $\tilde{T}_{\max}$ is the maximum thrust expressed in normalized units. To complete the model, the mass decreases proportionally to the delivered thrust

$$\dot{m}(t) = -\frac{\tilde{T}_{\max}}{I_{sp} g_0} \|u(t)\|, \tag{5}$$

where $I_{sp}$, the specific impulse, is thruster-dependent, and $g_0$ is the gravitational acceleration at Earth's sea level. For our numerical tests we use

$$T_{\max} = 22\,\text{N},\ I_{sp} = 230\,\text{s},\ g_0 = 9.80665\,\text{m/s}^2,\ m_0 = 350\,\text{kg}.$$

With our fixed control structure it is simple to change these parameters using a continuation to consider other thruster specifications *e.g.* a solar electric propulsion with smaller $T_{\max}$ but higher $I_{sp}$. Indeed, the main obstacle to a continuation that is based solely on the thruster parameters would be the possible change in the control structure since there is no smooth continuation path between a minimum fuel transfer with $n$ switchings and another with $n \pm 1$ switchings.

# 3 Problem Statement

The aim of this paper is to design a minimum fuel round-trip transfer to 2006 $RH_{120}$ from a hibernating orbit including a rendezvous period during which the spacecraft travels with the TCO. Since we want the round trip journey to account for various synchronization constraints, it becomes complex when written as a single optimal control problem. To avoid this issue and obtain more general results on each portion of the global transfer, we decouple the rendezvous and return transfers. After optimizing these two problems for departure time, rendezvous point, return time and transfer duration, it is straightforward to pair them with a lock-in phase between the spacecraft and the asteroid to complete the round trip. In this section we introduce the rendezvous and return transfers as optimal control problems and provide the necessary conditions for an optimal control strategy with its associated trajectory.

## *3.1 Approach Transfer*

The first component of the round trip journey is to enable the spacecraft to encounter 2006 $RH_{120}$. We fix the origin of our mission time frame, $t_c = 0$, as the asteroid's geocentric capture time, June 1st 2006 and introduce the following assumptions.

**Assumption 3.1** At $t_c$ the spacecraft is already hibernating on a CR3BP-periodic halo orbit around the Earth-Moon $L_2$ libration point with a $z$-excursion of 5000 km. We arbitrarily fix the position and velocity of the spacecraft at $t_c$ to be $q_{HaloL_2} \approx (1.119, 0, 0.013, 0, 0.180, 0)$ which corresponds to the highest $z$ point on the halo orbit. The spacecraft's starting location can be easily altered in future work.

We denote by $t_{\mathrm{start}}$ the departing time of the spacecraft from the hibernating orbit. The position and velocity $q_{\mathrm{start}}$ of the spacecraft at $t_{\mathrm{start}}$ are determined as the result of the CR3BP uncontrolled dynamic, see Eq. (1). We integrate from $q_{HaloL_2}$ at $t_c = 0$ to $t_{\mathrm{start}}$ to guarantee the spacecraft departs from its correct location on the $L_2$-halo periodic orbit. Our algorithm treats $t_{\mathrm{start}}$ as an optimization variable of the rendezvous problem and we will discretize the spacecraft's departure time to analyze its impact on the final mass. The detection time $t_d^{\mathrm{RH}}$ is a practical mission constraint and in Sect. 4 we discuss how our results provide information regarding ideal windows of detection for 2006 $RH_{120}$ corresponding to the best transfers. 2006 $RH_{120}$ was actually discovered on September 14th, 2006, about 3.5 months after it was captured, but we consider it as a parameter of the problem associated with the mission start time to gain insights on future studies with other TCOs.

The rendezvous point between the spacecraft and the asteroid is a position and velocity $q_f^{\mathrm{rdv}}$ on the 2006 $RH_{120}$ orbit corresponding to a time $t_{\mathrm{rdv}}^{\mathrm{RH}} > t_{\mathrm{start}}$. Our algorithm treats the rendezvous point as an optimization variable and includes a discretization of the 2006 $RH_{120}$ orbit to analyze the impact of the rendezvous point on the fuel consumption. We also require that $t_{\mathrm{rdv}}^{\mathrm{RH}} \leq$ July 31st 2007, *i.e.* that the rendezvous must take place before the asteroid leaves Earth's gravitational field.

To reduce the complexity of the optimization problem we fix the structure of the thrusts for the candidate trajectories to achieve optimality. Our choice is motivated by the desire to mimic a pseudo-realistic strategy with at most three impulsive boosts: one to depart from the halo orbit, a second to redirect the spacecraft to encounter 2006 $RH_{120}$ on its orbit, and a third for the rendezvous to match the position and velocity of the spacecraft to the asteroid. In other words, we impose a control strategy with at most three thrust arcs and two ballistic arcs. Prior numerical calculations have shown that this strategy provides fuel efficient transfers. We employ the same thrust structure for the return portion of the journey. To summarize:

**Assumption 3.2** For our transfers, we restrict the thrust strategy $u(\cdot) : [t_{\text{start}}, t_{\text{rdv}}^{\text{RH}}] \to \bar{B}(0, 1) \subset \mathbb{R}^3$, i.e. the control, to have a piecewise constant norm with at most three switchings:

$$\|u(t)\| = \begin{cases} 1 \text{ if } t \in [t_{\text{start}}, t_1] \cup [t_2, t_3] \cup [t_4, t_{\text{rdv}}^{\text{RH}}] \\ 0 \text{ if } t \in (t_1, t_2) \cup (t_3, t_4) \end{cases}, \tag{6}$$

where $t_1, t_2, t_3, t_4$ are called the switching times and satisfy $t_{\text{start}} < t_1 < t_2 \leq t_3 < t_4 < t_{\text{rdv}}^{\text{RH}}$. We denote by $\mathscr{U}$ the set of measurable functions $u : [t_{\text{start}}, t_{\text{rdv}}^{\text{RH}}] \to B^3(0, 1)$ satisfying (6) for some switching times $(t_{\text{start}}, t_1, t_2, t_3, t_4, t_{\text{rdv}}^{\text{RH}})$.

We denote by $\xi = (q, m)$ the state variables, and by $f(t, \xi, u)$ the controlled Sun perturbed CR3BP dynamics (3) including the mass evolution of the spacecraft (5). The optimal control problem can then be written as:

$$(OCP)^{\text{rdv}}_{t_{\text{start}}, t_{\text{rdv}}^{\text{RH}}} \begin{cases} \min_{t_1, t_2, t_3, t_4, u(\cdot)} \int_{t_{\text{start}}}^{t_{\text{rdv}}^{\text{RH}}} \|u(t)\| dt \\ s.t.\ \dot{\xi}(t) = f(t, \xi(t), u(t)),\ a.e.\ t \in [t_{\text{start}}, t_{\text{rdv}}^{\text{RH}}] \\ \xi(t_{\text{start}}) = (q_{\text{start}}, m_0) \\ q(t_{\text{rdv}}^{\text{RH}}) = q_{\text{rdv}}^{\text{RH}} \\ t_{\text{start}} < t_1 < t_2 \leq t_3 < t_4 < t_{\text{rdv}}^{\text{RH}} \\ u(.) \in \mathscr{U} \end{cases} \tag{7}$$

Since our optimization approach is variational, studying the impact of each of our variables on fuel consumption would produce a large number of local extrema in a direct optimization with respect to $(t_{\text{start}}, t_{\text{rdv}}^{\text{RH}})$ and $(t_1, t_2, t_3, t_4, u(\cdot))$. To address this issue, we discretize the set of departure times and durations of the approach transfer $(t_{\text{start}}, t_{\text{rdv}}^{\text{RH}})$ and solve $(OCP)^{\text{rdv}}_{t_{\text{start}}, t_{\text{rdv}}^{\text{RH}}}$ for the finite number of discretized combinations. This produces an approximation of the optimal transfer with respect to $(t_{\text{start}}, t_{\text{rdv}}^{\text{RH}})$ and $(t_1, t_2, t_3, t_4, u(\cdot))$ that becomes more accurate as the discretization on $(t_{\text{start}}, t_{\text{rdv}}^{\text{RH}})$ is refined. We use a 15 day discretization of $t_{\text{start}}$ from June 1st 2006 to 360 days later, and a 1 day discretization on $t_{\text{rdv}}^{\text{RH}}$ from June 1st 2006 to July 31st 2007, with the additional constraint that $t_{\text{rdv}}^{\text{RH}} > t_{\text{start}}$. Note that by fixing the departure time and rendezvous point $q_{\text{rdv}}^{\text{RH}}$, we fix the transfer duration. Finally, as it is unlikely that a short duration transfer would yield reasonable fuel consumption, we require that the transfer duration $t_{\text{rdv}}^{\text{RH}} - t_{\text{start}}$ be greater than 7 days. Our choice of discretization leads to 5975 different $(OCP)^{\text{rdv}}_{t_{\text{start}}, t_{\text{rdv}}^{\text{RH}}}$ to be solved.

To summarize our optimization algorithm, let us consider how we would select the best approach transfer after 2006 $RH_{120}$ was detected on September 14th 2006. First, the calculation of a high-accuracy TCO orbit sufficient for mission planning requires that the object be observed for many days, perhaps a few weeks, unless high-accuracy radar range and range-rate measurements can be obtained (in which case the orbit element accuracy collapses much faster). Thus, in practice, there is an additional constraint expressed as $t_{\text{start}} > t_d^{\text{RH}} + t_{calc}$ where $t_{calc}$ is the time required to calculate a sufficiently accurate TCO orbit. However, preliminary orbits obtained soon after detection are useful for producing preliminary mission scenarios and should not be neglected. For this reason, our algorithm ignores $t_{calc}$. Indeed, we imagine that if a spacecraft is actually hibernating in a $L_2$ halo orbit awaiting discovery of a suitable TCO then every effort will be made to obtain a high accuracy orbit as rapidly as possible.

**Step 1**: Solve $(OCP)^{\text{rdv}}_{t_{\text{start}}, t_{\text{rdv}}^{\text{RH}}}$ for all $t_{\text{start}}$ and $t_{\text{rdv}}^{\text{RH}}$ satisfying:

(i) $t_{\text{start}} \in [\![ t_c, t_c + 360 \text{ days} ]\!]$ with 15 day steps
(ii) $t_{\text{rdv}}^{\text{RH}} \in [\![ t_{\text{start}} + 7 \text{ days, July 31st 2007} ]\!]$ with 1 day steps.

**Step 2**: Select the $(OCP)^{\text{rdv}}_{t_{\text{start}}, t_{\text{rdv}}^{\text{RH}}}$ with the best final mass among those with $t_{\text{start}} \geq$ September 14th 2006.

The first step is performed without any consideration for the detection time and is only performed once. Step 2 is essentially instantaneous as it is only a simple analysis of the results of Step 1. Note that once Step 2 is performed it is always possible to locally refine the discretization of $t_{\text{start}}$ and $t_{\text{rdv}}^{\text{RH}}$ around the selected values in order to improve the final mass. This however, assumes that the chosen discretization is fine enough to already capture the best $(t_{\text{start}}, t_{\text{rdv}}^{\text{RH}})$.

## *3.2 Return Transfer*

After the approach transfer, the spacecraft will travel with the asteroid in a lock-in configuration. The optimal duration of the lock-in phase in terms of fuel efficiency, not science, is determined by the start time of the return transfer. In this section we focus on the third component of the round trip journey that optimizes the spacecraft's return trip from 2006 $RH_{120}$ to the Earth-Moon system. For simplicity, we select the final point of the journey as the Earth-Moon $L_2$ libration point rather than an $L_2$ halo orbit because the latter requires significantly higher number of calculations and the difference in fuel consumption would be minimal (but it should be a topic for further study). The position and velocity of the $L_2$ point are given by $q_{L_2} \approx (1.15569383, 0, 0, 0, 0, 0)$.

We denote by $q_{\text{start}}^{\text{RH}}$ the departure point from the 2006 $RH_{120}$ orbit at time $t_{\text{start}}^{\text{RH}}$. The spacecraft can depart the asteroid only after its rendezvous so that $t_{\text{start}}^{\text{RH}} > t_{\text{rdv}}^{\text{RH}}$. On the return trip portion of the journey we also require the control $u(.)$ to be composed of three thrust arcs and two ballistic ones, i.e. $u(.) \in \mathcal{U}$. The final time of the return transfer is denoted by $t_f$ and satisfies $t_f > t^{\text{RH}}_{\text{start}}$. The optimal control for the return transfer is now:

$$(OCP)^{\text{return}}_{t^{\text{RH}}_{\text{start}}, t_f} \begin{cases} \min_{t_1,t_2,t_3,t_4,u(\cdot)} \int_{t^{\text{RH}}_{\text{start}}}^{t_f} \|u(t)\| dt \\ s.t. \ \dot{\xi}(t) = f(t, \xi(t), u(t)), \ a.e. \ t \in [t^{\text{RH}}_{\text{start}}, t_f] \\ \xi(t^{\text{RH}}_{\text{start}}) = (q^{\text{RH}}_{\text{start}}, m^{\text{RH}}_0) \\ q(t_f) = q_{L_2} \\ t_{\text{start}} < t_1 < t_2 \leq t_3 < t_4 < t_f \\ u(.) \in \mathcal{U} \end{cases} \tag{8}$$

*Remark 3.1* The initial mass for the return trip, $m^{\text{RH}}_0$, is not necessarily identical to the final mass of the approach transfer, $m^{\text{rdv}}_f$. Depending on mission specifics $m^{\text{RH}}_0$ may be equal to, less than (if the mission leaves some equipment or consumes some fuel), or greater than (if the mission brings back samples for example) $m^{\text{rdv}}_f$. We don't expect $m^{\text{RH}}_0$ to play a large role in the fuel consumption, therefore, for simplicity we chose to set it to 300 kg, which is 50 kg less than the mass $m_0$ of the spacecraft when it departed its $L_2$ halo orbit.

As for the approach transfer, we discretize the optimization variables to study their impact on the fuel consumption. We also use a discretization of $(t^{\text{RH}}_{\text{start}}, t_f - t^{\text{RH}}_{\text{start}})$ and solve $(OCP)^{\text{return}}_{t^{\text{RH}}_{\text{start}}, t_f}$ for all the discretization pairs. The discretization on $t^{\text{RH}}_{\text{start}}$ is the same as on $t^{\text{RH}}_{\text{rdv}}$ with 1 day time steps from June 1st 2006 to July 31st 2007, while the discretization on the transfer duration $t_f - t^{\text{RH}}_{\text{start}}$ is in 30 day time steps from 30 to 360 days. This leads to 5112 different $(OCP)^{\text{return}}_{t^{\text{RH}}_{\text{start}}, t_f}$ combinations. In the final analysis, when constructing the complete round transfer, we will impose a constraint on the relation between the rendezvous and departure times but we treat them separately now to gain insight on the problem.

Figure 4 provides an overview of the round trip journey.

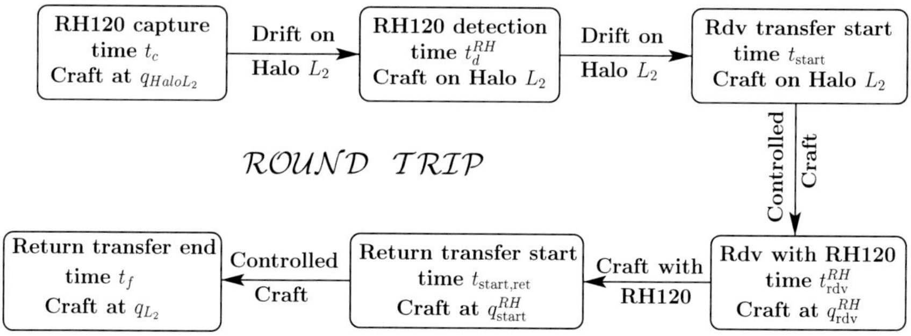


**Fig. 4** Chronology of the round trip journey

### 3.3 Necessary Conditions for Optimality

The maximum principle, see [13], provides first order necessary conditions for a control and associated trajectory to be optimal. In this section we apply the maximum principle to our optimization problems.

Let us first focus on the approach transfer. We denote by $\xi(t) = (q(t), m(t)) \in \mathbb{R}^6 \times \mathbb{R}_+$ the state of $(OCP)^{\text{rdv}}_{t_{\text{start}}, t^{\text{RH}}_{\text{rdv}}}$, with $q(t) = (r(t), v(t))$ the position and velocity of the vehicle, and $m(t)$ its mass, at time $t$. For $(OCP)^{\text{rdv}}_{t_{\text{start}}, t^{\text{RH}}_{\text{rdv}}}$, the maximum principle introduces an adjoint state $(p^0, p_\xi(\cdot))$ defined on $[t_{\text{start}}, t^{\text{RH}}_{\text{rdv}}]$ and the Hamiltonian,

$$H(t, \xi(t), p^0, p_\xi(t), u(t)) = p^0\|u(t)\| + \langle p_\xi(t), \dot{\xi}(t)\rangle, \text{ for a.e. } t \in [t_{\text{start}}, t^{\text{RH}}_{\text{rdv}}], \tag{9}$$

where $\langle , \rangle$ is the standard inner product. One of the conditions of the maximum principle is that the optimal control maximizes the Hamiltonian. This maximization requires that the optimal control be a multiple of the vector $p_v(\cdot)$ which translates into the following condition:

$$u(t) = \|u(t)\|\frac{p_v(t)}{\|p_v(t)\|}, \text{ for a.e. } t.$$

Without any constraint on the structure of the control the maximization of the Hamiltonian leads to the definition of the switching function $\psi$

$$\psi(t) = p^0 + \tilde{T}_{\max}\left(\frac{\|p_v(t)\|}{m(t)} - \frac{1}{I_{\text{sp}}g_0}p_m(t)\right), \tag{10}$$

where $\|u(t)\| = 1$ if $\psi(t) > 0$ or $\|u(t)\| = 0$ is $\psi(t) < 0$. However, since in $(OCP)^{\text{rdv}}_{t_{\text{start}}, t^{\text{RH}}_{\text{rdv}}}$ the control structure is constrained to have at most three thrust arcs, a rewriting of the optimal control problem following an approach similar to [14] implies that the switching function cancels at the constrained switching times but does not prescribe $\|u(t)\|$. The following theorem gives all the necessary conditions obtained from the maximum principle applied to $(OCP)^{\text{rdv}}_{t_{\text{start}}, t^{\text{RH}}_{\text{rdv}}}$.

**Theorem 3.1** *If* $(q(\cdot), m(\cdot), u(\cdot)) : [t_{\text{start}}, t^{\text{RH}}_{\text{rdv}}] \to \mathbb{R}^7 \times B^3(0, 1)$ *and* $(t_1, t_2, t_3, t_4) \in \mathbb{R}^4_+$ *is an optimal solution of* $(OCP)^{\text{rdv}}_{t_{\text{start}}, t^{\text{RH}}_{\text{rdv}}}$*, then there exists an absolutely continuous adjoint state* $(p^0, p_\xi(\cdot)) = (p^0, p_r(\cdot), p_v(\cdot), p_m(\cdot)) \in \mathbb{R}^- \times \mathbb{R}^7$ *defined on* $[t_{\text{start}}, t^{\text{RH}}_{\text{rdv}}]$ *and such that:*

*(a)* $(p^0, p_\xi(\cdot)) \neq 0, \ \forall t \in [t_{\text{start}}, t^{\text{RH}}_{\text{rdv}}]$*, and* $p^0 \leq 0$ *is constant.*
*(b) The state and adjoint state satisfy the Hamiltonian dynamics:*

$$\begin{aligned} \dot{\xi}(t) &= \frac{\partial H}{\partial p_\xi}(t, \xi(t), p^0, p_\xi(t), u(t)), \textit{ for a.e. } t \in [t_{\text{start}}, t^{\text{RH}}_{\text{rdv}}] \\ \dot{p}_\xi(t) &= -\frac{\partial H}{\partial \xi}(t, \xi(t), p^0, p_\xi(t), u(t)), \textit{ for a.e. } t \in [t_{\text{start}}, t^{\text{RH}}_{\text{rdv}}] \end{aligned} \tag{11}$$

*(c)*

$$u(t) = \frac{p_v(t)}{\|p_v(t)\|}, \ \forall t \in [t_{\text{start}}, t_1] \cup [t_2, t_3] \cup [t_4, t_{\text{rdv}}^{\text{RH}}] \tag{12}$$

*(d)* $\psi(t_1) = \psi(t_2) = \psi(t_3) = \psi(t_4) = 0$
*(e)* $p_m(t_{\text{rdv}}^{\text{RH}}) = 0$

Condition $(e)$ is the final transversality condition and comes from the fact that the final mass is free. In the case where we also allow a free initial time, $t_{\text{start}}$, we obtain an initial transversality condition of the form

$$\langle p_q(t_{\text{start}}), F_0^{\text{CR3BP}}(q(t_{\text{start}}))\rangle = 0,$$

where $F_0^{\text{CR3BP}}(\cdot)$ is the uncontrolled dynamics of the vehicle in the CR3BP model (without the Sun perturbation).

*Remark 3.2* Notice that transversality conditions at the rendezvous with asteroid 2006 $RH_{120}$ cannot be used because we do not have an analytic expression for its orbit. In case there is an analytic expression for the rendezvous orbit it would imply that the Hamiltonian must be zero at the rendezvous, and that $p_q(t_{\text{rdv}}^{\text{RH}})$ should be orthogonal to $\dot{q}_{\text{rdv}}^{\text{RH}}$.

*Remark 3.3* A state, control, and adjoint state $(\xi(\cdot), u(\cdot), p^0, p_\xi(\cdot))$ satisfying the conditions of Theorem 3.1 is called an extremal of $(OCP)^{\text{rdv}}_{t_{\text{start}}, t_{\text{rdv}}^{\text{RH}}}$. We assume that the extremals of $(OCP)^{\text{rdv}}_{t_{\text{start}}, t_{\text{rdv}}^{\text{RH}}}$ are normal, that is $p^0 \neq 0$.

For the return transfer, the maximum principle applied to $(OCP)^{\text{return}}_{t_{\text{start}}^{\text{RH}}, t_f}$ gives the same necessary conditions as in Theorem 3.1, with $t_{\text{start}}$ and $t_{\text{rdv}}^{\text{RH}}$ replaced by $t_{\text{start}}^{\text{RH}}$ and $t_f$.

## 3.4 Numerical Method

For our numerical calculations we assume that the extremals are normal, i.e. $p^0 \neq 0$ so we can normalize it to $-1$. A study of the existence of abnormal extremals is out of the scope of this paper.

Both optimal control problems are solved using a shooting method based on the necessary conditions. The shooting method consists in rewriting the necessary conditions of the maximum principle as the zero of a nonlinear function, namely the shooting function. Using the necessary conditions, in particular the Hamiltonian dynamics (11) and the maximization of the control (12), $(\xi(t), p_\xi(t))$ is completely defined by its initial value $(\xi_0, p_{\xi,0})$ at times $t_{\text{start}}$ (respectively $t_{\text{start}}^{\text{RH}}$ for the return transfer) and by the switching times $(t_1, t_2, t_3, t_4)$. Then, fixing $\xi_0$, we denote by $S$ the shooting function for the approach transfer:

$$S(p_{\xi,0}, t_1, t_2, t_3, t_4) = \begin{cases} q(t_{\text{rdv}}^{\text{RH}}) - q^{\text{rdv}} \\ \psi(t_i),\ i = 1, 2, 3, 4 \\ p_m(t_{\text{rdv}}^{\text{RH}}) \end{cases} \quad S \in \mathbb{R}^{11}. \tag{13}$$

For the return transfer we replace $t_{\text{rdv}}^{\text{RH}}$ by $t_f$, and $q^{\text{rdv}}$ by $q_{L_2}$. It follows that if we find $(p_{\xi,0}, t_1, t_2, t_3, t_4)$ such that $S(p_{\xi,0}, t_1, t_2, t_3, t_4) = 0 \in \mathbb{R}^{11}$, then the associated $(\xi(\cdot), u(\cdot), -1, p_\xi(\cdot))$ satisfies the necessary conditions of Theorem 3.1.

We computed the shooting function using the adaptative step integrator DOP853, see [15]. To find a zero of $S$ we used the quasi-Newton solver *HYBRD* of the Fortran *minpack* package. Since $S(\cdot)$ is nonlinear, Newton's method are very sensitive to the initial guess and seldom converge. To address the initialization sensitivity we used two initialization techniques as described below.

The first initialization technique is a direct approach, see [16], consisting of discretizing the state $\xi$ and control $u$ in order to rewrite the optimal control problem as a nonlinear parametric optimization problem (*NLP*). In (*NLP*) the dynamic is discretized using a fixed step fourth order Runge-Kutta scheme. The size of the (*NLP*) depends on the size of the discretization. The (*NLP*) is solved using the modeling language *Ampl*, see [17], and the optimization solver *IpOpt*, see [18]. Once a solution of the (*NLP*) is obtained we use the value of the Lagrange multipliers associated with the discretization of the dynamic at the initial time as our initial guess for $p_{\xi,0}$. The other unknowns are directly transcribable from (*NLP*). This approach cannot be used to solve our optimal control problems because, in order to have a sufficiently accurate solution, each (*NLP*) should be solved with a high-resolution time discretization which would yield execution times of a few hours. We thus use this initialization technique when the second one fails.

The second initialization technique is a continuation from the known solution of one optimal control problem to another. For instance, if we know the solution of a $(OCP)^{\text{rdv}}_{t_{\text{start}}, t_{\text{rdv}}^{\text{RH}}}$, then we can reasonably hope that in some, if not most, of the cases this solution is *connected* to the solution of a nearby $(OCP)^{\text{rdv}}_{t_{\text{start}}, t_{\text{rdv}}^{\text{RH}}+\delta t}$. To follow the connection between these two problems we could use elaborate continuation methods like in [8] or [19]. Here, we chose to use a linear prediction continuation which doesn't require the computation of the sensitivity of the shooting function but is nevertheless enough for our purpose. A solving with the continuation method usually takes a few seconds on a standard laptop.

Typically, the direct approach is used on one case and the continuation method enables us to solve tens or hundreds of other close cases. When the continuation method fails we then use the direct approach to once again initiate the continuation. To limit the number of local minima we add direct approach solvings and continuations from other neighbors to try to improve the solutions in terms of final mass. This *local minima trimming* is based on two heuristics. The first one is a selection of locally optimal cases with the assumption that the evolution of the final mass should be more or less continuous with respect to the 2006 $RH_{120}$ rendezvous point (for the rendezvous trip) or to the 2006 $RH_{120}$ departure point (for the return trip). For instance, for the forward trip, if two transfers with a comparable duration and neighbor rendezvous points exhibit a large final mass difference (say more than 10 kg), we employ a direct approach with the rendezvous point corresponding to the lower final mass. The second heuristic is a random selection of transfers and is used sparsely. This second round of calculations is essential and allowed us to greatly improve the solutions computed on the first solving round. The large number of optimal control problems we need to solve and the trimming of local minima requires several days of computation.

# 4 Numerical Results

In this section, we provide the best rendezvous and return transfers and the evolution of the criterion with respect to the discretization of the initial and final times. We then discuss how the results can be combined to design a global round trip mission.

Since our numerical approach is variational it is not possible to guarantee that the proposed trajectories are optimal despite the restriction of the control structure. Even the use of a (second order) sufficient condition, see [20], would only yield a proof of local optimality. It is thus likely that among the 5945 rendezvous trips and the 5112 return trips there are local minima that can be improved by finding other better local or, ideally, global minima.

## *4.1 Approach Transfer*

The best approach transfer under the imposed restricted thrust structure is represented on Figs. 5 and 6. Table 2 summarizes the main features of this transfer.

The departure time from the hibernating orbit is $t_{\text{start}} = 15$ days, which implies that the detection of the asteroid should occur before capture time to allow for an accurate calculation of 2006 $RH_{120}$'s orbit before the mission. The rendezvous between the spacecraft and 2006 $RH_{120}$ occurs 133 days later on October 27th 2006, 148 days after capture, i.e. $t_{\text{rdv}}^{\text{RH}} = 148$ days. The point on 2006 $RH_{120}$'s orbit at which rendezvous occurs is $q_f^{\text{rdv}} \approx (-1.958, 0.401, -3.992, 0.224, 1.728, -0.029)$. The best transfer rendezvous point is far from the Earth-Moon orbital plane and 5.08 LD from the $L_2$ libration point. A possible explanation is that the thrusters have a larger impact on the motion of the vehicle as the spacecraft's distance from the two primaries increases. It is also important to note that this rendezvous point is not simply the closest one in terms of distance or energy (see Fig. 7). There is first a rapid increase in the final mass for the rendezvous with 2006 $RH_{120}$ that occurs near capture time, the reason being that these transfers correspond to approach transfers with a short duration. The best transfers are obtained between 120 and 170 days, after which the final mass is roughly constant with a mean of 252.85 kg and a standard deviation of 8.21 kg. This behavior is good in terms of the design of a real mission as it provides flexibility for the spacecraft's departure time from its hibernating orbit.

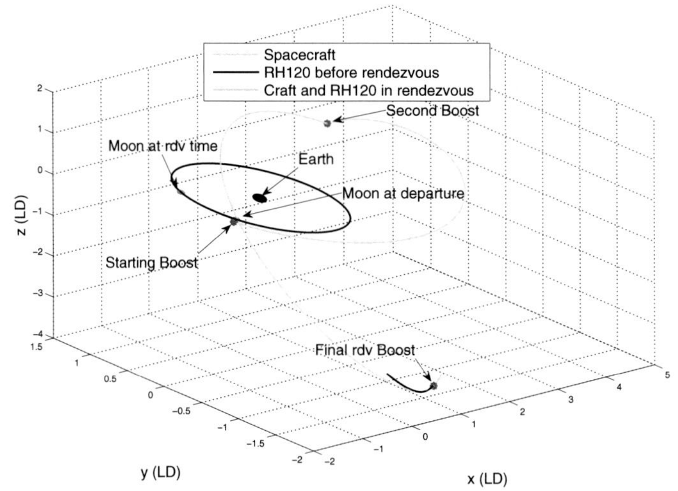


**Fig. 5** Best approach transfer to 2006 $RH_{120}$ in a geocentric inertial frame

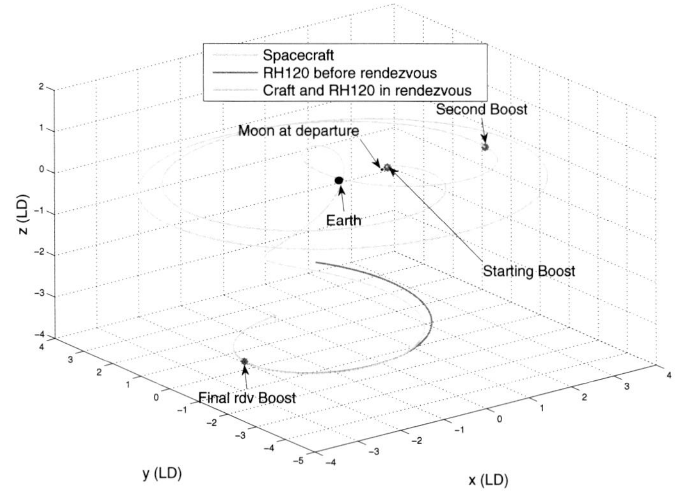


**Fig. 6** Best approach transfer to 2006 $RH_{120}$ in the CR3BP rotating frame

The final mass of the spaceraft for the best overall rendezvous is $m_f \approx 280.855$ kg corresponding to a $\Delta V \approx 496.43$ m/s where we compute $\Delta V$ such that $m_f = m_0 e^{-\frac{\Delta V}{I_{sp} g_0}}$. Note that the spacecraft performs only one revolution around the Earth in the inertial reference frame. We obtained, in the CR3BP model, a better transfer in [5] with a $\Delta V$ of 203.6 m/s but in this case the rendezvous would take place on June 26th 2006 and the duration would be about 415 days. This would require the unrealistic scenario that 2006 $RH_{120}$ be detected and spacecraft launched about 14 months before June 1st 2006.

**Table 2** Table summarizing the best approach transfer to 2006 $RH_{120}$

| Best rendezvous to 2006 $RH_{120}$ | | |
|---|---|---|
| Parameter | Symbol | Value |
| Departure date | $t_{\text{start}}$ | 06/16/2006 |
| Arrival date | $t_{\text{rdv}}^{\text{RH}}$ | 10/27/2006 |
| Final position | | $(-1.958, 0.401, -3.992)$ |
| Final velocity | | $(0.224, 1.728, -0.029)$ |
| Final mass | $m_f$ | 280.855 kg |
| Delta-V | $\Delta V$ | 496.43 m/s |
| Max dist. to earth | | 1714 mm (4.46 LD) |
| Min dist. to earth | | 366080 km (0.95 LD) |

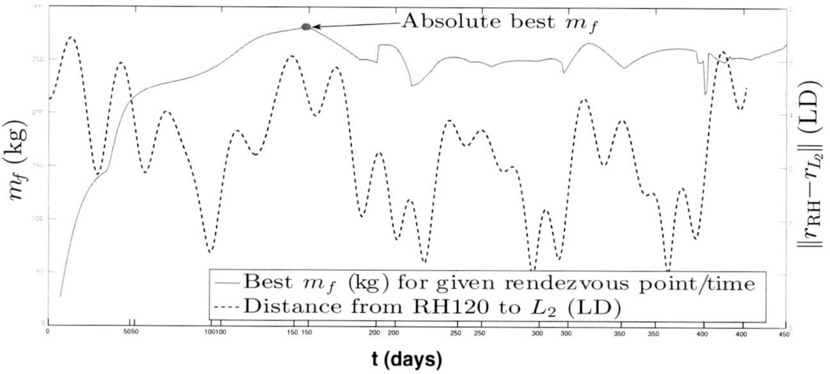


**Fig. 7** Best final mass (kg) among all $t_{\text{start}}$ with respect to the rendezvous time (*left y-axis*) and distance from the rendezvous point to $L_2$ (LD, *right y-axis*)

Finally, the norm of the control is shown on Fig. 8 with three thrust arcs lasting respectively 16.44 min, 1.62 h and 4.23 min and two ballistic arcs lasting 68.70 and 64.25 days. The second thrust takes place approximately in the middle of the transfer but this is typically not the case (see the best return transfer below). The position and velocity of the spacecraft at the beginning of the second thrust arc is $q(t_2) = (3.286, -0.141, -0.012, -0.476, -3.185, 0.012)$, at a EM barycenter distance of 3.29 LD or 1.26 million km.

Figure 9 gives the evolution of the final mass with respect to $t_{\text{rdv}}^{\text{RH}}$ and $t_{\text{start}}$. Notice that the scale for $t_{\text{start}}$ needs to be multiplied by 15 (the discretization rate) to justify the void region for which no approach transfer are associated. It also reflects the fact that $t_{\text{rdv}}^{\text{RH}} \geq t_{\text{start}} + 7\ days$.

Figure 10 is a selection of the evolution of the final mass and $\Delta V$ with respect to the rendezvous time $t_{\text{rdv}}^{\text{RH}}$ for various starting dates $t_{\text{start}}$, i.e. it provides cross sections through the 3D Fig. 9.

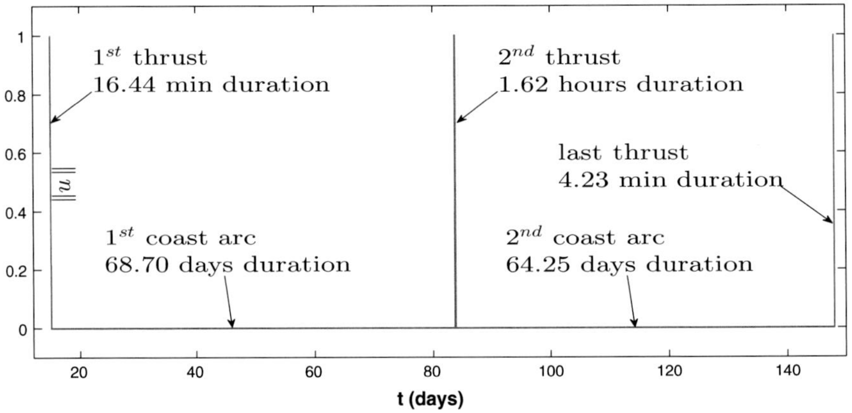


**Fig. 8** Norm of control for the best approach transfer to 2006 $RH_{120}$

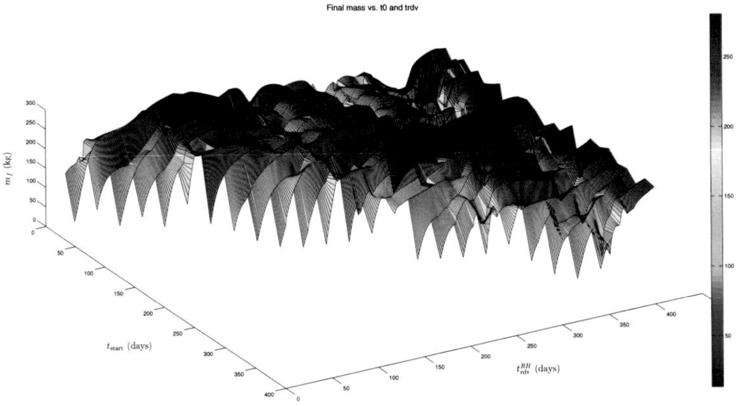


**Fig. 9** Evolution of final mass for $(OCP)^{\mathrm{rdv}}_{t_{\mathrm{start}}, t^{\mathrm{RH}}_{\mathrm{rdv}}}$ with respect to the rendezvous and starting dates

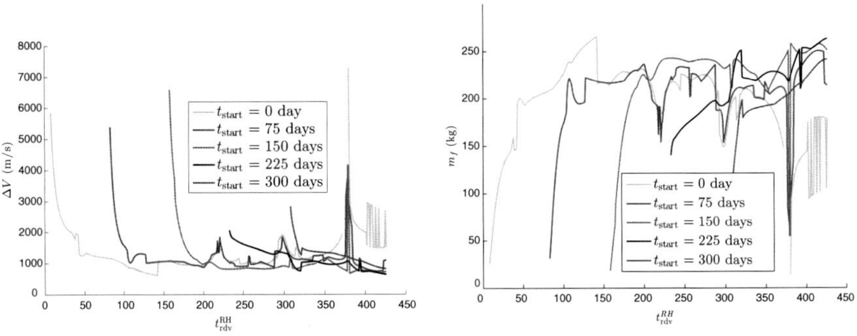


**Fig. 10** Evolution of $\Delta V$ (*left*) and final mass (*right*) for $(OCP)^{\mathrm{rdv}}_{t_{\mathrm{start}}, t^{\mathrm{RH}}_{\mathrm{rdv}}}$ with respect to the rendezvous date and various starting dates $t_{\mathrm{start}} \in \{0, 75, 150, 225, 300\}$ days

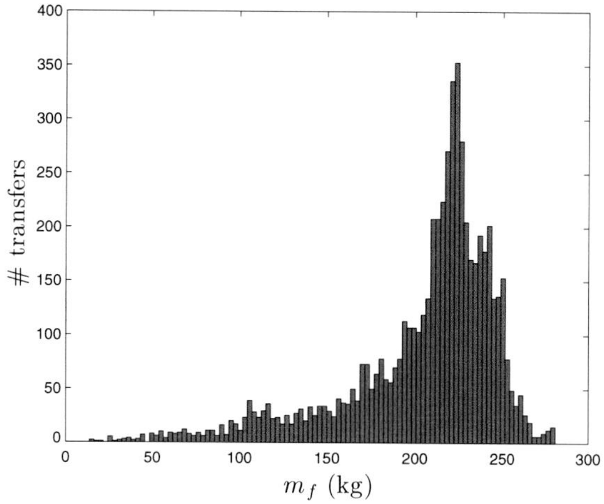


**Fig. 11** Number of approach transfers per final mass range. Most approach transfers provide a final mass above 200 kg with a peak at around 230 kg

Figures 9 and 10 show that there is first a gradual increase of the final mass with respect to the transfer duration $t_{\rm rdv}^{\rm RH} - t_{\rm start}$ which then stops after 30–120 days, depending on the starting date. This suggests that after a period of about 2 months the final mass is less sensitive to an increase in transfer duration and depends more heavily on the rendezvous point.

Figure 11 provides a histogram of the number of approach transfers corresponding to a given final mass range. 68.8% of transfers provide a final mass above 200 kg. This is promising for the mission design because it provides flexibility with respect to the departure time and transfer duration and because the less fuel that is used per mission the more additional TCOs that can be targeted.

Table 3 provides the best approach transfers for each of 25 departure times for the spacecraft from its hibernating location. Officially, 2006 $RH_{120}$ was discovered on September 14th 2006, 105 days after its capture by Earth's gravity, so that $t_d^{\rm RH} = 105$. Table 3, suggests that in this case the best departure time satisfying $t_{\rm start} \geq t_d^{\rm RH}$ is $t_{\rm start} = 180$ days after June 1st 2006. In this scenario the 75 days between the detection time and the departure of the spacecraft for the rendezvous mission ensure that the celestial mechanics computations required to predict 2006 $RH_{120}$'s orbit with enough accuracy can be completed. This approach transfer provides a final mass of 267.037 kg or, equivalently $\Delta V = 610.224$ m/s, and a rendezvous date 312 days after capture, on April 9th 2007. In particular, we will see that this approach transfer can be combined with the best return transfer provided in the following section. If practical considerations delay the spacecraft departure then Table 3 shows that it will have minimal impact on the mission's fuel consumption. For instance, for $t_{\rm start} = 285$ days the final mass is 266.525 kg, not even a one kilo difference from a departure 180 days after capture. However, this late rendezvous time might seriously compromise the efficiency of the return transfer. Clearly, an early detection of the TCO or timely departure of the spacecraft once the asteroid orbit has been determined is preferable for a fuel efficient round trip transfer.

**Table 3** Best rendezvous dates and final mass for the 25 different $t_{start}$

| $t_{start}$ (d) | $t_{rdv}^{RH}$ (d) | $m_f$ (kg) | $t_{start}$ (d) | $t_{rdv}^{RH}$ (d) | $m_f$ (kg) |
|---|---|---|---|---|---|
| 0 | 141 | 265.831 | 195 | 392 | 255.172 |
| 15 | 148 | 280.855 | 210 | 363 | 262.025 |
| 30 | 111 | 234.909 | 225 | 425 | 263.304 |
| 45 | 138 | 251.379 | 240 | 359 | 240.076 |
| 60 | 146 | 231.103 | 255 | 425 | 260.92 |
| 75 | 414 | 250.772 | 270 | 416 | 256.618 |
| 90 | 273 | 250.608 | 285 | 425 | 266.525 |
| 105 | 414 | 250.02 | 300 | 425 | 241.721 |
| 120 | 290 | 252.547 | 315 | 425 | 246.111 |
| 135 | 390 | 245.707 | 330 | 407 | 254.773 |
| 150 | 380 | 258.222 | 345 | 425 | 251.158 |
| 165 | 314 | 244.521 | 360 | 425 | 245.369 |
| 180 | 312 | 267.037 | | | |
| Avg | – | 253.091 | $\sigma$ | – | 11.136 |

The average final mass is 253 kg with a standard deviation of 11 kg

## 4.2 *Return Transfer from 2006 $RH_{120}$ to $L_2$*

To get a global idea of the impact of the choice of departure time from asteroid 2006 $RH_{120}$ and duration of the transfer we first study the return transfer as a completely decoupled problem from the approach transfer. In an unrealistic way we assume the spacecraft can depart 2006 $RH_{120}$ as soon as June 1st 2006 and we use a 1 day discretization of the 2006 $RH_{120}$ orbit. However, to keep the number of calculations under control we use a 30 days discretization for the transfer duration, $t_f - t_{start}^{RH}$, from 30 to 360 days, yielding a total of 5112 combinations for $(t_{start}^{RH}, t_f - t_{start}^{RH})$. Since the mass at departure from 2006 $RH_{120}$ is unknown beforehand, and in order to compare all the return trips, we choose arbitrarily to set the initial mass of the return trip to 300 kg, exactly 50 kg less than the initial mass of the approach transfer.

The best return transfer to $L_2$ under our thrust restrictions is shown in Fig. 12 and Table 4 summarizes its main features.

The starting date of the best return transfer is 365 days after June 1st 2006, June 1st 2007, and the transfer duration is 240 days. This departure date occurs shortly before 2006 $RH_{120}$ escapes Earth's gravity and after the best approach transfer which makes it an ideal candidate for a complete round trip. The final mass for this transfer is $m_f \approx 250.712$ kg, corresponding to $\Delta V \approx 404.815$ m/s, slightly better than the $\Delta V$ for the best approach transfer.

Figure 13 provides the norm of the control associated with the best return transfer. This thrust strategy has three thrust arcs lasting 2.15 min, 1.32 h and 3.06 min with two intervening ballistic arcs lasting 213.79 and 26.15 days respectively. Contrary

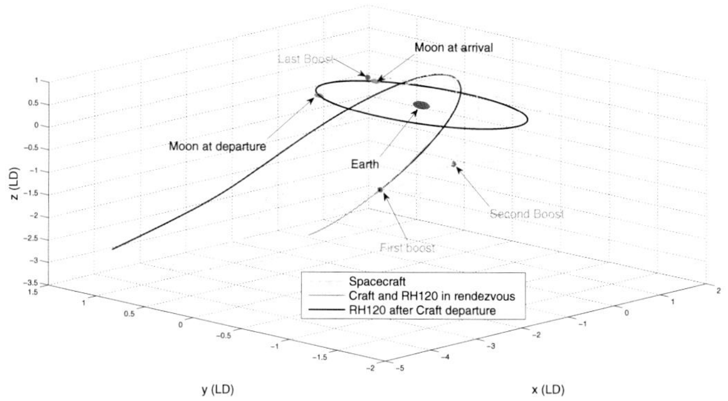


**Fig. 12** Best return transfer from 2006 $RH_{120}$ to Earth-Moon $L_2$ in a geocentric inertial frame

**Table 4** Table summarizing the best return transfer from 2006 $RH_{120}$ to $L_2$

| Best return trip from 2006 $RH_{120}$ | | |
|---|---|---|
| Parameter | Symbol | Value |
| Departure date | $t_{\text{start}}^{\text{RH}}$ | 06/01/2007 |
| Arrival date | $t_f$ | 01/27/2008 |
| Initial position | | (0.238, −0.598, −2.228) |
| Initial velocity | | (−0.947, −0.477, 0.496) |
| Final mass | $m_f$ | 250.712 kg |
| Delta-V | $\Delta V$ | 404.815 m/s |
| Max dist. to earth | | 2031 mm (5.28 LD) |
| Min dist. to earth | | 265520 km (0.69 LD) |

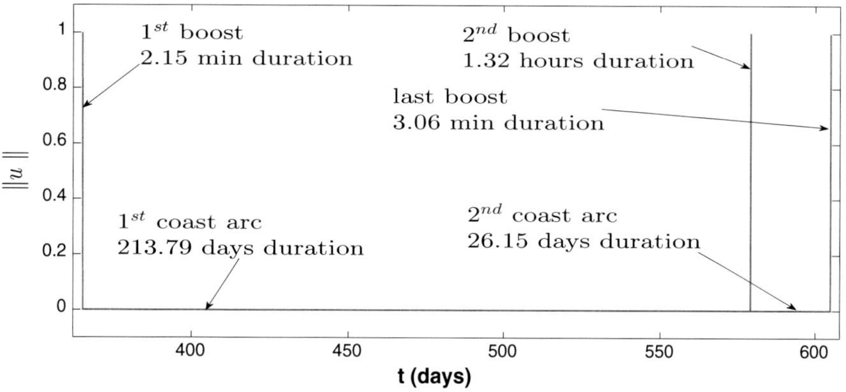


**Fig. 13** Norm of control for the best return transfer from 2006 $RH_{120}$

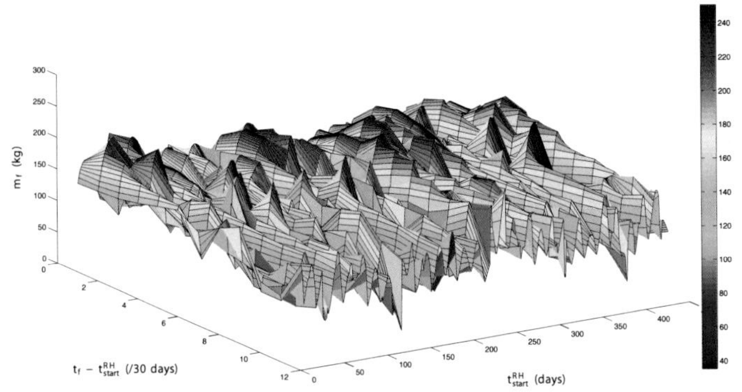


**Fig. 14** Evolution of final mass for $(OCP)^{\text{return}}_{t^{\text{RH}}_{\text{start}}, t_f}$ with respect to the starting date and transfer time

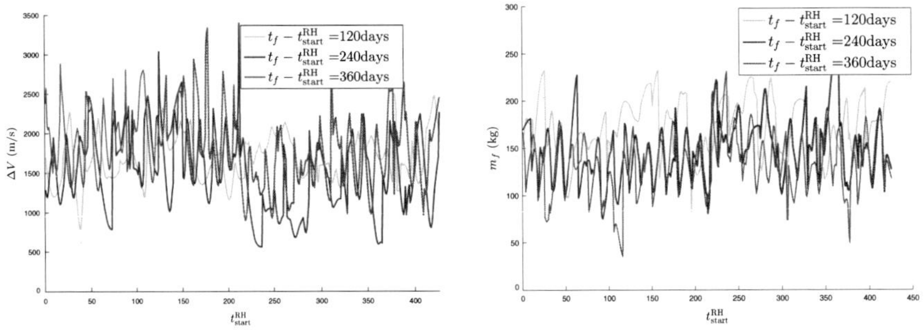


**Fig. 15** Evolution of $\Delta V$ (*left*) and final mass (*right*) for $(OCP)^{\text{return}}_{t^{\text{RH}}_{\text{start}}, t_f}$ with respect to the starting date and various transfer durations $t_f - t^{\text{RH}}_{\text{start}} \in \{120, 240, 360\}$ days

to the best approach transfer, the second thrust arc does not occur in the middle of the transfer but rather near the end. However, the second thrust arc is the longest one as was the case for the approach transfer.

Figure 14 illustrates the evolution of the final mass with respect to $t^{\text{RH}}_{\text{rdv}}$ for various choices of $t^{\text{RH}}_{\text{start}}$. For the return trip there is not a large variation in the final mass with respect to either the starting date or the transfer times.

Figure 15 provides a selection of the evolution of the final mass and $\Delta V$ with respect to the departure date $t^{\text{RH}}_{\text{start}}$ for various transfer durations $t_f - t^{\text{RH}}_{\text{start}}$, i.e. it presents several cross sectional views through 3D Fig. 14.

Based on the evolution of the final mass with respect to the transfer duration, allowing more time for the transfer does not always provide a more efficient return transfer. However, it is possible that the optimal control problem from a fixed duration to one with a maximum allowed duration would give better results. Indeed, for the return transfer, it would make sense to be more lax with respect to the transfer duration than for the synchronized approach transfer. This remark is partially illustrated by the results from [5] where the transfer duration was free, albeit those results are applicable only to a rendezvous type transfer and the CR3BP model.

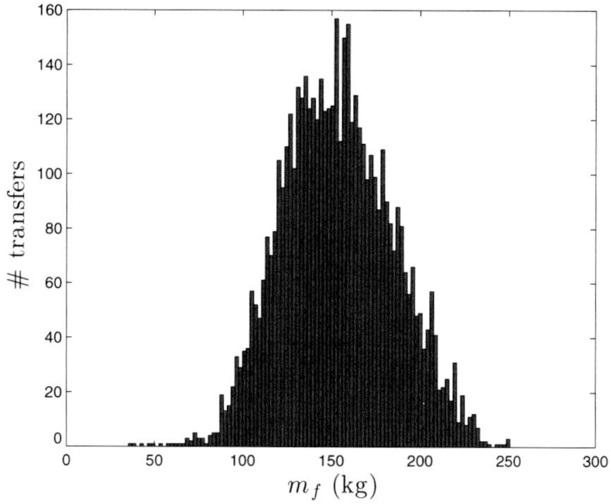


**Fig. 16** Number of return transfers per final mass range. Notice the shape resembles a normal distribution with a maximum at ∼170 kg

**Table 5** Best starting dates for the return trip for 12 different transfer durations $\Delta t_f = t^f - t_{\text{start}}^{\text{RH}}$

| $\Delta t^f$ (d) | $t_{\text{start}}^{\text{RH}}$ (d) | $m_f$ (kg) | $\Delta t^f$ (d) | $t_{\text{start}}^{\text{RH}}$ (d) | $m_f$ (kg) |
|---|---|---|---|---|---|
| 30 | 37 | 211.681 | 210 | 271 | 231.035 |
| 60 | 18 | 225.091 | 240 | 365 | 250.712 |
| 90 | 149 | 220.053 | 270 | 221 | 205.216 |
| 120 | 25 | 232.328 | 300 | 271 | 207.765 |
| 150 | 154 | 236.009 | 330 | 218 | 201.274 |
| 180 | 236 | 233.768 | 360 | 236 | 231.093 |

Mean value of final mass is 224 kg and standard deviation is 15 kg

Figure 16 provides the number of approach transfers corresponding to a given final mass range. Contrary to the approach transfers, it resembles a normal distribution with an average final mass of about 160 kg. This distribution does not provide as much flexibility as the approach transfers because in the latter case there are many more transfers with a final mass close to the best one.

Table 5 gives a quick overview of the best return trips for each transfer duration $\Delta t^f = t^f - t_{\text{start}}^{\text{RH}}$. The best transfers require durations between 120 and 240 days and extending the duration beyond 240 days does not provide more fuel efficiency. Except for the return transfers lasting less than 150 days, they all depart 2006 $RH_{120}$ at a late date. This is desirable because it is unrealistic to expect that the approach transfer arrives at 2006 $RH_{120}$early.

## *4.3 Complete Round Trip Mission*

In this section we combine an approach transfer with a return transfer in a realistic way to design a round trip mission to 2006 $RH_{120}$. To do so, we need to account for some practical constraints such as the fact that the return transfer must start after the completion of the approach transfer, that is $t_{\text{rdv}}^{\text{RH}} < t_{\text{start}}^{\text{RH}}$. This means that the spacecraft

**Table 6** Parameters of the best return transfer from 2006 $RH_{120}$ to $L_2$ after pairing with the approach transfer (so $m_0 = 280.855$ kg)

| Best return trip from 2006 $RH_{120}$ for the round trip mission | | |
|---|---|---|
| Parameter | Symbol | Value |
| Stay on 2006 $RH_{120}$ | $t^{RH}_{start} - t^{RH}_{rdv}$ | 217 days |
| Departure date | $t^{RH}_{start}$ | 06/01/2007 |
| Arrival date | $t_f$ | 01/27/2008 |
| Initial position | | (0.238, −0.598, −2.228) |
| Initial velocity | | (−0.947, −0.477, 0.496) |
| Final mass | $m_f$ | 234.713 kg |
| Delta-V | $\Delta V$ | 404.814 m/s |
| Max dist. to earth | | 2031 mm (5.28 LD) |
| Min dist. to earth | | 265519 km (0.69 LD) |

stays with 2006 $RH_{120}$ for $t^{RH}_{start} - t^{RH}_{rdv}$ days. We prefer to think of the lock-in duration between the spacecraft and 2006 $RH_{120}$ to be a consequence of our calculation rather than a fixed value by the user. Our calculations determine what the ideal lock-in duration should be and the only remaining constraint is to ensure that it corresponds to a realistic duration for the science component of the mission.

From our prior calculations, the best rendezvous and return transfers satisfy the time constraint so we only need to modify the initial mass of the return transfer to match our desired scenario. We chose to simply impose that the mass at the end of the approach transfer equals the mass at the departure of the return transfer, in other words there is no loss or addition of mass during the lock-in phase. This is an arbitrary choice, and we could for instance have decided that some equipment was left on the asteroid or some material collected form the asteroid that would alter the departure mass. Based on our choice, the return transfer must start with an initial mass of 280.855 kg instead of the 300 kg prescribed previously. This modification is addressed easily through a continuation on the previous best return transfer. It provides a return transfer that is nearly the same as the one with the higher mass. Table 6 provides the main features of the modified return trip while Table 2 remains applicable to the approach transfer. Figure 17 shows the entire round trip transfer in a geocentric inertial reference frame.

As mentioned in Sect. 4.1, the best round trip transfer requires that 2006 $RH_{120}$ be detected at, or almost immediately after, capture which is not an ideal scenario especially given the fact that 2006 $RH_{120}$ was actually detected 105 days after June 1st 2006. This suggests that additional scenarios should be considered. Moreover, the round trip transfer should also allow the spacecraft ample time to perform its science mission at the TCO. If we denote by $\delta t_{mission}$ the minimum time the spacecraft needs to remain at 2006 $RH_{120}$ this constraint can be expressed as $t^{RH}_{start} \geq t^{RH}_{rdv} + \delta t_{mission}$. Table 7 gives a sample of the best round trip transfers for various $t^{RH}_{d}$ and $\delta t_{mission}$. Since the best return transfer departs 1 year after June 1st 2006 it can be used in almost all scenarios except the last one when the rendezvous portion ends 395 days

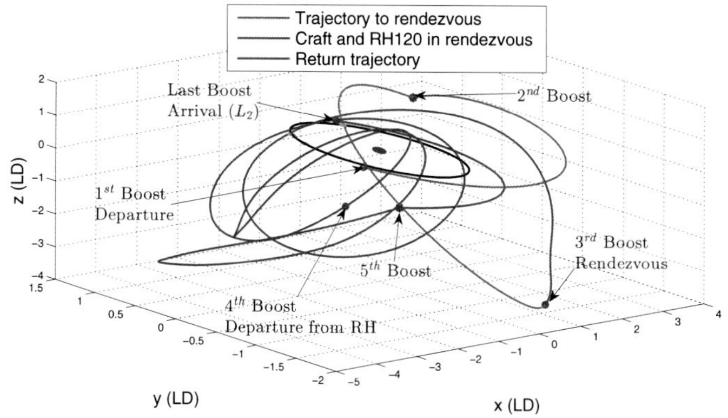


**Fig. 17** Best round trip transfer from Earth-Moon $L_2$ to 2006 $RH_{120}$ in a geocentric inertial frame

**Table 7** The best round trips with detection and mission duration constraints, $m_0 = 350\,\text{kg}$, and $m_0^{RH} = 300\,\text{kg}$

| $t_d^{RH}$ | $\delta t_{mission}$ | Approach transfer | | | Return transfer | | | Round trip | |
|---|---|---|---|---|---|---|---|---|---|
| | | $t_{start}$ | $t_{rdv}^{RH}$ | $\Delta V$ (m/s) | $t_{start}^{RH}$ | $t_f - t_{start}^{RH}$ | $\Delta V$ (m/s) | Total $\delta V$ (m/s) | Duration |
| 0 | 30 | 15 | 148 | 496.43 | 365 | 240 | 404.82 | 901.25 | 373 |
| 30 | 30 | 180 | 312 | 610.22 | 365 | 240 | 404.82 | 1.01504 | 372 |
| 30 | 60 | 180 | 305 | 684.66 | 365 | 240 | 404.82 | 1089.48 | 365 |
| 210 | 30 | 210 | 312 | 732.23 | 365 | 240 | 404.82 | 1137.05 | 342 |
| 210 | 60 | 210 | 305 | 809.61 | 365 | 240 | 404.82 | 1214.43 | 335 |
| 240 | 30 | 255 | 319 | 892.82 | 365 | 240 | 404.82 | 1297.63 | 304 |
| 240 | 60 | 240 | 305 | 1010.88 | 365 | 240 | 404.82 | 1415.70 | 305 |
| 270 | 30 | 270 | 335 | 1034.27 | 365 | 240 | 404.82 | 1439.09 | 305 |
| 300 | 30 | 330 | 395 | 936.80 | 425 | 120 | 704.06 | 1640.85 | 185 |

All times expressed in days

after June 1st 2006. For instance, if we assume that the detection occurs 210 days after June 1st 2006, and that we need only 30 days for the lock-in phase, we can select 102 day duration approach transfer to combine with the best overall return transfer. However, if we impose a 60 day lock-in constraint for the spacecraft and the asteroid we need to select an approach transfer duration of 95 days to reach 2006 $RH_{120}$. In general, the longer the lock-in phase the more expensive the round trip transfer. Another way to look at our calculations would be to design efficient round trip transfers and deduce from these data the ideal windows for detection and lock-in phases. This would provide additional information for the overall design of a TCO mission.

**Acknowledgments** This research is partially supported by the National Science Foundation (NSF) Division of Mathematical Sciences, award #DMS-1109937 and by the NASA, proposal *Institute for the Science of Exploration Targets* from the program Solar System Exploration Research Virtual Institute.

# References


1. Granvik M, Vauballion J, Jedicke R (2012) The population of natural Earth satellites. Icarus 218(1):262–277
2. Bolin B, Jedicke R, Granvik M, Brown P, Howell E, Nolan MC, Jenniskens P, Chyba M, Patterson G, Wainscoat R (2014) Detecting Earth's temporarily captured natural satellites. Icarus 241:280–297
3. Chyba M, Patterson G, Picot G, Granvik M, Jedicke R, Vaubaillon J (2014) Designing rendezvous missions with mini-moons using geometric optimal control. J Ind Manage Optim 10(2):477–501
4. Chyba M, Patterson G, Picot G, Granvik M, Jedicke R, Vaubaillon J (2014) Time-minimal orbital transfers to temporarily-captured natural Earth satellites. In: PROMS series: advances of optimization and control with applications. Springer
5. Brelsford S, Chyba M, Haberkorn T, Patterson G Rendezvous missions to temporarily-captured near Earth asteroids. http://arxiv.org/abs/1508.00738
6. Russell C, Angelopoulos V (2013) The ARTEMIS Mission (Google eBook). Springer (112 pp)
7. Sweetser TH, Broschart SB, Angelopoulos V, Whiffen GJ, Folta DC, Chung M-K, Hatch SJ, Woodard MA (2011) ARTEMIS mission design. Space Sci Rev 165(1–4):27–57
8. Gergaud J, Haberkorn T, Martinon P (2004) Low thrust minimum-fuel orbital transfer: a homotopic appoach. J Guidance Control Dyn 27(6): 1046–1060
9. Larson W, Wertz J (1999) Space Mission Analysis and Design, vol. 8, p. 969, 3rd edn. Space Technology Library. ISBN 978-1881883104
10. Fedorets G, Granvik M, Jones L, Jedicke R (2015) Discovering asteroids temporarily captured by the Earth with LSST. IAU Gen Assembly 22:57052
11. Koon WS, Lo MW, Marsden JE, Ross SD (2011) Dynamical Systems, the Three-Body Problem and Space Mission Design. Springer, New York
12. Mingotti G, Topputo F, Bernelli-Zazzera F (2007) A method to design sun-perturbed earth-to-moon low-thrust transfers with ballistic capture. In: XIX Congresso Nazionale AIDAA, vol. 17–21. Forle, Italia, Sept 2007
13. Pontryagin LS, Boltyanskii VG, Gamkrelidze RV, Mishchenko EF (1962) The Mathematical Theory of Optimal Processes. Wiley, New York
14. Dmitruk AV, Kaganovich AM (2008) The hybrid maximum principle is a consequence of Pontryagin maximum principle. Syst Control Lett 57(11):964–970
15. Hairer E, Norsett SP, Wanner G (1993) Solving ordinary differential equations I. Nonstiff problems. In: Springer Series in Computational Mathematics, 2nd edn. Springer
16. Betts JT (2001) Practical Methods for Optimal Control Using Nonlinear Programming. Society for Industrial and Applied Mathematics
17. Fourer R, Gay DM, Kernighan BW (1993) AMPL: A Modeling Language for Mathematical Programming. Duxbury Press, Brooks-Cole Publishing Company
18. Waechter A, Biegler LT (2006) On the implementation of an interior-point filter-line search algorithm for large-scale nonlinear programming. Research Report RC 23149. IBM T.J. Watson Research Center, Yorktown, New York
19. Caillau JB, Cots O, Gergaud J (2012) Differential continuation for regular optimal control problems. Optim Methods Softw 27:177–196
20. Caillau JB, Chen Z, Chitour Y (2015) $L^1$-minimization of mechanical systems. https://hal.archives-ouvertes.fr/hal-01136676

# Low-Thrust Transfers Between Libration Point Orbits Without Explicit Use of Manifolds

**Richard Epenoy**

**Abstract** In this paper, we investigate the numerical computation of minimum-energy low-thrust transfers between Libration point orbits in the Circular Restricted Three-Body Problem. We develop a three-step methodology based on optimal control theory, indirect shooting methods and variational equations without using information from invariant manifolds. Numerical results are provided in the case of transfers between Lyapunov orbits around $L_1$ and $L_2$ in the Earth-Moon system demonstrating the efficiency of the developed approach for different values of the transfer duration leading to trajectories with one or two revolutions around the Moon.

**Keywords** Three-body problem · Libration Point Orbits · Minimum-energy control · Indirect shooting methods · Continuation techniques

## 1 Introduction

Libration point orbits, a subset of unstable periodic orbits in the three-body problem, have attracted a major interest over the last three decades. The launch of the first spacecraft to a Libration Point Orbit (LPO) in the Sun-Earth system, ISEE-3, dates back to the late 70s [1]. Since then, the design of trajectories based on three-body dynamics has undergone a significant evolution due to missions to the Libration points of the Sun-Earth and Earth-Moon systems. There are currently several missions operating in LPOs, such as SOHO [2] and HERSCHEL/PLANCK [3]. Upcoming missions, such as ARTEMIS [4] and JWST [5] will use Earth-Moon and Sun-Earth LPOs, respectively. LPOs have also been proposed for a wide range of missions, such as lunar navigation and communication relays [6–8].

Efficient techniques have been developed by different authors [9–11] to yield zero cost transfers between LPOs of the same energy using invariant manifolds and dynamical systems theory. In fact, the cost of such transfers is not exactly zero in

R. Epenoy (✉)
Centre National d'Etudes Spatiales,
18 Avenue Edouard Belin, 31401 Toulouse, France
e-mail: Richard.Epenoy@cnes.fr



B. Bonnard and M. Chyba (eds.), *Recent Advances in Celestial and Space Mechanics*, Mathematics for Industry 23, DOI 10.1007/978-3-319-27464-5_5

finite time scale because small thrusts are required, first to inject the spacecraft on the unstable manifold of the departure orbit, second to insert it into the target orbit from the stable manifold of this last one. Some authors [12–15] constructed transfers between unstable periodic orbits of differing energies using invariant manifolds and impulsive maneuvers. These maneuvers are, in general, determined by using primer vector theory [16, 17]. The use of low-thrust propulsion in a $n$-body problem was also investigated to design Earth-Mars transfers [18, 19], Earth-Venus transfers [20], trajectories to the Moon [21], transfers between planetary moons [22], and trajectories involving LPOs [23–27]. Some authors used direct methods [28, 29] to solve the corresponding optimal control problems [18, 21]. In other approaches, as in [19] or in [22, 23], the structure of the solution is assumed a priori and defined in terms of a few number of parameters. These last ones are derived by solving a nonlinear programming problem. Independently of the solution method used, all these papers take advantage of invariant manifolds, when they are useful, to yield propellant-efficient solutions. This often means that, even when the solution is not designed a priori by taking into account invariant manifolds as in [19] or in [22, 23], knowledge from invariant manifolds is explicitly incorporated in the solution process. Direct methods [28, 29] are known to be better suited than indirect methods to take into account prior knowledge about the solution. They are able, for example, to include coast arcs along the invariant manifolds as it is done in [24].

Concerning transfers involving LPOs, a recent study [23] determines the direction of the thrust acceleration in such a way as to decrease (or increase) the Jacobi constant (energy), leading to an anti-tangential steering (or tangential steering) control law. Once a target value of the Jacobi constant is reached, a coast arc is added until the spacecraft encounters a given Poincaré section where the insertion into a target manifold is searched. In [24–26], a spacecraft with a Variable Specific Impulse propulsion system is considered. Transfers between LPOs are investigated in [24]. The associated optimal control problem is solved thanks to a multiple shooting algorithm with analytical gradients, combined with direct optimization techniques [28, 29] to include coast times along the invariant manifolds when required. In [25–27], transfers from an Earth parking orbit to different kinds of LPOs are designed by means of direct multiple shooting techniques. These efficient approaches are based on the existence of a predefined reference coast arc computed as a trajectory on the invariant manifold associated with the final periodic orbit.

This investigation focuses on low-thrust minimum-energy transfers between Lyapunov orbits around $L_1$ and $L_2$ in the Earth-Moon planar restricted three-body problem. These departure and arrival planar periodic orbits are computed using Lindstedt-Poincaré techniques [30]. It is clear from [9–11] that low-energy low-thrust transfers may exist for particular values of the transfer duration when the two orbits share the same Jacobi constant. Trying to determine the associated optimal control laws in a direct way, by using indirect shooting methods, appears to be very difficult or even impossible from a medium value of the transfer duration. This is due to different reasons. Numerical difficulties regarding the computation of the shooting function and its Jacobian are a first reason. The existence of local optima with higher energy values than that of the low-energy solution is a second source of difficulty, but most of all, the two time-scale structure characterizing the low-energy solution is the central issue that will be addressed in this paper. Indeed, this particular structure makes the problem ill-conditioned and requires the development of a specific solution method.

On the other hand, when the energies of the initial and target orbits are different, the use of invariant manifolds offers little advantage in terms of reduction of the transfer cost. In this case, it will be shown that the associated optimal control laws can be computed in a direct way by means of indirect shooting methods. This is due to the fact that the solution does not exhibit a two-time-scale structure as in the case of orbits with the same energy.

This paper focuses on the development of a three-step methodology for computing low-thrust low-energy trajectories between Lyapunov orbits of the same energy without explicit use of invariant manifolds. Contrarily to the approach developed in [24], coast times along the invariant manifolds are not enforced. More precisely, a feasible control with quadratic-zero-quadratic time structure connecting the departure and arrival orbits is first determined. Then an optimal control problem whose solution is equal to this feasible control is built. In a second step, this problem is embedded in a family of problems depending on a parameter $\varepsilon$. For each problem, the departure location from the first orbit and the arrival location at the target orbit are fixed to the non-optimal values associated with the feasible control. These problems are solved by continuation on $\varepsilon$ until a locally energy-optimal trajectory connecting the two Lyapunov orbits is obtained. Each problem is solved thanks to an indirect single shooting method. The Jacobian of the shooting function is computed using variational equations. Finally, in the last step of the method, the low-energy solution is obtained by determining the optimal value of the departure location from the initial orbit and that of the arrival location at the target orbit. This is achieved by means of a gradient algorithm.

This paper is organized as follows. In Sect. 2 the low-thrust minimum-energy transfer problem is formulated and the associated necessary optimality conditions are derived. The main features of the method are developed in Sect. 3. In Sect. 4 the efficiency of the method is illustrated through numerical results obtained in the case of transfers between Lyapunov orbits of the same energy around $L_1$ and $L_2$ in the Earth-Moon planar restricted three-body problem. Results related to transfers between Lyapunov orbits of differing energies around the same collinear points are also provided. Conclusions are drawn in Sect. 5.

## 2 Problem Statement

### *2.1 Equations of Motion*

The dynamic model is based on the Planar Circular Restricted Three Body Problem (PCR3BP) [31] with the Earth as one primary and the Moon as the second. The equations of motion are constructed within the context of a rotating reference frame.

The $x$-axis extends from the barycenter of the Earth-Moon system to the Moon and the $y$-axis completes the right-hand coordinate frame. Moreover, a set of nondimensional units is chosen such that the unit of distance is the distance between the two primaries, the unit of mass is the sum of the primaries' masses, and the unit of time is such that the angular velocity of the primaries around their barycenter is unitary. Thus, the Moon has mass $\mu$ and is fixed at coordinates $(1-\mu, 0)$ while the Earth has mass $(1-\mu)$ and is fixed at coordinates $(-\mu, 0)$. The mass parameter $\mu$ is defined as,

$$\mu = \frac{M_m}{M_e + M_m} = 0.0121506683 \tag{1}$$

where $M_e$ and $M_m$ are the masses of the Earth and the Moon.

The coordinates of the spacecraft in the rotating frame are indicated with $(x, y)$ for the positions and $(v_x, v_y)$ for the relevant velocities. Thus, the planar equations of motion for the spacecraft are given by [31]:

$$\begin{cases} \dot{x} = v_x \\ \dot{y} = v_y \\ \dot{v}_x = x + 2v_y - \frac{(1-\mu)(x+\mu)}{r_1^3} - \frac{\mu(x+\mu-1)}{r_2^3} + u_1 \\ \dot{v}_y = y - 2v_x - \frac{(1-\mu)y}{r_1^3} - \frac{\mu y}{r_2^3} + u_2 \end{cases} \tag{2}$$

where the dots indicate the nondimensional time derivatives relative to an observer in the rotating frame, and where $r_1$ and $r_2$ are the distances from the Earth and the Moon respectively:

$$r_1 = \sqrt{(x+\mu)^2 + y^2} \tag{3}$$

$$r_2 = \sqrt{(x+\mu-1)^2 + y^2} \tag{4}$$

The control variables $u_1$ and $u_2$ denote the spacecraft's accelerations in the rotating frame.

In addition, let $J$ denote the following quantity:

$$J\left(x, y, v_x, v_y\right) = 2\Omega\left(x, y\right) - v_x^2 - v_y^2 \tag{5}$$

where,

$$\Omega\left(x, y\right) = \frac{x^2 + y^2}{2} + \frac{1-\mu}{r_1} + \frac{\mu}{r_2} + \frac{\mu\left(1-\mu\right)}{2} \tag{6}$$

When $u_1$ and $u_2$ are identically zero, $J$ remains constant along the trajectory and is known as the Jacobi constant, or Jacobi integral of the motion.

In what follows, Eq. (2) will be written under the following compact form:

$$\dot{\boldsymbol{\xi}} = \varphi\left(\boldsymbol{\xi}, \boldsymbol{u}\right) \tag{7}$$

where,

$$\boldsymbol{\xi} = \left\{ x \; y \; v_x \; v_y \right\}^T \quad \boldsymbol{u} = \left\{ u_1 \; u_2 \right\}^T \tag{8}$$

## 2.2 Optimal Control Problem

The minimum-energy optimal control problem to be solved can be written as follows:

Find

$$\left\{ \bar{\boldsymbol{u}}, \bar{\tau}_0, \bar{\tau}_f \right\} = arg \min_{\boldsymbol{u}, \tau_0, \tau_f} K\left(\boldsymbol{u}, \tau_0, \tau_f\right) = \frac{1}{2} \int_{t_0}^{t_f} \|\boldsymbol{u}\|^2 dt \tag{9a}$$

such that

$$\begin{aligned} &\dot{\boldsymbol{\xi}} = \varphi\left(\boldsymbol{\xi}, \boldsymbol{u}\right) \\ &\boldsymbol{\xi}\left(t_0\right) - \boldsymbol{\xi}_I\left(\tau_0\right) = \boldsymbol{0} \\ &\boldsymbol{\xi}\left(t_f\right) - \boldsymbol{\xi}_T\left(\tau_f\right) = \boldsymbol{0} \end{aligned} \tag{9b}$$

where $\|\boldsymbol{u}\| = \sqrt{u_1^2 + u_2^2}$ denotes the Euclidian norm of vector $\boldsymbol{u}$.

The initial and final dates $t_0$ and $t_f$ are fixed. As the problem defined in Eqs. (9a) and (9b) is autonomous, $t_0$ will be fixed from now on to $t_0 = 0$. The initial and target states on the Lyapunov orbits are denoted by $\boldsymbol{\xi}_I\left(\tau_0\right)$ and $\boldsymbol{\xi}_T\left(\tau_f\right)$, respectively, and are computed by means of Lindstedt-Poincaré techniques [30]. More precisely, $\tau_0$ and $\tau_f$ are nondimensional times that determine the departure location from the initial orbit and the arrival location at the target orbit. These parameters have to be optimized at the same time as the control $\boldsymbol{u}$.

## 2.3 Necessary Optimality Conditions

### 2.3.1 Derivation of Maximum Principle

The optimality conditions for Eqs. (9a) and (9b) can be established using Pontryagin's Maximum Principle (PMP) [32, 33].

The Hamiltonian of Eqs. (9a) and (9b) is given by:

$$H = \frac{1}{2} \|\mathbf{u}\|^2 + \boldsymbol{\lambda}^T \varphi\left(\boldsymbol{\xi}, \mathbf{u}\right) \tag{10}$$

where $\boldsymbol{\lambda}$ is a 4-dimension costate vector. The optimal controls are obtained by minimizing the Hamiltonian with respect to $\boldsymbol{u}$ so that,

$$\boldsymbol{u} = -\left\{ \lambda_3 \; \lambda_4 \right\}^T \tag{11}$$

According to the PMP, the costates equations are given by:

$$\dot{\boldsymbol{\lambda}} = -\nabla_{\xi} H = -\frac{\partial \varphi}{\partial \boldsymbol{\xi}} (\boldsymbol{\xi}, \boldsymbol{u})^T \boldsymbol{\lambda} \tag{12}$$

Finally, the transversality conditions associated with the initial and final boundary conditions can be written:

$$\boldsymbol{\lambda} (t_0)^T \frac{d\boldsymbol{\xi}_I (\tau_0)}{d\tau_0} = \boldsymbol{\lambda} (t_0)^T \varphi \left(\boldsymbol{\xi}_I (\tau_0), \mathbf{0}\right) = 0 \tag{13}$$

$$\boldsymbol{\lambda} \left(t_f\right)^T \frac{d\boldsymbol{\xi}_T \left(\tau_f\right)}{d\tau_f} = \boldsymbol{\lambda} \left(t_f\right)^T \varphi \left(\boldsymbol{\xi}_T \left(\tau_f\right), \mathbf{0}\right) = 0 \tag{14}$$

In Eqs. (13) and (14), the derivatives of $\boldsymbol{\xi}_I (\tau_0)$ and $\boldsymbol{\xi}_T \left(\tau_f\right)$ with respect to $\tau_0$ and $\tau_f$, respectively, reduce to nondimensional time derivatives from the definition of $\tau_0$ and $\tau_f$.

Finally, solving the Two-Point Boundary Value problem (TPBVP) arising from the PMP is equivalent to finding the 6-dimension unknown vector $z = \left\{\boldsymbol{\lambda} (t_0)^T, \tau_0, \tau_f\right\}^T$ such that Eqs. (7) and (11)–(14) and the boundary conditions in Eq. (9b) are satisfied. The corresponding shooting function, denoted by $F$, associates to $z$ the following value:

$$F(z) = \begin{bmatrix} \boldsymbol{\xi} \left(t_f\right) - \boldsymbol{\xi}_T \left(\tau_f\right) \\ \boldsymbol{\lambda} (t_0)^T \varphi \left(\boldsymbol{\xi}_I (\tau_0), \mathbf{0}\right) \\ \boldsymbol{\lambda} \left(t_f\right)^T \varphi \left(\boldsymbol{\xi}_T \left(\tau_f\right), \mathbf{0}\right) \end{bmatrix} \tag{15}$$

where $\boldsymbol{\xi} \left(t_f\right)$ is obtained by using Eqs. (7), (11) and (12), with the boundary condition at $t_0$ given in Eq. (9b), and where $\boldsymbol{\lambda} \left(t_f\right)$ is computed by integrating Eq. (12), with $\boldsymbol{u}$ given in Eq. (11). Thus, solving the TPBVP is equivalent to finding a zero of function $F$.

In what follows, the combined state and costate 8-dimension vector will be denoted as:

$$\boldsymbol{\eta} = \begin{Bmatrix} \boldsymbol{\xi} \\ \boldsymbol{\lambda} \end{Bmatrix} \tag{16}$$

and Eqs. (7) and (12) (with $\boldsymbol{u}$ given in Eq. (11)) will be written under the following compact form:

$$\dot{\boldsymbol{\eta}} = \psi (\boldsymbol{\eta}) \tag{17}$$

### 2.3.2 Computing Jacobian of Shooting Function

In order to accurately compute the Jacobian of Eq. (15), variational equations are used. First, the gradients of $\boldsymbol{\eta} \left(t_f\right)$ with respect to $z_i (i = 1, \ldots, 5)$ are computed as follows.

Consider the following extended system of 48 differential equations, integrated over $[t_0, t_f]$:

$$\begin{aligned} &\dot{\boldsymbol{\eta}} = \psi\left(\boldsymbol{\eta}\right) \\ &\dot{\boldsymbol{\alpha}}_i = \frac{\partial \psi}{\partial \boldsymbol{\eta}}\left(\boldsymbol{\eta}\right)\boldsymbol{\alpha}_i \quad (i = 1, .., 5) \\ &\boldsymbol{\eta}\left(t_0\right) = \left\{\boldsymbol{\xi}_I\left(\tau_0\right)^T \; z_1 \; z_2 \; z_3 \; z_4\right\}^T \\ &\boldsymbol{\alpha}_i\left(t_0\right) = \left\{0\,0\,0\,0\;\boldsymbol{\delta}_i^T\right\}^T \quad (i = 1, ..., 4) \\ &\boldsymbol{\alpha}_5\left(t_0\right) = \left\{\varphi\left(\boldsymbol{\xi}_I\left(\tau_0\right), \mathbf{0}\right)^T \; 0\,0\,0\,0\right\}^T \end{aligned} \tag{18}$$

where $\boldsymbol{\delta}_i$ $(i = 1, \ldots, 4)$ is a 4-dimension column vector with all entries equal to zero except the $i$-th entry which is equal to one.

Then the following holds:

$$\nabla_{z_i}\boldsymbol{\eta}\left(t_f\right) = \boldsymbol{\alpha}_i\left(t_f\right) (i = 1, \ldots, 5) \tag{19}$$

Now, consider the component-wise nondimensional time derivative of $\varphi\left(\boldsymbol{\xi}, \mathbf{0}\right)$ defined as:

$$\varphi^1\left(\boldsymbol{\xi}\right) = \left\{\dot{v}_x \; \dot{v}_y \; \ddot{v}_x \; \ddot{v}_y\right\}^T \tag{20}$$

where $\dot{v}_x$ and $\dot{v}_y$ are given in Eq. (2) (with $u_1 = u_2 = 0$), and where $\ddot{v}_x$ and $\ddot{v}_y$ are obtained by differentiating the last two equations in Eq. (2) (for $u_1 = u_2 = 0$) with respect to the nondimensional time, leading to:

$$\begin{aligned} \ddot{v}_x = 2y - 3v_x &- \frac{(1-\mu)(2y+v_x)}{r_1^3} - \frac{\mu(2y+v_x)}{r_2^3} \\ &+ \frac{3(1-\mu)(x+\mu)\left[v_x(x+\mu)+v_y y\right]}{r_1^5} + \frac{3\mu(x-1+\mu)\left[v_x(x-1+\mu)+v_y y\right]}{r_2^5} \end{aligned} \tag{21}$$

$$\begin{aligned} \ddot{v}_y = -2x - 3v_y &+ \frac{(1-\mu)\left[2(x+\mu)-v_y\right]}{r_1^3} + \frac{\mu\left[2(x-1+\mu)-v_y\right]}{r_2^3} \\ &+ \frac{3(1-\mu)y\left[v_x(x+\mu)+v_y y\right]}{r_1^5} + \frac{3\mu y\left[v_x(x-1+\mu)+v_y y\right]}{r_2^5} \end{aligned} \tag{22}$$

Then consider the $8 \times 5$ matrix $\mathbf{A(t)}$ whose columns are the column vectors $\boldsymbol{\alpha}_i$ $(i = 1, \ldots, 5)$ evaluated at the adimensional time $t$:

$$\mathbf{A}\left(t\right) = \left[\boldsymbol{\alpha}_1\left(t\right) \boldsymbol{\alpha}_2\left(t\right) \boldsymbol{\alpha}_3\left(t\right) \boldsymbol{\alpha}_4\left(t\right) \boldsymbol{\alpha}_5\left(t\right)\right] = \begin{bmatrix} \mathbf{A}_1\left(t\right) \\ \mathbf{A}_2\left(t\right) \end{bmatrix} \tag{23}$$

where $\mathbf{A}_1\left(t\right)$ and $\mathbf{A}_2\left(t\right)$ denote the $4 \times 5$ upper and lower submatrices of $\mathbf{A}\left(t\right)$, respectively.

Thus, the Jacobian of Eq. (15) can be written as follows:

$$\frac{\partial F}{\partial z}\left(z\right) = \begin{bmatrix} \mathbf{A}_1\left(t_f\right) & -\varphi\left(\boldsymbol{\xi}_T\left(\tau_f\right), \mathbf{0}\right) \\ \varphi\left(\boldsymbol{\xi}_I\left(\tau_0\right), \mathbf{0}\right)^T \quad \boldsymbol{\lambda}\left(t_0\right)^T \varphi^1\left(\boldsymbol{\xi}_I\left(\tau_0\right)\right) & 0 \\ \varphi\left(\boldsymbol{\xi}_T\left(\tau_f\right), \mathbf{0}\right)^T \mathbf{A}_2\left(t_f\right) & \boldsymbol{\lambda}\left(t_f\right)^T \varphi^1\left(\boldsymbol{\xi}_T\left(\tau_f\right)\right) \end{bmatrix} \tag{24}$$

# 3 New Approach for Computing Low-Energy Transfers

## 3.1 Introduction

It will be shown in Sect. 4, that Eqs. (9a) and (9b) can be solved in a direct way by means of shooting methods when the transfer duration is relatively short. However, for medium to large transfer durations, Eqs. (9a) and (9b) cannot be solved without the use of a specific methodology in the particular case where the initial and target orbits are of the same energy.

In [34], a method has been developed for solving optimal control problems over large time intervals by connecting the solutions of two infinite time problems. These two problems are solved thanks to Hamilton-Jacobi-Bellman theory [33], limiting the applicability of [34] to problems with a very small number of state variables due to the well-known curse of dimensionality phenomenon. In [35, 36], the solution of long time horizon hyper-sensitive optimal control problems is approximated by decomposing the Hamiltonian vector field into its stable and unstable components. This is achieved thanks to the use of a local eigenvalue decomposition of the Jacobian of the vector field. In [35, 36], as in [34], the authors show that the optimal solution exhibits a particular structure made of an initial boundary-layer segment, an equilibrium segment, and a terminal boundary-layer segment. Notice that the same research group developed another approach, based on Lyapunov exponents, for solving hyper-sensitive problems [37, 38].

The approximate techniques proposed in [35–38] are strongly inspired by the singular perturbation methodology developed in [39]. They are not straightforward to implement and require that the problem fulfills specific properties. Nevertheless, in accordance with [34], they highlight the very particular structure of the solution of hyper-sensitive optimal control problems whose dynamics equations are characterized by the presence of two time scales. This three-segment structure, that is also described qualitatively as 'take-off, cruise, and landing' in [36], indicates that the controls are close to zero during the intermediate 'cruise' (or equilibrium) segment. The 'take-off' segment is determined by the initial conditions, the state dynamics, and the goal of reaching the 'cruise' segment in forward time. The 'landing' segment is determined by the terminal conditions, the state dynamics, and the goal of reaching the 'cruise' segment in backward time. Finally, the 'cruise' segment is determined by the cost function and the state dynamics, while it is almost independent of the boundary conditions.

Low-energy transfers between Lyapunov orbits of the same energy exhibit this particular three-segment pattern. In this case, the equilibrium segment corresponds to a part of the trajectory where the spacecraft is close to an invariant manifold. Nevertheless, this three-segment structure is shared by the solutions of a wide class of optimal control problems with long time horizon.

In Sect. 3.2, a three-step methodology based on the knowledge of this particular structure is developed. An initial feasible control with an intermediate coast arc is sought. Contrarily to [24], a coast arc is enforced on the feasible control but not on the optimal control corresponding to the low-energy solution.

## 3.2 *Three-Step Solution Method*

### 3.2.1 First Step: Initial Feasible Solution

The aim of this first step of the method is to find a control law with the three-segment structure highlighted in Sect. 3.1, that is feasible for Eqs. (9a) and (9b). This control, denoted by $\boldsymbol{u}^0$, is assumed here to have the following elementary quadratic-zero-quadratic time-shape:

$$u_i^0(t) = \begin{cases} a_i (t - t_1)^2 & t \in [t_0, t_1] \\ 0 & t \in [t_1, t_2] \quad (i = 1, \ldots, 2) \\ b_i (t - t_2)^2 & t \in \left[t_2, t_f\right] \end{cases} \tag{25}$$

where $t_1$ and $t_2$ are two switching adimensional times.

Let $w$ denote the following 8-dimension unknown vector

$$\boldsymbol{w} = \left\{a_1, a_2, b_1, b_2, t_1, t_2, \tau_0^0, \tau_f^0\right\}^T$$

where $\tau_0^0$ and $\tau_f^0$ determine the departure location from the initial orbit and the arrival location at the target orbit.

Hence, finding a feasible control $\boldsymbol{u}^0$ of the form given in Eq. (25) is equivalent to finding a zero-cost solution of the following optimization problem:

Find

$$\bar{\boldsymbol{w}} = \arg \min_{\boldsymbol{w}} \; G(\boldsymbol{w}) = \frac{1}{2} \left\| \boldsymbol{\xi}_T \left(\tau_f^0\right) \right\|^2 \tag{26a}$$

such that

$$\begin{aligned} &\dot{\boldsymbol{\xi}} = \varphi\left(\boldsymbol{\xi}, \boldsymbol{u}^0\right) \\ &\boldsymbol{\xi}(t_0) - \boldsymbol{\xi}_I\left(\tau_0^0\right) = \boldsymbol{0} \end{aligned} \tag{26b}$$

where $\boldsymbol{u}^0$ depends on $\boldsymbol{w}$ through Eq. (25).

The derivative-free Nelder-Mead simplex method [40] is used for solving Eqs. (26a) and (26b). The method is initialized by a set of randomly generated values of $\boldsymbol{w}$. The solutions obtained that satisfy the condition $G(\bar{\boldsymbol{w}}) \leq \sigma$, where $\sigma$ is a given small threshold, are classified according to the value of the performance index $K\left(\boldsymbol{u}^0, \tau_0^0, \tau_f^0\right)$ defined in Eq. (9a).

Finally, among these solution candidates, the solution $\bar{\boldsymbol{w}}$ associated with the lowest value of $K$ is selected.

### 3.2.2 Second Step: Suboptimal Solution

Using the maximum principle, it is straightforward to show that the solution $\left\{\boldsymbol{u}^0, \tau_0^0, \tau_f^0\right\}$ obtained in Sect. 3.2.1 satisfies:

$$\boldsymbol{u}^0 = \arg\min_{\boldsymbol{u}} K_0(\boldsymbol{u}) = \frac{1}{2}\int_{t_0}^{t_f} \left\|\boldsymbol{u} - \boldsymbol{u}^0\right\|^2 dt \tag{27a}$$

such that

$$\begin{aligned} &\dot{\boldsymbol{\xi}} = \varphi(\boldsymbol{\xi}, \boldsymbol{u}) \\ &\boldsymbol{\xi}(t_0) - \boldsymbol{\xi}_I\left(\tau_0^0\right) = \mathbf{0} \\ &\boldsymbol{\xi}\left(t_f\right) - \boldsymbol{\xi}_T\left(\tau_f^0\right) = \mathbf{0} \end{aligned} \tag{27b}$$

and that the optimal costate vector in Eqs. (27a) and (27b) is identically zero on $[t_0, t_f]$.

Indeed, the Hamiltonian of Eqs. (27a) and (27b) is given by:

$$H = \frac{1}{2}\left\|\boldsymbol{u} - \boldsymbol{u}^0\right\|^2 + \boldsymbol{\lambda}^T \varphi(\boldsymbol{\xi}, \boldsymbol{u}) \tag{28}$$

leading to the following optimal control vector:

$$\boldsymbol{u} = \boldsymbol{u}^0 - \left\{\lambda_3\ \lambda_4\right\}^T \tag{29}$$

In addition, the costates equations for Eqs. (27a) and (27b) are the same as those given in Eq. (12).

Thus, $\left\{\boldsymbol{u} = \boldsymbol{u}^0, \boldsymbol{\lambda} = \boldsymbol{\lambda}^0 = \mathbf{0}\right\}$ is a trivial solution of the necessary optimality conditions given by Eqs. (12), (27b) and (29).

Consider now the following family of optimal control problems, parametrized by $\varepsilon \in [0, 1]$:

Find

$$\boldsymbol{u}^\varepsilon = \arg\min_{\boldsymbol{u}} K_\varepsilon(\boldsymbol{u}) = \frac{1}{2}\int_{t_0}^{t_f} \left\{(1-\varepsilon)\left\|\boldsymbol{u} - \boldsymbol{u}^0\right\|^2 + \varepsilon\left\|\boldsymbol{u}\right\|^2\right\} dt \tag{30a}$$

such that

$$\begin{aligned} &\dot{\boldsymbol{\xi}} = \varphi(\boldsymbol{\xi}, \boldsymbol{u}) \\ &\boldsymbol{\xi}(t_0) - \boldsymbol{\xi}_I\left(\tau_0^0\right) = \mathbf{0} \\ &\boldsymbol{\xi}\left(t_f\right) - \boldsymbol{\xi}_T\left(\tau_f^0\right) = \mathbf{0} \end{aligned} \tag{30b}$$

For $\varepsilon = 0$, Eqs. (30a) and (30b) is identical to Eqs. (27a) and (27b) and its solution is simply $\boldsymbol{u}^0$.

Starting from this initial solution, Eqs. (30a) and (30b) has to be solved by continuation on $\varepsilon$ until the value $\varepsilon = 1$ is reached. For this final value of $\varepsilon$, Eqs. (30a) and (30b) is a minimum-energy problem with fixed departure location from the initial Lyapunov orbit and fixed arrival location at the target Lyapunov orbit. Thus, for $\varepsilon = 1$, the solution of Eqs. (30a) and (30b) is suboptimal compared with that of Eqs. (9a) and (9b). During the continuation process, small steps in $\varepsilon$ must be taken in order not to switch to another local optimum with a higher energy level. For each value of $\varepsilon$, Eqs. (30a) and (30b) is solved by means of an indirect single shooting method.

The Hamiltonian of Eqs. (30a) and (30b) is given by:

$$H = \frac{1}{2}\left\{(1-\varepsilon)\left\|\boldsymbol{u}-\boldsymbol{u}^0\right\|^2 + \varepsilon\left\|\boldsymbol{u}\right\|^2\right\} + \boldsymbol{\lambda}^T\varphi\left(\boldsymbol{\xi},\boldsymbol{u}\right) \tag{31}$$

leading to the following optimal control vector:

$$\boldsymbol{u}^{\varepsilon} = (1-\varepsilon)\,\boldsymbol{u}^0 - \left\{\lambda_3\ \lambda_4\right\}^T \tag{32}$$

and the costates equations for Eqs. (30a) and (30b) are the same as those given in Eq. (12).

The Jacobian of the associated shooting function is computed using variational equations as described in Sect. 2.3.2. The difference with the calculations developed in Sect. 2.3.2, is that the number of unknowns of the shooting function is only four here, because $\tau_0$ and $\tau_f$ are fixed to the values $\tau_0^0$ and $\tau_f^0$, respectively. Thus, the unknown vector reduces to $z = \boldsymbol{\lambda}\left(t_0\right)$ and the shooting function becomes:

$$F\left(z\right) = \boldsymbol{\xi}\left(t_f\right) - \boldsymbol{\xi}_T\left(\tau_f\right) \tag{33}$$

In addition, the controls are given here by Eq. (32) instead of Eq. (11) as in Sect. 2.3.2. With these differences, the Jacobian of the shooting function is equal now to the following $4 \times 4$ matrix:

$$\frac{\partial F}{\partial z}\left(z\right) = \left[\boldsymbol{\alpha}_1\left(t_f\right)\ \boldsymbol{\alpha}_2\left(t_f\right)\ \boldsymbol{\alpha}_3\left(t_f\right)\ \boldsymbol{\alpha}_4\left(t_f\right)\right] \tag{34}$$

and Eq. (18) reduces to a system of 40 differential equations, instead of 48 as in Sect. 2.3.2, due to the fact that the equations associated with $\boldsymbol{\alpha}_5$ are dropped.

Finally, notice here that the same kind of continuation on the control variable, based on a family of problems similar to Eqs. (30a) and (30b), has been applied to a space shuttle reentry problem in [41].

### 3.2.3 Third Step: Low-Energy Solution

Once the solution of Eqs. (30a) and (30b) has been computed for $\varepsilon = 1$, the departure location from the initial Lyapunov orbit and the arrival location at the target Lyapunov orbit have to be optimized in order to yield the low-energy trajectory, solution of Eqs. (9a) and (9b). At this step, one may consider solving Eqs. (9a) and (9b) by means of the single shooting method starting from the solution of Eqs. (30a) and (30b) (for $\varepsilon = 1$). Unfortunately, this approach turns out to be tricky from the numerical point of view. To reduce the numerical difficulties, consider solving a sequence of problems with a lower number of unknown variables, typically four instead of six.

With this aim consider a problem similar to of Eqs. (9a) and (9b), but where $\tau_0$ and $\tau_f$ are fixed parameters:

Find

$$\boldsymbol{u}_{\tau_0,\tau_f} = \arg\min_{\boldsymbol{u}} K\left(\boldsymbol{u}, \tau_0, \tau_f\right) = \frac{1}{2}\int_{t_0}^{t_f} \|\boldsymbol{u}\|^2 dt \tag{35a}$$

such that

$$\begin{aligned} &\dot{\boldsymbol{\xi}} = \varphi\left(\boldsymbol{\xi}, \boldsymbol{u}\right) \\ &\boldsymbol{\xi}\left(t_0\right) - \boldsymbol{\xi}_I\left(\tau_0\right) = \boldsymbol{0} \\ &\boldsymbol{\xi}\left(t_f\right) - \boldsymbol{\xi}_T\left(\tau_f\right) = \boldsymbol{0} \end{aligned} \tag{35b}$$

First, it is clear that the solution of Eqs. (35a) and (35b) for $\tau_0 = \tau_0^0$ and $\tau_f = \tau_f^0$ is equal to that of Eqs. (30a) and (30b) for $\varepsilon = 1$; that is, $\boldsymbol{u}_{\tau_0^0,\tau_f^0} = \boldsymbol{u}^1$. Now, consider that the optimal cost function in Eq. (35a) is a function of the two parameters $\tau_0$ and $\tau_f$:

$$K\left(\boldsymbol{u}_{\tau_0,\tau_f}, \tau_0, \tau_f\right) = L\left(\tau_0, \tau_f\right) \tag{36}$$

Hence, solving Eqs. (9a) and (9b) is equivalent to solving the following problem:

$$\left\{\bar{\tau}_0, \bar{\tau}_f\right\} = \arg\min_{\tau_0,\tau_f} L\left(\tau_0, \tau_f\right) \tag{37}$$

and the associated optimal control in Eqs. (9a) and (9b) is simply given by $\bar{\boldsymbol{u}} = \boldsymbol{u}_{\bar{\tau}_0,\bar{\tau}_f}$.

Consider the use of a gradient algorithm for solving Eq. (37). First, the gradient of function $L$ defined in Eq. (36) follows from applying the maximum principle to Eqs. (35a) and (35b). The following holds:

$$\nabla L\left(\tau_0, \tau_f\right) = \begin{bmatrix} \boldsymbol{\lambda}_{\tau_0,\tau_f}\left(t_0\right)^T \varphi\left(\boldsymbol{\xi}_I\left(\tau_0\right), \boldsymbol{0}\right) \\ -\boldsymbol{\lambda}_{\tau_0,\tau_f}\left(t_f\right)^T \varphi\left(\boldsymbol{\xi}_T\left(\tau_f\right), \boldsymbol{0}\right) \end{bmatrix} \tag{38}$$

where $\boldsymbol{\lambda}_{\tau_0,\tau_f}$ is the optimal costate vector for Eqs. (35a) and (35b).

Thus, solving Eq. (37) by means of a gradient algorithm leads to the solution of a sequence of problems Eqs. (35a) and (35b), with varying values of $\tau_0$ and $\tau_f$. Here again, as in the case of Eqs. (30a), (30b) and (35a), (35b) is solved by means of an indirect single shooting method and the Jacobian of the shooting function is computed using variational equations (see Sect. 2.3.2).

At step $k = 0$, the algorithm starts with $\tau_0 = \tau_0^0$ and $\tau_f = \tau_f^0$. At step $k$, the current values are denoted by $\tau_0^k$ and $\tau_f^k$. Then the values at step $(k+1)$ are computed as follows:

$$\begin{Bmatrix} \tau_0^{k+1} \\ \tau_f^{k+1} \end{Bmatrix} = \begin{Bmatrix} \tau_0^{k} \\ \tau_f^{k} \end{Bmatrix} - \beta \nabla L \left( \tau_0^k, \tau_f^k \right) \tag{39}$$

where $\beta$ is a small constant step size.

The gradient algorithm stops at a given step $N$ when the following condition is satisfied:

$$\parallel \nabla L \left( \tau_0^N, \tau_f^N \right) \parallel \leq \nu \tag{40}$$

for a given threshold $\nu > 0$.

# 4 Numerical Results

## 4.1 Orbits with Same Energy

Consider a transfer from a Lyapunov orbit around $L_1$ to a Lyapunov orbit around $L_2$ with the same energy in the Earth-Moon three-body problem. The characteristics of these two orbits are summarized in Table 1.

For the above value of the Jacobi constant, it has been shown in [42] that heteroclinic connections exist between the two Lyapunov orbits, i.e., zero-cost transfers (in infinite time scale), with one or two revolutions around the Moon. This implies that low-energy solutions exist for Eqs. (9a) and (9b), for appropriately chosen transfer durations.

### 4.1.1 Example 1: Small Transfer Duration

Consider the solution of the minimum-energy problem Eqs. (9a) and (9b) with a transfer duration fixed to $T_f = 12$ days.

**Table 1** Characteristics of the two Lyapunov orbits

| Lagrange point | Jacobi constant | Adimensional x-amplitude | Adimensional period |
|---|---|---|---|
| $L_1$ | 3.1780 | 0.13515959512207 | 2.776024944790715 |
| $L_2$ | 3.1780 | 0.10041124020000 | 3.385292341037150 |

After time scaling, the corresponding value of $t_f$ is equal to:

$$t_f = 2\pi \frac{T_f}{P_M} \tag{41}$$

where $P_M = 27.321577$ days is the orbital period of the Moon, leading here to $t_f = 2.759659$.

For this small value of $t_f$, Eqs. (9a) and (9b) can be solved in a direct way by means of a single shooting method coupled with the use of variational equations for computing the Jacobian of the shooting function (see Sect. 2.3).

Starting from randomly generated initial guesses, a zero of the shooting function Eq. (15) is searched by means of Powell's hybrid method [43] implemented in routine C05PCF from NAG FORTRAN Library [44]. The differential system Eq. (18) is integrated numerically using an explicit Runge-Kutta method of order eight [45]. A set of solutions, corresponding to different values of the performance index $K$ in Eq. (9a), is found. Some of these solutions with their associated values of $z$ are listed in Table 2.

Solution 1 is associated with the lowest value of $K$. The corresponding control history and the trajectory obtained are presented in Figs. 1 and 2, respectively. The

**Table 2** Some low-cost solutions of the 12-day transfer

| Solution # | Performance index K | z |
|---|---|---|
| 1 | $3.6513857715175225 \times 10^{-3}$ | $\{-0.16848021336495497,$<br>$-9.670309150887528 \times 10^{-3},$<br>$-0.04760618141086829,$<br>$-0.060647942633170776,$<br>$1.6548647700861398,$<br>$3.0315500530220513\}^T$ |
| 2 | 0.023206311154351245 | $\{-0.1504418369886553,$<br>$-0.10857363835560588,$<br>$-0.037298142153247795,$<br>$-0.10521925343747937,$<br>$1.6826178039929275,$<br>$1.2114995052368988\}^T$ |
| 3 | 0.05611351253574562 | $\{0.6259367206833839,$<br>$0.12804101240764947,$<br>$0.25137878106392086,$<br>$-0.1464292499136169,$<br>$2.579354774482469,$<br>$2.601598366193081\}^T$ |
| 4 | 0.062989036389057575 | $\{0.28624444620589095,$<br>$-0.4493787016358408,$<br>$0.06923526173484933,$<br>$-0.13431844202420811,$<br>$0.34082463069298635,$<br>$2.90468607327521590\}^T$ |

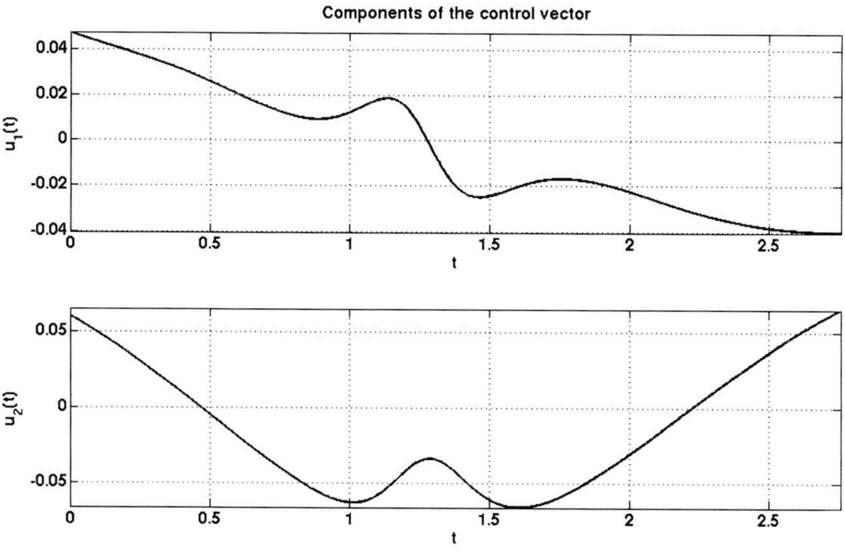


**Fig. 1** Control history for the 12-day transfer: solution 1

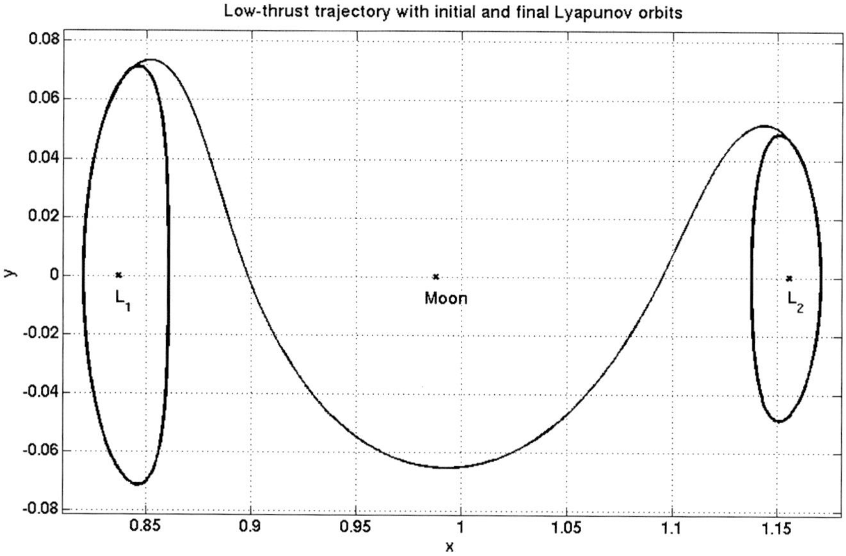


**Fig. 2** Optimal trajectory for the 12-day transfer: solution 1

control laws and the trajectories associated with solutions 2, 3 and 4 are given in Figs. 3, 4, 5, 6, 7 and 8. Solutions 2, 3 and 4 seem clearly non-optimal from these figures due to either a departure from the initial orbit in opposite direction to the

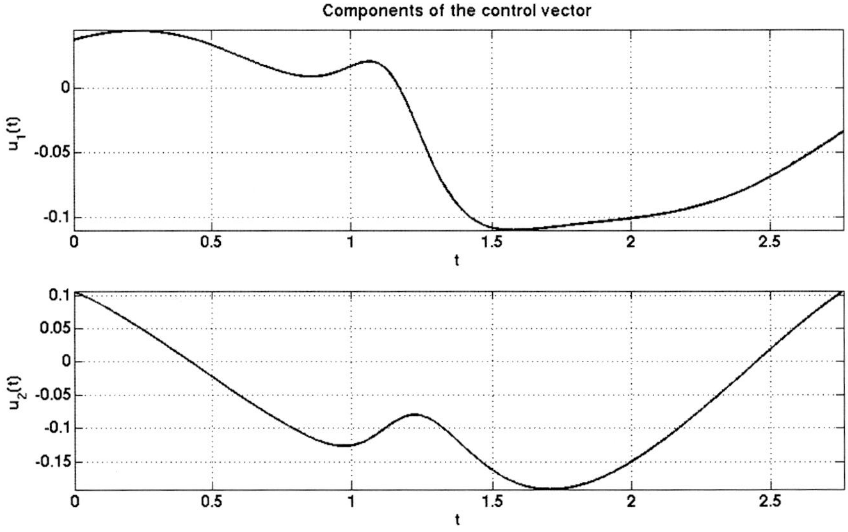


**Fig. 3** Control history for the 12-day transfer: solution 2

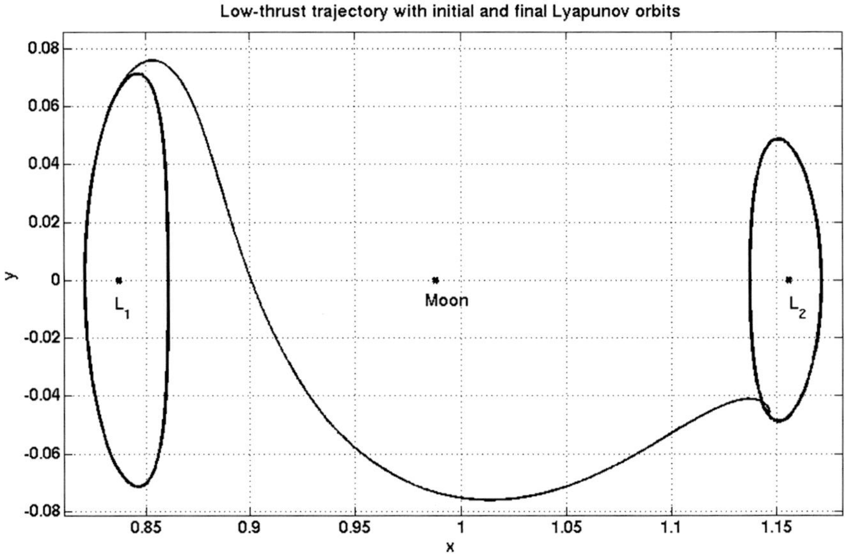


**Fig. 4** Optimal trajectory for the 12-day transfer: solution 2

target orbit (as in solutions 3 and 4), or a backward injection into the target orbit, i.e., an injection from above the target orbit in direction of the Moon (as in solution 2).

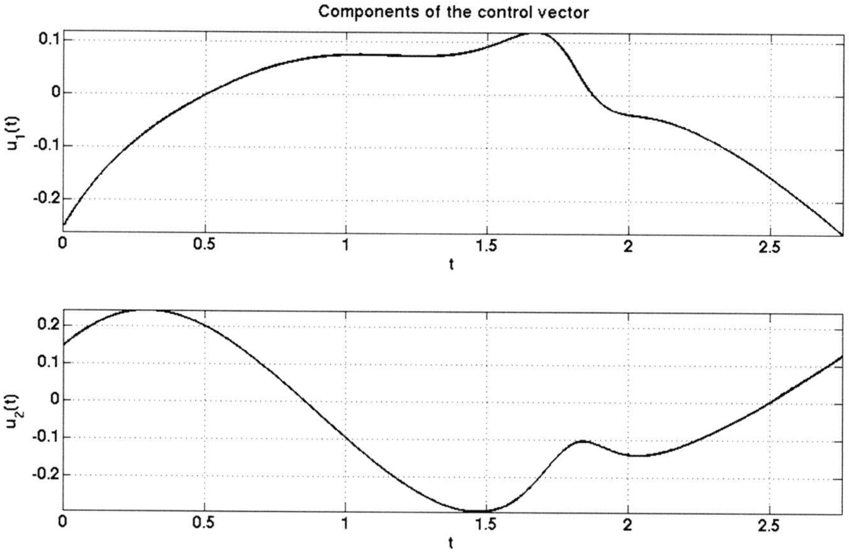


**Fig. 5** Control history for the 12-day transfer: solution 3

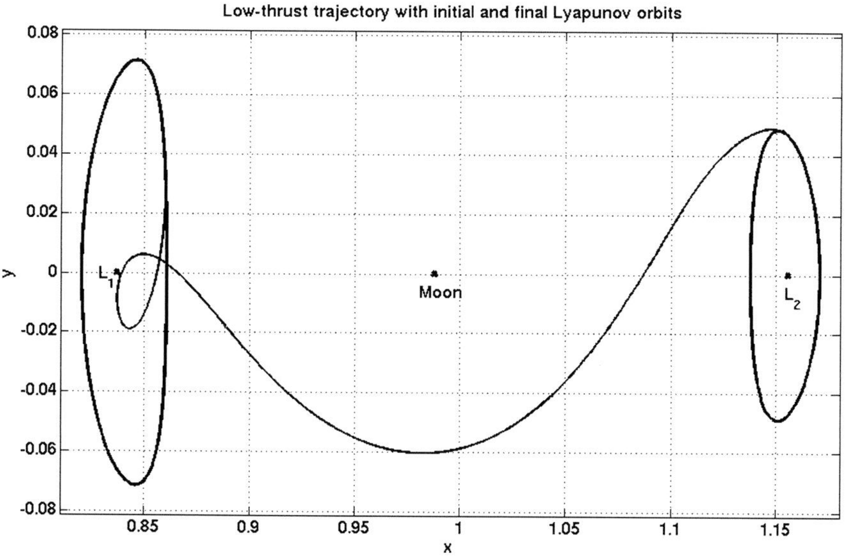


**Fig. 6** Optimal trajectory for the 12-day transfer: solution 3

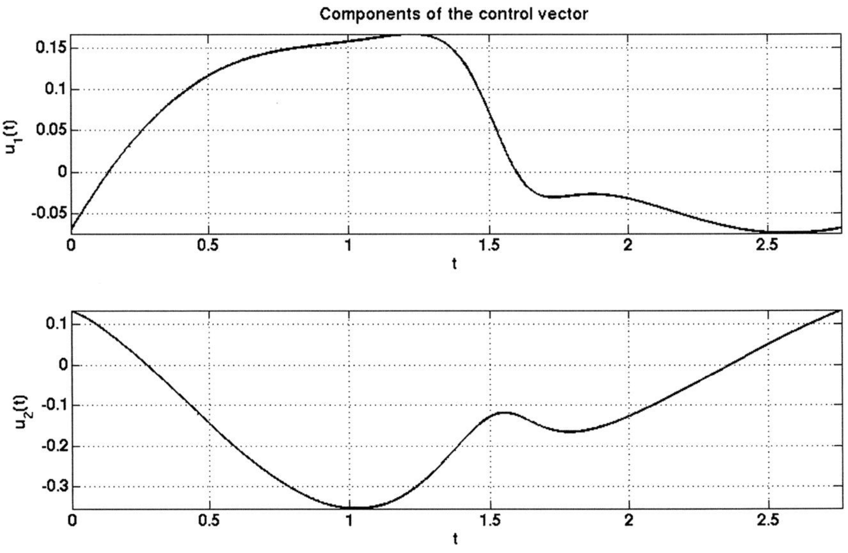


**Fig. 7** Control history for the 12-day transfer: solution 4

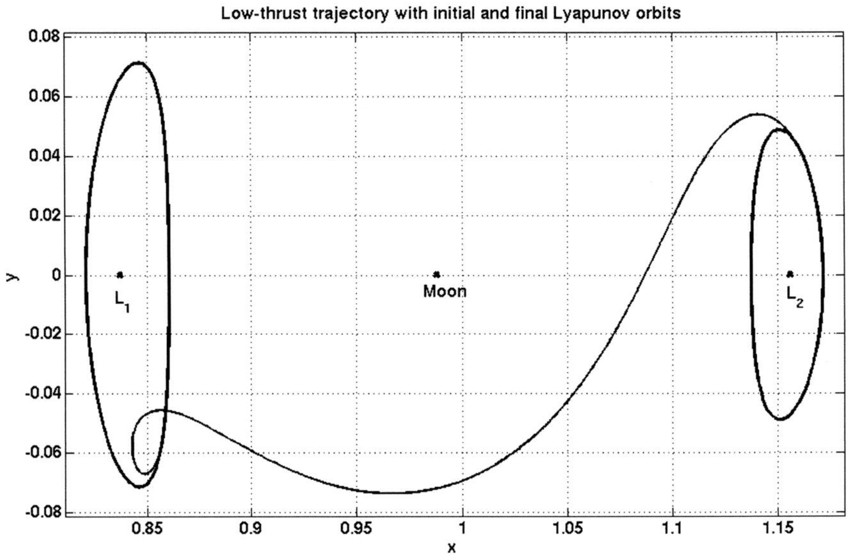


**Fig. 8** Optimal trajectory for the 12-day transfer: solution 4

### 4.1.2 Example 2: Medium Transfer Duration

Consider now the solution of the minimum-energy problem Eqs. (9a) and (9b) with a transfer duration fixed to $T_f = 29$ days. The corresponding value of $t_f$ is derived from Eq. (41), leading to $t_f = 6.669175$.

In this case, trying to solve Eqs. (9a) and (9b) in a direct way by means of a single shooting method fails whatever the number of randomly generated values of vector $z$ used for initializing Powell's hybrid Method. The shooting function in Eq. (15) turns out to be ill-conditioned, inducing a very small convergence radius for Powell's method. As the solution of Eqs. (9a) and (9b) is expected to exhibit the three-segment structure identified in Sect. 3, Eqs. (9a) and (9b) is going to be solved by means of the three-step methodology developed in Sect. 3.2.

In the first step, an initial feasible control $\boldsymbol{u}^0$ is sought by determining a zero-cost solution of Eqs. (26a) and (26b) thanks to the Nelder-Mead simplex method. The zero-cost solution associated with the lowest value of the performance index $K$ defined in Eq. (9a), corresponds here to the following value of the unknown vector $\boldsymbol{w}$:

$$\bar{\boldsymbol{w}} = \begin{Bmatrix} -0.047953478770013 69 \\ 0.09068237475600469 \\ -1.938474730931543 \times 10^{-3} \\ -3.5800293854974633 \times 10^{-4} \\ 0.3650177920519315 \\ 4.8481805396773705 \\ 0.3556394515840182 \\ 0.725904853320663 \end{Bmatrix} \tag{42}$$

and satisfies:

$$G(\bar{\boldsymbol{w}}) = 1.0964331333708144 \times 10^{-11}$$

$$K\left(\mathbf{u}^0, \tau_0^0, \tau_f^0\right) = 1.462274559923132 \times 10^{-5}$$

The feasible control $\boldsymbol{u}^0$ appears in Fig. 9. The associated trajectory in Fig. 10 indicates that the spacecraft completes one revolution around the Moon.

In the second step of the method, Eqs. (30a) and (30b) is solved by continuation on $\varepsilon$ until the target value $\varepsilon = 1$ is reached. This continuation process requires 671 iterations.

For $\varepsilon = 1$, the zero of the shooting function obtained when solving Eqs. (30a) and (30b) by means of the single shooting method is given as follows:

$$z = \boldsymbol{\lambda}^1(t_0) = \begin{Bmatrix} -4.044584655813334 \times 10^{-3} \\ -1.1751010990337217 \times 10^{-3} \\ -8.026696676597337 \times 10^{-4} \\ -1.0910975885703793 \times 10^{-3} \end{Bmatrix} \tag{43}$$

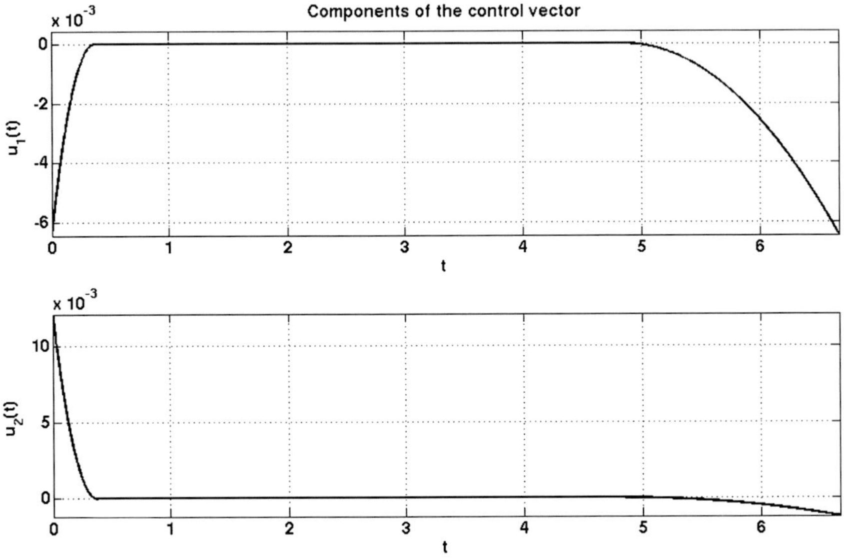


**Fig. 9** Feasible controls for the 29-day transfer

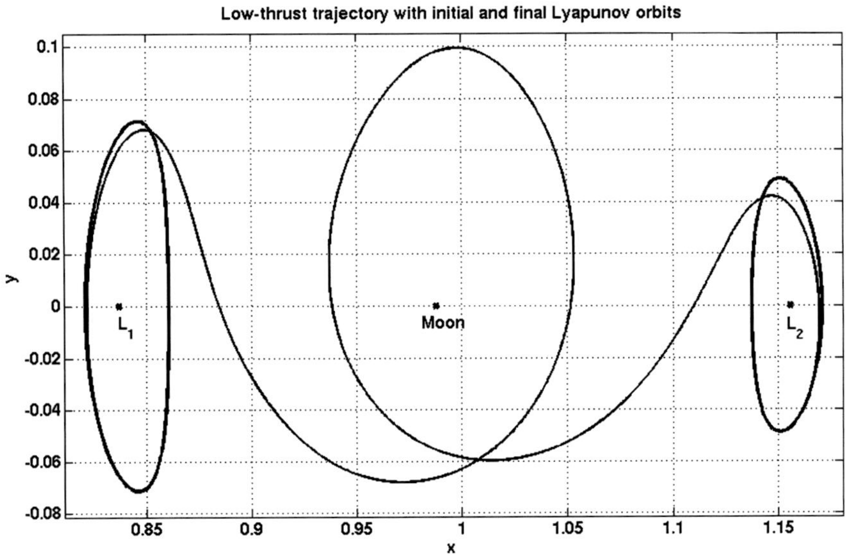


**Fig. 10** Feasible trajectory for the 29-day transfer

and the value of the performance index,

$$K_1\left(\boldsymbol{u}^1\right) = K\left(\boldsymbol{u}^1, \tau_0^0, \tau_f^0\right) = 4.760839250309593 \times 10^{-6}$$

is logically lower than the value obtained for the feasible control $\boldsymbol{u}^0$. Remember that $\tau_0^0$ and $\tau_f^0$ are the non-optimal departure location from the initial orbit and arrival location at the target orbit associated with the feasible solution. Thus, they are equal to the two last components of $\bar{\boldsymbol{w}}$ in Eq. (42):

$$\tau_0^0 = 0.3556394515840182 \quad \tau_f^0 = 0.725904853320663 \tag{44}$$

The control $\boldsymbol{u}^1 = \boldsymbol{u}_{\tau_0^0, \tau_f^0}$ appears in Fig. 11. The associated trajectory in Fig. 12 is similar to that given in Fig. 10. This trajectory seems to follow the paths of the invariant manifolds even if it is not the low-energy trajectory. Indeed, the control law in Fig. 11 does not clearly exhibit an intermediate coast arc, as expected for the low-energy control.

Finally, in the third step of the method, $\tau_0$ and $\tau_f$ are optimized thanks to the gradient algorithm described in Sect. 3.2.3. The convergence is slow and is achieved after $N = 1152$ iterations.

The optimal values obtained at the convergence of the gradient algorithm are relatively different to those given in Eq. (44):

$$\bar{\tau}_0 = \tau_0^N = 0.6681909682675602 \quad \bar{\tau}_f = \tau_f^N = 1.0156271587811724 \tag{45}$$

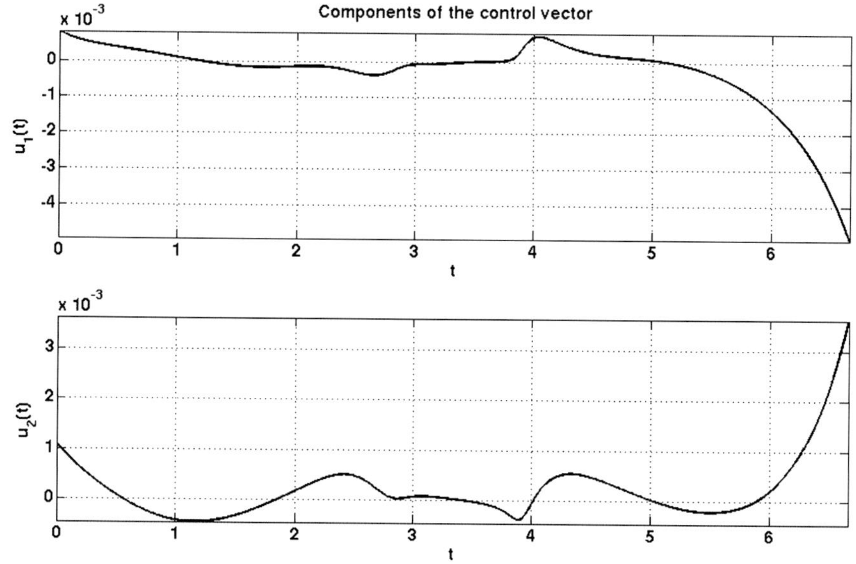


**Fig. 11** Control history obtained at the end of step 2 for the 29-day transfer

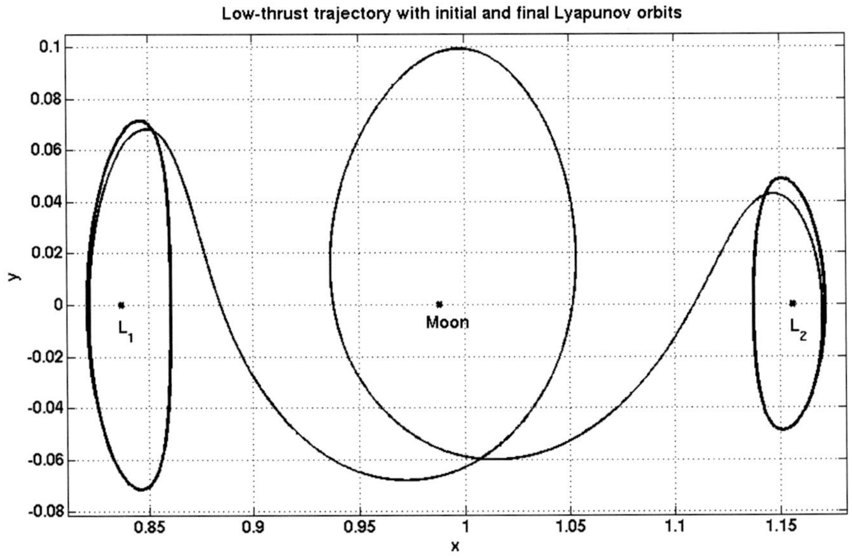


**Fig. 12** Trajectory obtained at the end of step 2 for the 29-day transfer

and the initial costate vector for Eqs. (35a) and (35b) associated with these values is given as follows:

$$\boldsymbol{\lambda}_{\tau_0^N,\tau_f^N}(t_0) = \begin{Bmatrix} -9.87126560148959 \times 10^{-3} \\ -7.613251807328916 \times 10^{-4} \\ -2.625874439708917 \times 10^{-3} \\ -1.728436523309004 \times 10^{-3} \end{Bmatrix} \tag{46}$$

The gradient algorithm has stopped with the condition: $\left\| \nabla L\left(\tau_0^N, \tau_f^N\right) \right\| = 9.990404225024985 \times 10^{-11}$.

As expected, the value of the performance index is lower than the suboptimal value obtained at the end of step 2:

$$K\left(\bar{\boldsymbol{u}}, \bar{\tau}_0, \bar{\tau}_f\right) = K\left(\boldsymbol{u}_{\tau_0^N,\tau_f^N}, \tau_0^N, \tau_f^N\right) = 2.014050772103423 \times 10^{-6} \tag{47}$$

Finally, the low-energy optimal controls appear in Fig. 13. The associated trajectory in Fig. 14 is very close to the trajectories obtained in Figs. 10 and 12. It is apparent in Fig. 13 that an intermediate arc exists where the control values are close to zero. The low-energy solution obtained seems consistent with the one revolution heteroclinic connection identified in [42].

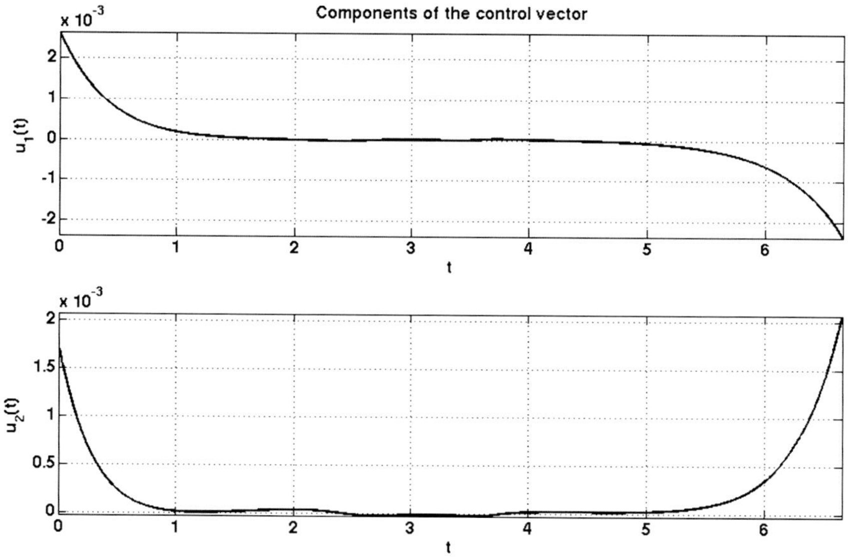


**Fig. 13** Control history for the 29-day transfer: low-energy solution

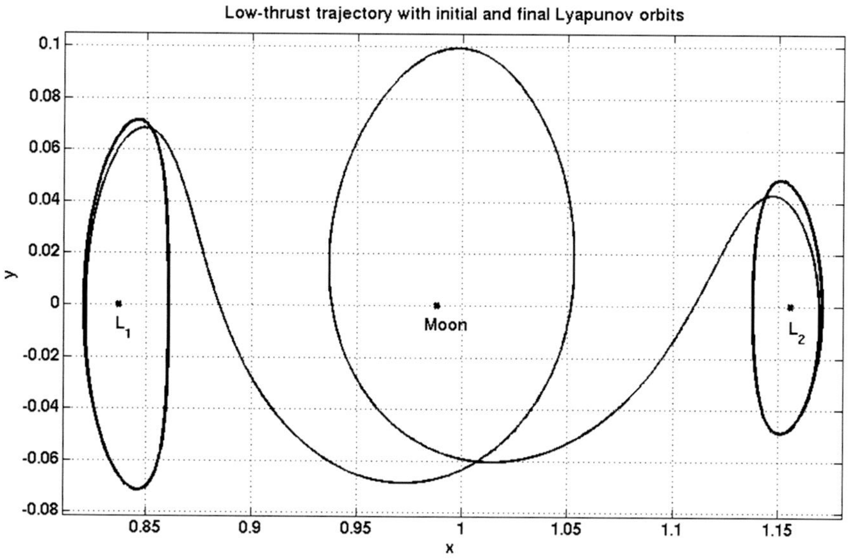


**Fig. 14** Trajectory obtained for the 29-day transfer: low-energy solution

### 4.1.3 Example 3: Long Transfer Duration

Starting from $T_f = 29$ days, Eqs. (9a) and (9b) has been solved for increasing values of the transfer duration $T_f$. It appears that the performance index $K$ is not monotonically decreasing with respect to $T_f$. However, a new low-energy transfer can be obtained by adding to $T_f$ =29 days about 15 days, which corresponds to slightly more than half the orbital period of the Moon. As an example, consider here the solution of Eqs. (9a) and (9b) with a transfer duration fixed to $T_f = 44$ days. The corresponding value of $t_f$ is derived from Eq. (41), leading to $t_f = 10.11874803$.

As in Sect. 4.1.2, trying to solve Eqs. (9a) and (9b) in a direct way by means of a single shooting method fails. Thus, Eqs. (9a) and (9b) is going to be solved by means of the three-step methodology developed in Sect. 3.2.

The optimal unknown vector $\bar{\boldsymbol{w}}$ obtained at the end of the first step of the method determines the initial feasible control $\boldsymbol{u}^0$ and its associated departure location from the initial orbit and arrival location at the target orbit. This vector takes the following value:

$$\bar{\boldsymbol{w}} = \begin{Bmatrix} -3.2801942129081313 \times 10^{-4} \\ 2.4746357420605364 \times 10^{-3} \\ -4.7359757243175854 \times 10^{-4} \\ -7.563535857307221 \times 10^{-7} \\ 0.3261051145072389 \\ 7.366376891137603 \\ 1.6321834714495793 \\ 0.677910923194809 \end{Bmatrix} \tag{48}$$

and satisfies:

$$G(\bar{\boldsymbol{w}}) = 6.252715130123296 \times 10^{-9}$$

$$K\left(\boldsymbol{u}^0, \tau_0^0, \tau_f^0\right) = 3.5451665346382667 \times 10^{-6}$$

The feasible control $\boldsymbol{u}^0$ appears in Fig. 15. The associated trajectory in Fig. 16 shows that the spacecraft completes two revolutions around the Moon.

In the second step of the method, Eqs. (30a) and (30b) is solved by continuation on $\varepsilon$ until the target value $\varepsilon = 1$ is reached. The continuation process requires 5498 iterations in the present case.

For $\varepsilon = 1$, the zero of the shooting function obtained when solving Eqs. (30a) and (30b) by means of the single shooting method is given as follows:

$$z = \boldsymbol{\lambda}^1(t_0) = \begin{Bmatrix} -5.9314173145129833 \times 10^{-5} \\ 2.2895029759933761 \times 10^{-5} \\ -2.2007589927380855 \times 10^{-5} \\ 5.542793032328769 \times 10^{-6} \end{Bmatrix} \tag{49}$$

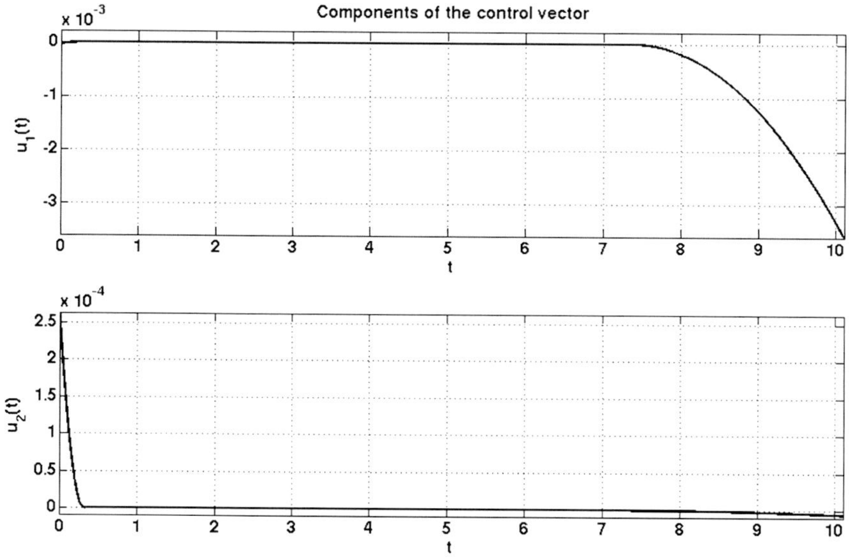


**Fig. 15** Feasible controls for the 44-day transfer

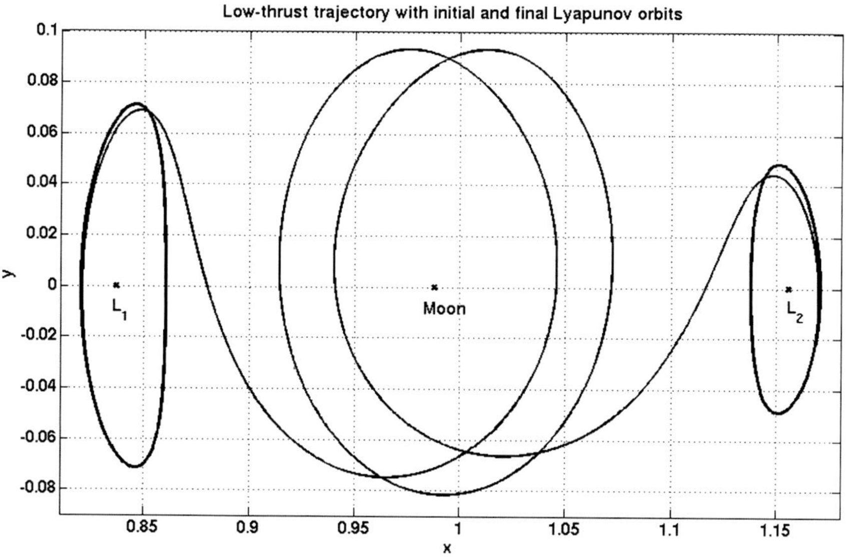


**Fig. 16** Feasible trajectory for the 44-day transfer

and the value of the performance index,

$$K_1\left(\boldsymbol{u}^1\right) = K\left(\boldsymbol{u}^1, \tau_0^0, \tau_f^0\right) = 2.2893636712440527 \times 10^{-6}$$

is logically lower than the value obtained for the feasible control $\boldsymbol{u}^0$. As in Sect. 4.1.2, $\tau_0^0$ and $\tau_f^0$ are fixed here to the non-optimal values associated with the feasible solution, i.e., they are equal to the two last components of $\bar{\boldsymbol{w}}$ in Eq. (48):

$$\tau_0^0 = 1.6321834714495793 \quad \tau_f^0 = 0.677910923194809 \tag{50}$$

The control $\boldsymbol{u}^1 = \boldsymbol{u}_{\tau_0^0, \tau_f^0}$ appears in Fig. 17. The associated trajectory in Fig. 18 is similar to that given in Fig. 16 and it seems to follow the paths of the invariant manifolds even if it is not the low-energy trajectory. Indeed, as in Sect. 4.1.2, the control law in Fig. 17 does not exhibit the 'take-off, cruise, and landing' pattern that is expected for the low-energy control.

Finally, in the third step of the method, $\tau_0$ and $\tau_f$ are optimized thanks to the gradient algorithm described in Sect. 3.2.3. The convergence is very slow in this case and is achieved after approximately $N = 6000$ iterations. The optimal values obtained at the convergence of the gradient algorithm are different to those given in Eq. (50):

$$\bar{\tau}_0 = \tau_0^N = 2.768373811637286 \quad \bar{\tau}_f = \tau_f^N = 1.812860236777545 \tag{51}$$

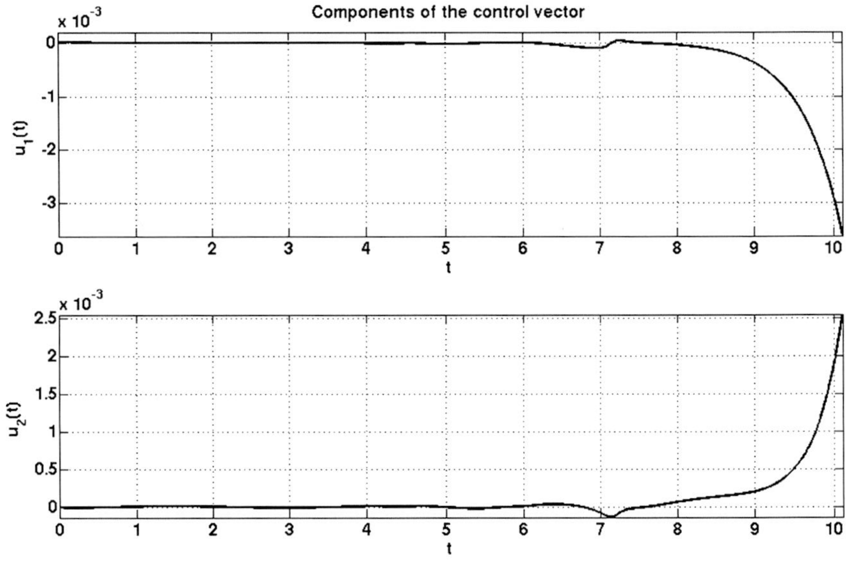


**Fig. 17** Control history obtained at the end of step 2 for the 44-day transfer

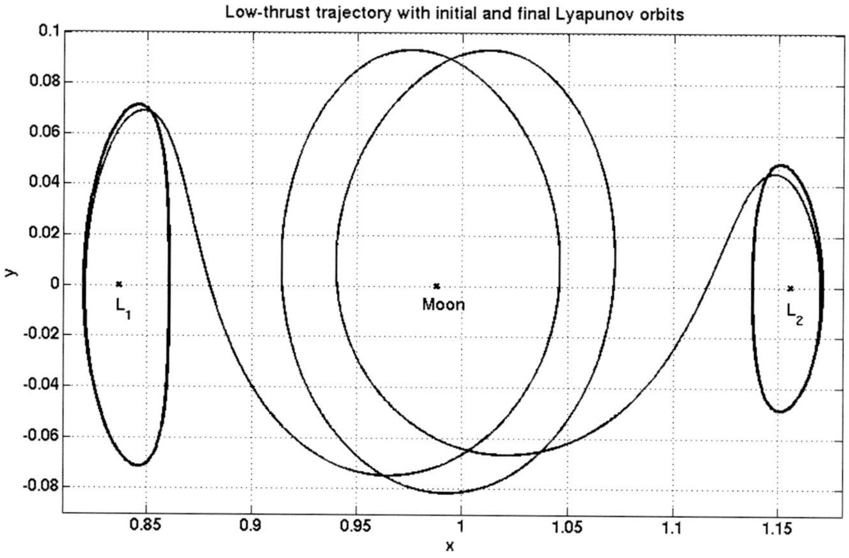


**Fig. 18** Trajectory obtained at the end of step 2 for the 44-day transfer

and the initial costate vector for Eqs. (35a) and (35b) associated with these values of $\tau_0$ and $\tau_f$ is given as follows:

$$\boldsymbol{\lambda}_{\tau_0^N,\tau_f^N}(t_0) = \begin{Bmatrix} -1.3984950996650867 \times 10^{-3} \\ 5.983979354559943 \times 10^{-5} \\ -2.853017218738709 \times 10^{-4} \\ -2.1552968007331502 \times 10^{-4} \end{Bmatrix} \tag{52}$$

The gradient algorithm has stopped with the following condition:

$$\| \nabla L \left(\tau_0^N, \tau_f^N\right) \| = 1.5595453701161325 \times 10^{-9} \tag{53}$$

As expected, the value of the performance index is lower than the value obtained at the end of step 2, but also lower than the optimal value obtained in Sect. 4.1.2 for the 29-day transfer:

$$K\left(\bar{\boldsymbol{u}}, \bar{\tau}_0, \bar{\tau}_f\right) = K\left(\boldsymbol{u}_{\tau_0^N,\tau_f^N}, \tau_0^N, \tau_f^N\right) = 2.5439143520205065 \times 10^{-8} \tag{54}$$

Finally, the low-energy optimal controls appear in Fig. 19 and the associated trajectory is given in Fig. 20. It is apparent in Fig. 19 that the control history exhibits an intermediate arc where the control values are close to zero. The low-energy solution obtained seems consistent with the two-revolution heteroclinic connection identified in [42].

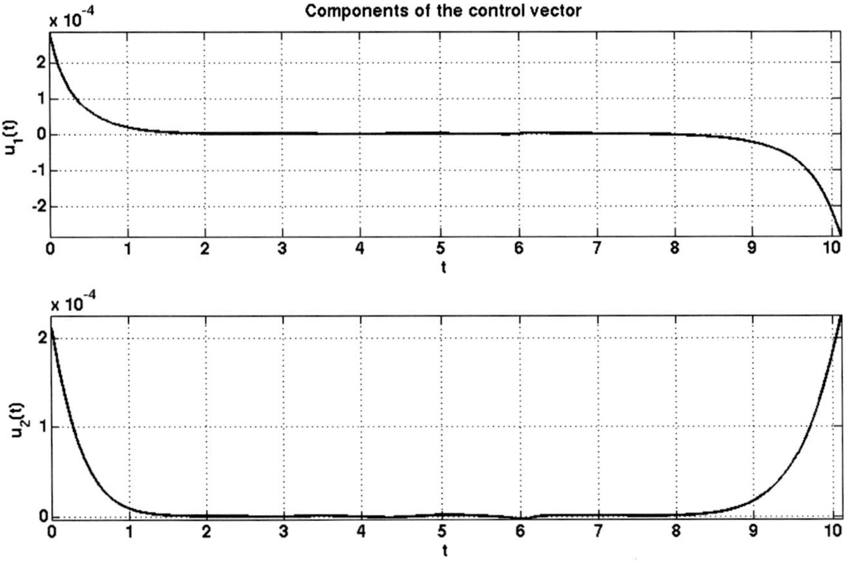


**Fig. 19** Control history for the 44-day transfer: low-energy solution

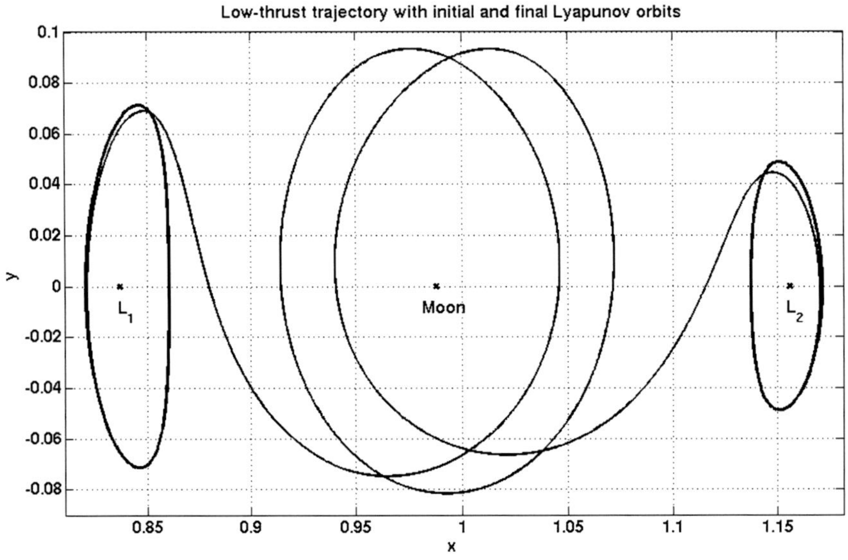


**Fig. 20** Trajectory obtained for the 44-day transfer: low-energy solution

**Table 3** Characteristics of the two Lyapunov orbits

| Lagrange point | Jacobi constant | Adimensional x-amplitude | Adimensional period |
|---|---|---|---|
| $L_1$ | 3.1960 | 0.05743698800000 | 2.707058919534849 |
| $L_2$ | 3.1780 | 0.10041124020000 | 3.385292341037150 |

## 4.2 *Orbits with Different Energies*

Consider, in the Earth-Moon three-body problem, a transfer from a Lyapunov orbit around $L_1$ to a Lyapunov orbit around $L_2$ with a different energy value. The characteristics of these two orbits are summarized in Table 3.

Notice that the Lyapunov orbit around $L_2$ is the same as in Sect. 4.1.

### 4.2.1 Example 4: Small Transfer Duration

Consider the solution of the minimum-energy problem Eqs. (9a) and (9b) with a transfer duration fixed to $T_f = 12$ days.

After time scaling, the corresponding value of $t_f$ is equal to $t_f = 2.759659$, as in Sect. 4.1.1.

For this small value of $t_f$, Eqs. (9a) and (9b) can be solved in a direct way by means of a single shooting method coupled with the use of variational equations for computing the Jacobian of the shooting function (see Sect. 2.3).

A set of solutions, corresponding to different values of the performance index $K$ in Eq. (9a), is found. The solution associated with the lowest value of $K$ corresponds to the following value of $z$:

$$z = \begin{Bmatrix} 0.01750290259912734 \\ -0.08553617948 10282 \\ 0.04158014770312709 \\ -0.1178985752629 1395 \\ 1.436303398216749 \\ 2.8645102420051462 \end{Bmatrix} \tag{55}$$

and leads to the following value of the performance index:

$$K\left(\bar{\boldsymbol{u}}, \bar{\tau}_0, \bar{\tau}_f\right) = 0.01174648595972109 15$$

The optimal control $\bar{\boldsymbol{u}}$ appears in Fig. 21 and the associated trajectory is depicted in Fig. 22.

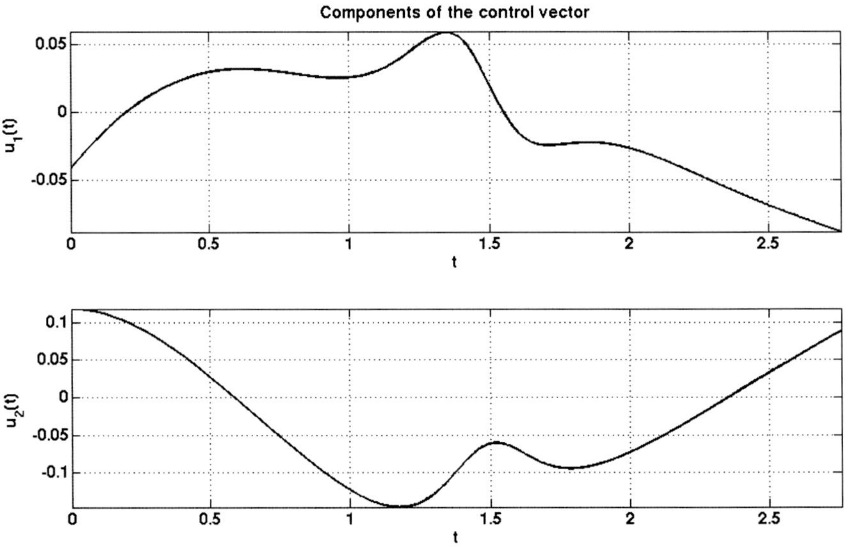


**Fig. 21** Control history for the 12-day transfer, orbits with differing energies: best solution

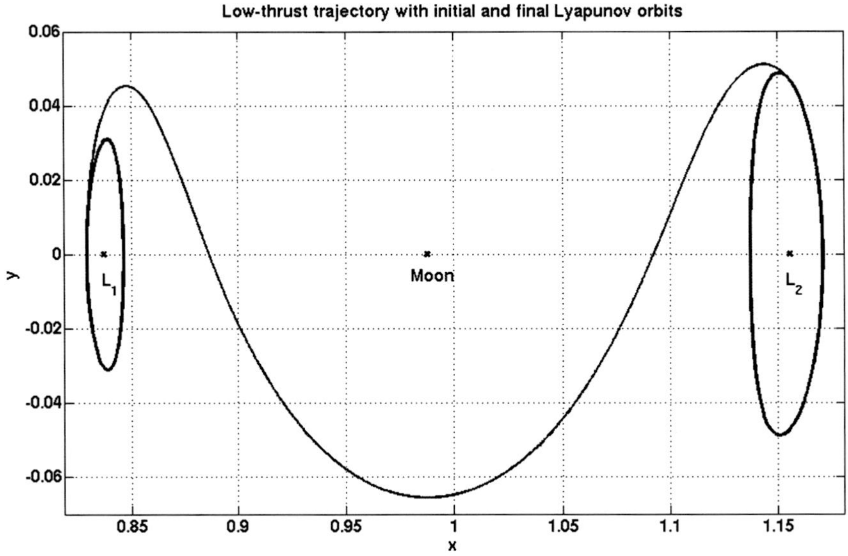


**Fig. 22** Optimal trajectory for the 12-day transfer, orbits with differing energies: best solution

For the same value of $t_f$, Eqs. (9a) and (9b) admits a local solution associated with following value of $z$:

$$\mathbf{z} = \begin{Bmatrix} -0.7737797925937719 \\ 0.1606956304951557 5 \\ -0.2103803873049919 \\ -0.0976142159055366 4 \\ 1.7491003325068834 \\ 2.7238355290200546 \end{Bmatrix} \tag{56}$$

The corresponding value of the performance index is only slightly greater than that obtained above:

$$K\left(\boldsymbol{u}, \tau_0, \tau_f\right) = 0.011993102276349195 \tag{57}$$

But what is interesting is that the control history that appears in Fig. 23, and the associated trajectory depicted in Fig. 24 are very different from their homologous in Figs. 21 and 22. In particular, the local solution corresponds to a transfer with one revolution around the Moon (see Fig. 24) contrarily to the best solution that corresponds to a direct transfer (see Fig. 22).

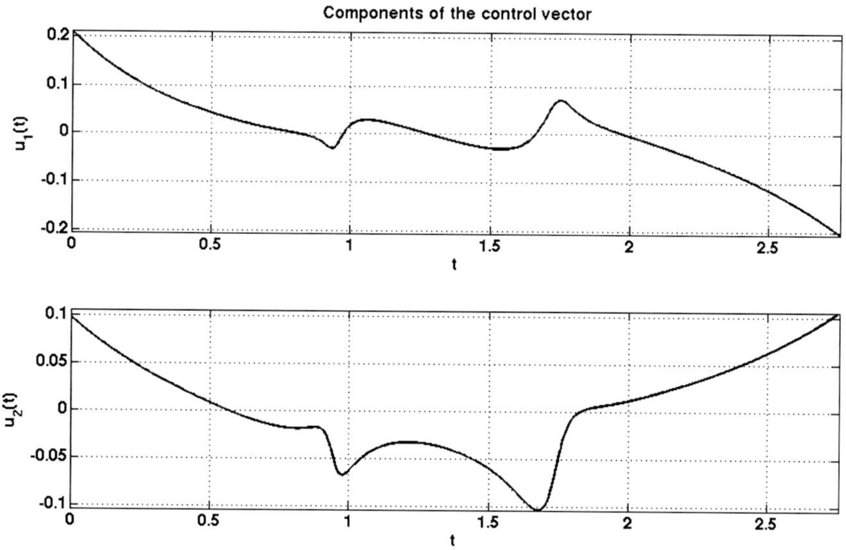


**Fig. 23** Control history for the 12-day transfer, orbits with differing energies: local solution

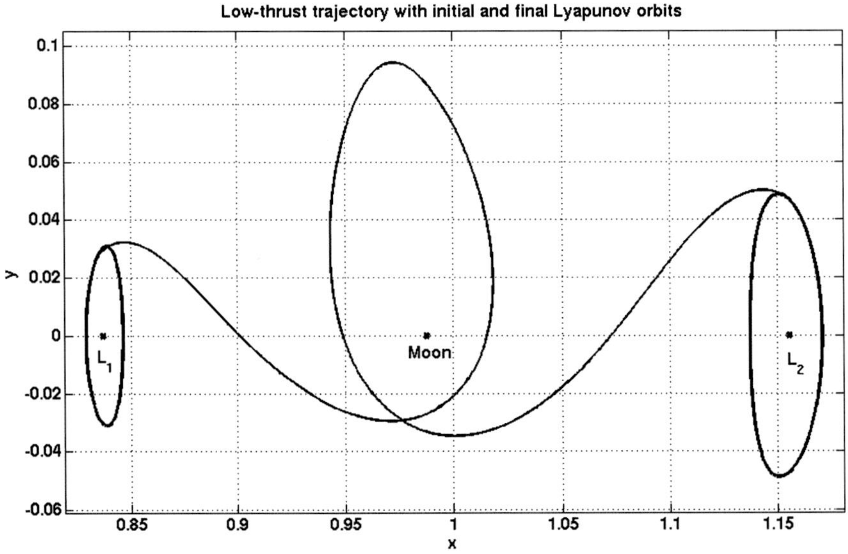


**Fig. 24** Optimal trajectory for the 12-day transfer, orbits with differing energies: local solution

### 4.2.2 Example 5: Medium Transfer Duration

Consider now the solution of the minimum-energy problem Eqs. (9a) and (9b) with a transfer duration fixed to $T_f = 29$ days. The corresponding value of $t_f$ is derived from Eq. (41), leading to $t_f = 6.669175$ as in Sect. 4.1.2.

Starting from randomly generated initial guesses, a zero of the shooting function Eq. (15) can be obtained in this case. The corresponding value of $z$ is given hereafter:

$$z = \begin{bmatrix} -0.046250694624153133 \\ 8.516613126702917 \times 10^{-3} \\ -0.012189208433373446 \\ -6.1754497660578 \times 10^{-3} \\ 1.1692039130673885 \\ 0.321159265519571 \end{bmatrix} \tag{58}$$

and is associated with the following value of the performance index:

$$K\left(\bar{\boldsymbol{u}}, \bar{\tau}_0, \bar{\tau}_f\right) = 1.3400494672052268 \times 10^{-4} \tag{59}$$

The optimal control $\bar{\boldsymbol{u}}$ appears in Fig. 25 and does not exhibit at all the 'take-off, cruise, and landing' pattern as in the case of transfers between orbits of the same energy (see Fig. 13 for example). The optimal trajectory is given in Fig. 26 and indicates that the spacecraft completes two revolutions around the Moon.

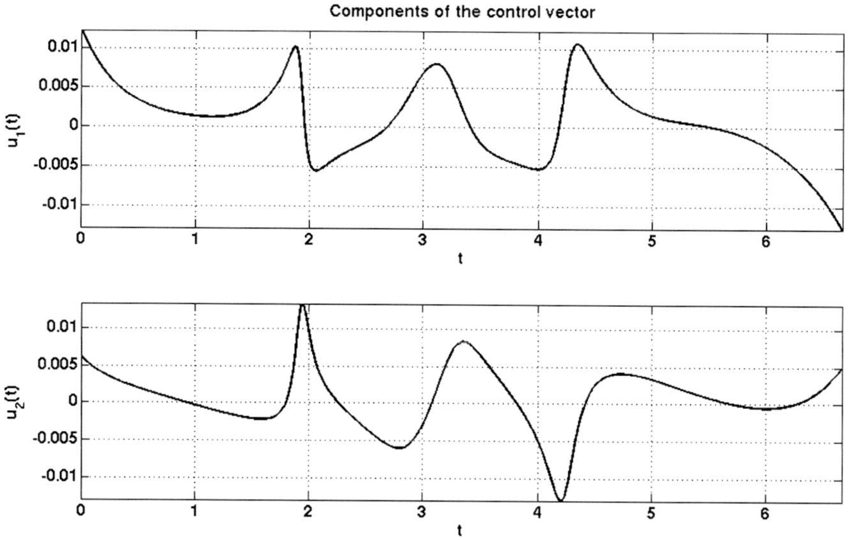


**Fig. 25** Control history for the 29-day transfer, orbits with differing energies

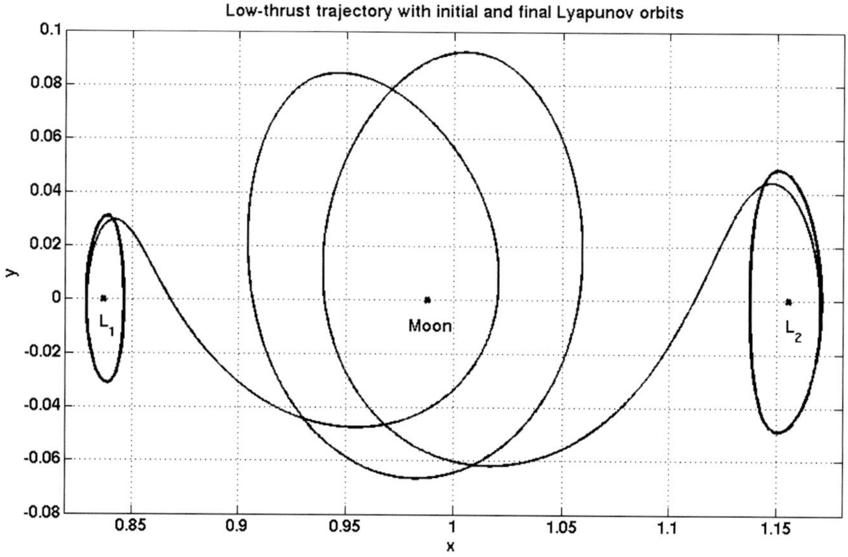


**Fig. 26** Optimal trajectory for the 29-day transfer, orbits with differing energies

# 5 Conclusion

In this investigation, the computation of low-thrust transfers between Lyapunov orbits is achieved via indirect optimal control. Combining single shooting methods, continuation techniques, and variational equations, this approach appears to be an efficient alternative to direct methods as demonstrated through numerical results.

Moreover, a three-step methodology has been developed to determine low-energy transfers between Lyapunov orbits of the same energy around the collinear points of the Earth-Moon system. The proposed methodology cope with the two time-scale behaviour of the problem without enforcing coast arcs along invariant manifolds contrarily to existing approaches. Thus, no prior knowledge of the manifolds is incorporated in the solution process.

The scope of application of this method needs to be extended by taking into account a real propulsion model for the spacecraft including mass variation. It should also be extended to be able to handle three-dimension transfers between LPOs in the Circular Restricted Three-Body Problem. Finally, this three-step methodology could also be applied to the determination of low-energy Earth-Moon transfers in the Sun-Earth-Moon system, by considering a Bicircular Four-Body dynamics model. Indeed, the use of this model is known to be useful to reduce the cost of transfers from a low Earth orbit to the Moon's vicinity.

**Acknowledgments** The greatest thanks to my CNES colleague Elisabet Canalias for fruitful discussions and advices. I would like also to express my gratitude to Josep Masdemont and Gerard Gómez from the University of Barcelona for the FORTRAN codes implementing Lindstedt-Poincaré techniques, provided under a CNES contract in 2008.

# References

1. Farquhar RW, Muhonen DP, Newman CR, Heuberger HS (1980) Trajectories and orbital maneuvers for the first libration-point satellite. J Guidance Control Dyn 3(6):549–554
2. Rodriguez-Canabal J, Hechler M (1989) Orbital aspects of the SOHO mission design. In: Teles, J (ed) AAS/NASA International Symposium, Greenbelt, MD, April 24–27 1989. Advances in the Astronautical Sciences, vol 69, pp 347–357. Univelt, San Diego, CA
3. Doyle D, Pilbratt G, Tauber J (2009) The herschel and planck space telescopes. Proc IEEE 97(8):1403–1411. doi:10.1109/JPROC.2009.2017106
4. Broschart SB, Sweetser TH, Angelopoulos V, Folta C, Woodard MA (2012) ARTEMIS lunar orbit insertion and science orbit design through 2013. In: Astrodynamics specialists conference, Girdwood AK, 2 Aug 2011 Advances in the Astronautical Sciences Series, vol 142. Univelt, CA
5. Gardner JP, Mather JC, Clampin M, Doyon R, Greenhouse MA (2006) The James web telescope. Space Sci Rev 123(4):485–606. doi:10.1007/s11214-006-8315-7
6. Farquhar RW, Kamel AA (1973) Quasi-periodic orbits about the translunar liberation point. Celest Mech Dyn Astron 7(4):458–473. doi:10.1007/BF01227511
7. Grebow DJ, Ozimek MT, Howell KC (2008) Multibody orbit architectures for lunar South pole coverage. J Spacecr Rockets 45(2):344–358. doi:10.2514/1.28738
8. Hill K, Parker J, Born GH, Demandante N (2012) A lunar $L_2$ navigation, commnication, and gravity mission. In: AIAA/AAS astrodynamics specialist conference and Exhibit, Aug. 2006,

Keystone, CO, Paper AIAA 2006–6662. http://ccar.colorado.edu/geryon/papers/Conference/AIAA-06-6662.pdf(2006). Accessed 20 Feb 2012

9. Koon WS, Lo MW, Marsden JE, Ross SD (2000) Heteroclinic connections between periodic orbits and resonance transitions in celestial mechanics. Chaos 10(4):427–469. doi:10.1063/1.166509
10. Gómez G, Masdemont J (2000) Some zero cost transfers between libration point orbits. In: Kluever CA, Neta B, Hall CD, Hanson JM (eds) AAS/AIAA Spaceflight Mechanics Meeting, Clearwater, Florida, Jan 2000, Paper AAS 00–177. Advances in the Astronautical Sciences Series, vol. 105. Univelt, CA
11. Gómez G, Koon WS, Marsden JE, Masdemont J, Ross SD (2004) Connecting orbits and invariant manifolds in the spatial restricted three-body problem. Nonlinearity 17(5):1571–1606. doi:10.1088/0951-7715/17/5/002
12. Davis KE, Anderson RL, Scheeres DJ, Born GH (2011) Optimal transfers between unstable periodic orbits using invariant manifolds. Celest Mech Dyn Astron 109(3):241–264. doi:10.1007/s10569-010-9327-x
13. Tantardini M, Fantino E, Ren Y, Pergola P, Gómez G (2010) Spacecraft trajectories to the $L_3$ point of the Sun-Earth three-body problem. Celest Mech Dyn Astron 108(3):215–232. doi:10.1007/s10569-010-9299-x
14. Howell KC, Hiday-Johnston LA (1994) Time-free transfers between libration-point orbits in the elliptic restricted problem. Acta Astronaut 32(4):245–254. doi:10.1016/0094-5765(94)90077-9
15. Nakamiya M, Yamakawa H, Scheeres DJ, Yoshikawa M (2010) Interplanetary transfers between halo orbits: connectivity between escape and capture trajectories. J Guidance Control Dyn 33(3):803–813. doi:10.2514/1.46446
16. Lawden DF (1953) Minimal rocket trajectories. J Am Rocket Soc 23(6):360–367
17. Lawden DF (1963) Optimal trajectories for space navigation. Butterworths & Co Publishers, London, pp 1–126
18. Mingotti G, Topputo F, Bernelli-Zazzera F (2011) Earth-Mars transfers with ballistic escape and low-thrust capture. Celest Mech Dyn Astron 110(2):169–188. doi:10.1007/s10569-011-9343-5
19. Pergola P, Geurts K, Casaregola C, Andrenucci M (2009) Earth-Mars halo to halo low thrust manifold transfers. Celest Mech Dyn Astron 105(1–3):19–32. doi:10.1007/s10569-009-9205-6
20. Dellnitz M, Junge O, Post M, Thiere B (2006) On target for Venus–set oriented computation of energy efficient low thrust trajectories. Celest Mech Dyn Astron 95(1–4):357–370. doi:10.1007/s10569-006-9008-y
21. Mingotti G, Topputo F, Bernelli-Zazzera F (2012) Efficient invariant-manifold, low-thrust planar trajectories to the Moon. Commun Nonlinear Sci Numer Simul 17(2):817–831. doi:10.1016/j.cnsns.2011.06.033
22. Tanaka K, Kawagushi J (2011) Low-thrust transfer between Jovian Moons using manifolds. In: Jah MK, Gua Y, Bowes AL, Lai PC (eds) AAS/AIAA Spaceflight Mechanics Meeting, New Orleans LA, Feb 13–17, 2011, Paper AAS 11–235. Advances in the Astronautical Sciences Series, vol 140. Univelt, CA (2011)
23. Ren Y, Pergola P, Fantino E, Thiere B (2012) Optimal low-thrust transfers between libration point orbits. Celest Mech Dyn Astron 112(1):1–21. doi:10.1007/s10569-011-9382-y
24. Stuart JR, Ozimek MT, Howell KC (2010) Optimal, low-thrust, path-constrained transfers between libration point orbits using invariant manifolds. In: Proceedings of the AIAA/AAS Astrodynamics specialists conference, Toronto, Canada, Aug 2–5, 2010, Paper AIAA 10–7831. https://engineering.purdue.edu/people/kathleen.howell.1/Publications/conferences/StuOziHow_10.pdf. Accessed 20 Feb 2012
25. Ozimek MT, Howell KC (2010) Low-thrust transfers in the Earth-Moon system, including applications to libration point orbits. J Guidance Control Dyn 33(2):533–549. doi:10.2514/1.43179

26. Senent J, Ocampo C, Capella A (2005) Low-thrust variable-specific-impulse transfers and guidance to unstable periodic orbits. J Guidance Control Dyn 28(2):280–290. doi:10.2514/1.6398
27. Mingotti G, Topputo F, Bernelli-Zazzera F (2007) Combined optimal low-thrust and stable-manifold trajectories to the Earth-Moon halo orbits. Am Inst Phys Conf Proc 886:100–112. doi:10.1063/1.2710047
28. Betts JT (1998) Survey of numerical methods for trajectory optimization. J Guidance Control Dyn 21(2):193–207
29. Conway BA (2012) A survey of methods available for the numerical optimization of continuous dynamic systems. J Optim Theory Appl 152(2):271–306. doi:10.1007/s10957-011-9918-z
30. Masdemont J (2005) High order expansions of invariant manifolds of libration point orbits with applications to mission design. Dyn Syst 20(1):59–113. doi:10.1080/14689360412331304291
31. Szebehely VG (1967) Theory of orbits–the restricted problem of three bodies, pp. 8–100. Academic Press Inc., Harcourt Brace Jovanovich Publishers, Orlando, Florida
32. Pontryagin L (1961) Optimal regulation processes. Am Math Soc Transl 18:17–66
33. Bryson AE, Ho YC (1975) Applied optimal control. Hemisphere Publishing Corporation, New York, pp 42–125
34. Anderson BD, Kokotovic PV (1987) Optimal control problems over large time intervals. Automatica 23(3):355–363. doi:10.1016/0005-1098(87)90008-2
35. Rao AV, Mease KD (1999) Dichotomic basis approach to solving hyper-sensitive optimal control problems. Automatica 35(4):633–642. doi:10.1016/S0005-1098(98)00161-7
36. Rao AV, Mease KD (2000) Eigenvector approximate dichotomic basis method for solving hyper-sensitive optimal control problems. Optimal Control Appl Methods 21(1):1–19. doi:10.1002/(SICI)1099-1514(200001/02)21:1<1:AID-OCA646>3.0.CO;2-V
37. Mease KD, Bharadwaj S, Iravanchy S (2003) Timescale analysis for nonlinear dynamical systems. J Guidance Control Dyn 26(2):318–330
38. Bharadwaj, S., Mease, K.D.: A new invariance property of Lyapunov characteristic directions. In: Proceedings of the American Control Conference, vol. 6, pp. 3800–3804. American Automatic Control Council, Evanston, IL (1999)
39. Ardema MD (1983) Solution algorithms for non-linear singularly perturbed optimal control problems. Optimal Control Appl. Methods 4(4):283–302. doi:10.1002/oca.4660040403
40. Nelder JA, Mead R (1965) A simplex method for function minimization. Comput. J. 7(4):308–313. doi:10.1093/comjnl/7.4.308
41. Graichen K, Petit N (2008) Constructive methods for initialization and handling mixed state-input constraints in optimal control. J. Guidance, Control Dyn. 31(5):1334–1343. doi:10.2514/1.33870
42. Canalias E, Masdemont J (2006) Homoclinic and heteroclinic transfer trajectories between Lyapunov orbits in the Sun-Earth and Earth-Moon systems. Discret. Contin. Dyn. Syst.–Ser. A 14(2):261–279. doi:10.3934/dcds.2006.14.261
43. Powell MJD (1970) A hybrid method for nonlinear equations. In: Rabinowitz P (ed) Numerical Methods for Nonlinear Algebraic Equations. Gordon and Breach, New York, pp 87–114
44. NAG Fortran Library (2009) Mark 22. The Numerical Algorithms Group Ltd, Oxford, UK
45. Hairer E, NØrsett SP, Wanner G (1987) Solving ordinary differential equations I. Nonstiff Prob lems. Springer Series in Computational Mathematics, vol 8, pp 173–185. Springer, Berlin

# Time-Minimum Control of the Restricted Elliptic Three-Body Problem Applied to Space Transfer

**Monique Chyba, Geoff Patterson and Gautier Picot**

**Abstract** In this chapter, we investigate time minimal transfers in the elliptic restricted 3-body problem. We study the controllability of the problem and show that it is small-time locally controllable at the equilibrium points. We present results about the structure of the extremal trajectories, based on a previous study of the time minimum control of the circular restricted 3-body problem. We use indirect numerical methods in optimal control to simulate time-minimizing space transfers using the elliptic model from the geostationary orbit to the equilibrium points $L_1$ and $L_2$ in the Earth-Moon system, as well as a rendezvous mission with a near-Earth asteroid.

**Keywords** Optimal control theory · Astrodynamics · Near Earth Orbiter

## 1 Introduction

The general three-body problem models the motion of three bodies under their mutual gravitational fields. This classic problem of celestial mechanics [31, 47] has aroused the curiosity of mathematicians for more than three hundred years, since its formulation at the end of the seventeenth century by Isaac Newton [43]. A standard simplification of the general problem consists of considering the motion of a massless body subjected to the gravitational attraction of two main bodies moving in a circular motion around their center of mass. This is the *spatial circular restricted three-body problem* [48]. When the motion of the massless body is restricted to the plane defined by the motion of the two main bodies, the problem is referred to as

M. Chyba · G. Patterson · G. Picot (✉)
Department of Mathematics, University of Hawaii at Manoa,
2565 McCarthy Mall, Honolulu, HI 96822, USA
e-mail: gautier@math.hawaii.edu

M. Chyba
e-mail: mchyba@math.hawaii.edu

G. Patterson
e-mail: geoff@math.hawaii.edu



B. Bonnard and M. Chyba (eds.), *Recent Advances in Celestial and Space Mechanics*,
Mathematics for Industry 23, DOI 10.1007/978-3-319-27464-5_6

the *planar circular restricted three-body problem*. This problem has been addressed extensively from the geometrical and dynamical systems point of view. In particular, the structure of invariant manifolds in the vicinity of the colinear equilibrium points [32, 33] or complex fractal regions of unstable and chaotic motion in space [8] have been used to design space missions with low energy cost. Recently, optimal control approaches, inspired by founding studies on the Kepler problem [10, 18, 29, 30], have led to new techniques for determining low-thrust space transfers in the Earth-Moon system. In [19, 45], indirect methods of optimal control are used to compute numerical time-minimal and energy-minimal trajectories of the circular restricted three-body problem. These computations provided numerical simulations of low-thrust space transfers from the geostationary orbit to a parking orbit around the Moon [45] and rendezvous missions with near-Earth asteroids temporarily captured by the gravitational field of the Earth [19]. In a contemporary chapter of Caillau and Daoud the authors study the controllability properties of the time-minimum control of the restricted three-body problem and provide an analysis of the structure of the time-minimizing controls [17].

The goal of this chapter is to generalize the results presented in [17] from the circular to the *elliptic restricted three-body problem* [48]. In this context, the two main bodies are assumed to move on elliptic orbits about their center of mass and the problem is reduced to the circular one when the eccentricity of the orbit is assumed to be zero. The main difference that arises when considering the elliptic case is that the mechanical potential of the problem is non-autonomous. As a consequence, there is no first integral which increases the complexity of the problem. Numerous in-depth studies on the dynamics of the elliptic problem have been carried out to improve the understanding of this model. In the 1960s, research has been conducted on the stability of the triangular equilibrium points and the integrals of motion for orbits with small eccentricities near the two main bodies of the problem [20, 23]. In a more recent past, a canonical transformation based on the Deprit-Hori method of Lie transforms has been applied to normalize the system dynamics about the circular case and one of the triangular points [25, 26]. Resonances and Nekhoroshev stability around triangular points have been analyzed as well [27, 41]. The dynamical properties of the elliptic restricted three-body problem have been applied to space mechanics. Among the greatest examples of such applications, we can mention low-fuel spacecraft missions trajectories constructed using the Lagrangian coherent structures in the problem [28] or moderate $\Delta v$ Earth-Mars transfers designed using ballistic capture [49]. Techniques from control theory have also been developped to derive quasi-periodic, peridodic and small-Halo orbits around the collinear equilibrium points, stabilize the motion on libration orbits [6, 35] and to investigate solar sail equilibria [7] in the elliptic restricted three-body problem.

This paper examines the structure of the time-minimal trajectories of the elliptic restricted three-body problem, defined as the solutions of a non-autonomous optimal control problem. It is organized as follows. In the first section, we derive the Hamiltonian forms of the controlled equations of both the circular and the elliptic restricted three-body problems to emphasize the intrinsic similarities and differences between these two models. In the second section, we emphasize the non-existence of a first integral as an obstacle to generalize the result of controllability previously established for the circular restricted three-body problem [17] to the elliptic restricted three-body problem. Nevertheless, we obtain a result of local controllability at the equilibrium points of the problem. In the third section, we apply the Pontryagin Maximum Principle and deduce necessary first-order optimality conditions for the time-minimum controls of the elliptic problem. We then study the structure of these time-minimum controls. In particular, we demonstrate the reason that the geometric control analysis performed in [17], based on the bi-input control affine system form of the equations, still holds in the elliptic case. In the fourth section, we introduce a shooting method which we use to compute numerical time-minimizing solutions of the elliptic problem. To overcome the challenges to initialize the algorithm, we use a continuation method [13, 15, 29, 30]. Our algorithm also verifies second-order conditions based on the notion of conjugate points [11] and allow us to generate time-minimum transfers from the geostationary orbit to the collinear equilibrium points $L_1$ and $L_2$ of the elliptic restricted three-body problem for different values of the eccentricity of the orbits of the main bodies. We also simulate a rendezvous mission with a temporarily captured near-Earth asteroid, namely 2006RH$_{120}$.

## 2 The Controlled Planar Elliptic Restricted Three-Body Problem

The planar elliptic restricted three-body problem is the simplest generalization of the classic planar circular restricted three-body problem [48], derived from the Newton's law of universal gravitation [43].

### *2.1 Controlled Equations of the Planar Circular Restricted Three-Body Problem*

The planar circular restricted three-body problem describes the motion of a massless body $M$ evolving in the the orbital plane of two main bodies called the primaries with constant mass $M_1$ and $M_2$ where $M_1 > M_2$, and circularly orbiting at constant angular velocity 1 around their center of mass $G$ under the influence of their mutual gravitational attraction [48]. In this problem, the distance between the two primaries is constant and can be normalized to 1. By defining the mass ratio $\mu = \frac{M_2}{M_1+M_2}$ and using a rotating frame centered at $G$ whose axis of abscissa is set as the line joining the primaries, the location of $M_1$ and $M_2$ can respectively be fixed to $(-\mu, 0)$ and

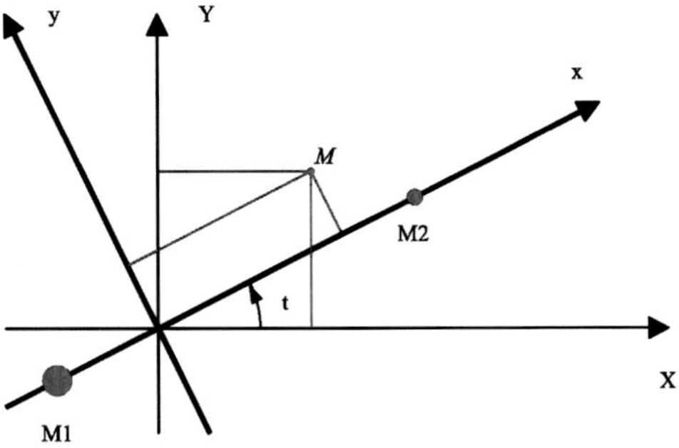


**Fig. 1** Representation of the primaries $M_1$ and $M_2$ of the planar circular restricted three-body problem in both the synodic frame $(G, X, Y)$ and the rotating frame $(G, x, y)$

$(1-\mu, 0)$, see Fig. 1. Denoting $(q_1(t), q_2(t))$ as the coordinates of $M$ in the rotating frame at time $t$, the equations of motion of $M$ are

$$\begin{cases} \ddot{q}_1(t) - 2\dot{q}_2(t) = \dfrac{\partial V}{\partial q_1}(q_1(t), q_2(t)) \\ \ddot{q}_2(t) + 2\dot{q}_1(t) = \dfrac{\partial V}{\partial q_2}(q_1(t), q_2(t)) \end{cases} \tag{1}$$

where

$$V(q_1, q_2) = \frac{q_1^2 + q_2^2}{2} + \frac{1-\mu}{\rho_1} + \frac{\mu}{\rho_2} + \frac{\mu(1-\mu)}{2} \tag{2}$$

so $-V$ is the mechanical potential of the problem and

$$\rho_1 = \sqrt{((q_1+\mu)^2 + q_2^2)},\ \rho_2 = \sqrt{((q_1 - 1 + \mu)^2 + q_2^2)}$$

are respectively the distances from $M$ to $M_1$ and $M_2$. Using the so-called Legendre transformation $p = (p_1, p_2) = (\dot{q}_1 - q_2, \dot{q}_1 + q_2)$, these equations can be written as a Hamiltonian system associated with the Hamiltonian function

$$H(q, p) = \frac{1}{2}\|p\|^2 + p_1 q_2 - p_2 q_1 - \frac{1-\mu}{\rho_1} - \frac{\mu}{\rho_2} + \frac{\mu(1-\mu)}{2}.$$

The function $H$ is a first integral of motion, called the Jacobian energy, which equals the total energy of the system. Thus we can deduce the five phase portraits for the topology of the possible region of motion, known as the Hill region [48]. The equilibrium points of the problem, defined as the critical points of the potential $-V$, divide into two categories: the collinear points $L_1$, $L_2$ and $L_3$ are located on the horizontal axis $y = 0$ joining the primaries and the equilateral points $L_4$ and $L_5$ are located at the vertices of the two equilateral triangles in the plane of motion sharing the same base given by the segment linking the primaries. We can show, using Arnold's stability theorem [5], that the collinear points are unstable whereas the equilateral points are stable when $\mu < \frac{1}{2}(1 - \frac{\sqrt{69}}{9})$. In the Earth-Moon system, whose mass ratio $\mu = 0.0121536$, the colinear points are then stable. The controlled planar restricted three-body problem is simply derived from (1) and is written as

$$\begin{cases} \ddot{q}_1(t) - 2\dot{q}_2(t) = \dfrac{\partial V}{\partial q_1}(q_1(t), q_2(t)) + u_1(t) \\ \ddot{q}_2(t) + 2\dot{q}_1(t) = \dfrac{\partial V}{\partial q_2}(q_1(t), q_2(t)) + u_2(t) \end{cases} \tag{3}$$

where the control $u(t) = (u_1(t), u_2(t))$ is a bounded measurable function valued in $\mathbf{R}^2$ and defined on an interval $[0, t(u)] \subset \mathbf{R}^+$. We say that $u$ is an admissible control if there exists a solution $q(t) = (q_1(t), q_2(t))$ to (3), called a trajectory associated with $u$, defined on $[0, t(u)]$. With the change of variables

$$x_1 = q_1, x_2 = q_2, x_3 = \dot{q}_1, x_4 = \dot{q}_2, \tag{4}$$

we can rewrite (3) as a first-order differential system

$$\begin{cases} \dot{x}_1(t) = x_3(t) \\ \dot{x}_2(t) = x_4(t) \\ \dot{x}_3(t) = 2x_4(t) + \dfrac{\partial V}{\partial x_1}(x_1(t), x_2(t)) + u_1(t) \\ \dot{x}_4(t) = -2x_3(t) + \dfrac{\partial V}{\partial x_2}(x_1(t), x_2(t)) + u_2(t). \end{cases} \tag{5}$$

Setting $x = (x_1, x_2, x_3, x_4)$, this system can be written as a so-called bi-input controlled system

$$\dot{x}(t) = F_0(x(t)) + u_1(t)F_1(x(t)) + u_2(t)F_2(x(t)) \tag{6}$$

where the vector fields $F_0$, $F_1$ and $F_2$ are

$$F_0(x) = \begin{pmatrix} x_3 \\ x_4 \\ 2x_4 + \frac{\partial V}{\partial x_1}(x_1, x_2) \\ -2x_3 + \frac{\partial V}{\partial x_2}(x_1, x_2) \end{pmatrix}, F_1(x) = \frac{\partial}{\partial x_3}, F_2(x) = \frac{\partial}{\partial x_4}.$$

## 2.2 Controlled Equations of the Planar Elliptic Restricted Three-Body Problem

The most natural generalization of the planar circular restricted three-body problem consists of assuming that the two primaries move on elliptic orbits

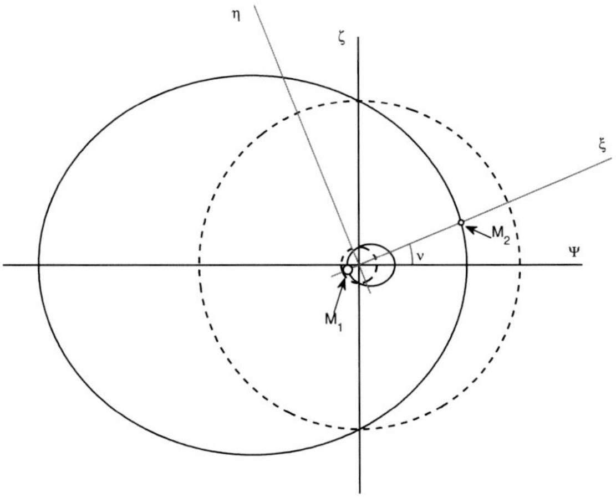


**Fig. 2** Representation of the primaries $M_1$ and $M_2$ of the planar elliptic restricted three-body problem in both the fixed frame $(G, \Psi, \zeta)$ and the rotating frame $(G, \xi, \eta)$

[25, 26, 28, 35, 48]. The smallest primary $M_2$ is orbiting the largest primary $M_1$ within an elliptic orbit with eccentricity $0 < e < 1$ and semimajor axis $a$ which fits the two-body problem [31]. We denote by $\nu(t)$ the true anomaly of $M_2$, defined as the angular time-dependent parameter given by the angle that the direction of the periapsis of the ellipse makes with the position of $M_2$ along the ellipse at time $t$. In this context, the instantaneous distance $\rho$ between the two primaries is a function of the true anomaly (Fig. 2)

$$\rho(\nu) = \frac{a(1-e^2)}{1+e\cos(\nu)}.$$

Furthermore, according to the principle of conservation of the angular momentum [43], the dynamics of true anomaly satisfies

$$\dot{\nu} = k(M_1+M_2)^{\frac{1}{2}} \frac{(1+e\cos(\nu))^2}{(a(1-e^2))^{\frac{3}{2}}} \tag{7}$$

where $k$ is the universal gravitational constant. The above equation provides a relation between the true anomaly and the time. By choosing the origin of the coordinate system at the center of mass $G$ of the two primaries and the axis of abscissa as the direction of the periapsis of the ellipse, we define an inertial, barycentric coordinate frame in which, up to appropriate normalizations of the units, the primaries $M_1$ and $M_2$ describe ellipses and have respective positions $(\Psi_1, \zeta_1)$ and $(\Psi_2, \zeta_2)$ parametrized by

$$(\Psi_1(\nu), \zeta_1(\nu)) = \left(\frac{-\mu}{1+\cos(\nu)}\cos(\nu), \frac{-\mu}{1+\cos(\nu)}\sin(\nu)\right)$$
$$(\Psi_2(\nu), \zeta_2(\nu)) = \left(\frac{1-\mu}{1+\cos(\nu)}\cos(\nu), \frac{1-\mu}{1+\cos(\nu)}\sin(\nu)\right)$$

where $\mu$ is the mass ratio defined in Sect. 2.1. By considering $\nu$ as the independent variable and introducing a non-uniformly rotating pulsating coordinate system, the respective positions of $M_1$ and $M_2$ can be fixed to $(-\mu, 0)$ and $(1-\mu, 0)$. In this system, the coordinates $(\xi(\nu), \eta(\nu))$ of the massless body $M$ satisfy the equations of motion

$$\begin{cases} \dfrac{d^2\xi}{d\nu^2}(\nu) - 2\dfrac{d\eta}{d\nu}(\nu) = \dfrac{\partial\omega}{\partial\xi}(\xi(\nu), \eta(\nu), \nu) \\ \dfrac{d^2\eta}{d\nu^2}(\nu) + 2\dfrac{d\xi}{d\nu}(\nu) = \dfrac{\partial\omega}{\partial\eta}(\xi(\nu), \eta(\nu), \nu) \end{cases} \tag{8}$$

where

$$\omega(\xi, \eta, \nu) = \frac{1}{1 + e\cos\nu} V(\xi, \eta) \tag{9}$$

so $-\omega$ is the non-autonomous mechanical potential of the problem. We remark that the equations of the circular restricted problem Fig. 1 correspond the equations of the elliptic problem (8) when $e = 0$. Using a similar Legendre transformation $q = (q_1, q_2) = (\xi, \eta), p = (p_1, p_2) = (\dot{q}_1 - q_2, \dot{q}_1 + q_2)$ as in Sect. 2.1, we can rewrite this dynamics as an Hamiltonian system through the new non-autonomous Hamiltonian function

$$H_e(q, p, \nu) = \frac{1}{2}\|p\|^2 + p_1 q_2 - p_2 q_1 - \frac{1}{1 + e\cos(\nu)}\left(\frac{1-\mu}{\rho_1} + \frac{\mu}{\rho_2} - \frac{\mu(1-\mu)}{2}\right)$$

where $\rho_1$ and $\rho_2$ are still respectively the distances from $M$ to $M_1$ and $M_2$. The function $\omega$ being a multiple of the function $V$, the elliptic restricted three-body problem has the exact same five equilibrium points as the circular restricted three-body problem. Previous studies about their stability showed that the three collinear points are unstable [48], whereas the equilateral points are linearly stable, provided that both the mass ratio $\mu$ and the eccentricity $e$ are appropriately chosen. However, the Hamiltonian function $H_e$ being non-autonomous, it is no longer a first integral of motion. As a consequence, we can not define any possible region of motion such as the Hill region (Fig. 3).

The controlled equations of the planar elliptic restricted three-body problem is derived similarly to (3) and is written

$$\begin{cases} \dfrac{d^2\xi}{d\nu^2}(\nu) - 2\dfrac{d\eta}{d\nu}(\nu) = \dfrac{\partial\omega}{\partial\xi}(\xi(\nu), \eta(\nu), \nu) + u_1(\nu) \\ \dfrac{d^2\eta}{d\nu^2}(\nu) + 2\dfrac{d\xi}{d\nu}(\nu) = \dfrac{\partial\omega}{\partial\eta}(\xi(\nu), \eta(\nu), \nu) + u_2(\nu). \end{cases} \tag{10}$$

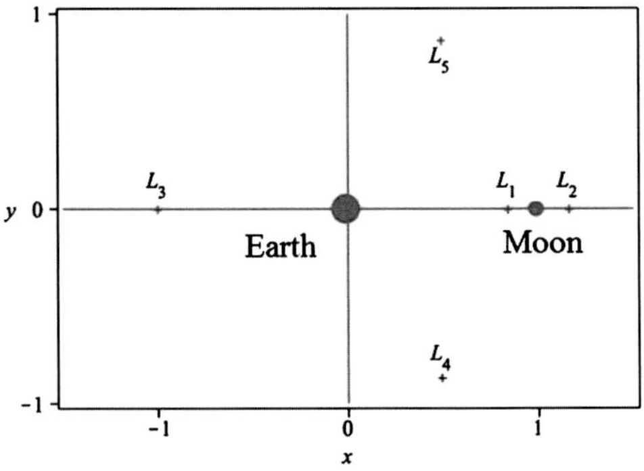


**Fig. 3** Locations of the equilibrium points of the Earth-Moon system, depicted in the rotating pulsating frame of the planar elliptic restricted 3-body problem. The locations are the same as in the planar circular restricted 3-body problem

where the control $u = (u_1, u_2)$ is a function of the independent variable $\nu$. Defining $x = (x_1, x_2, x_3, x_4) \in \mathbf{R}^4$, where

$$x_1 = \xi, x_2 = \eta, x_3 = \frac{\partial \xi}{\partial \nu}, x_4 = \frac{\partial \eta}{\partial \nu}, \tag{11}$$

we get the first-order differential system

$$\begin{cases} \dfrac{dx_1}{d\nu} = x_3 \\ \dfrac{dx_2}{d\nu} = x_4 \\ \dfrac{dx_3}{d\nu} = 2x_4 + \dfrac{\partial \omega}{\partial x_1}(x_1, x_2, \nu) + u_1 \\ \dfrac{dx_3}{d\nu} = -2x_3 + \dfrac{\partial \omega}{\partial x_2}(x_1, x_2, \nu) + u_2. \end{cases} \tag{12}$$

which we write as a non-autonomous bi-input controlled system

$$\frac{dx}{d\nu}(\nu) = F_0(\nu, x(\nu)) + u_1(\nu)F_1(x(\nu)) + u_2(\nu)F_2(x(\nu)) \tag{13}$$

where the non-autonomous drift vector field, $F_0$, is

$$F_0(\nu, x) = \begin{pmatrix} x_3 \\ x_4 \\ 2x_4 + \frac{\partial \omega}{\partial x_1}(x_1, x_2, \nu) \\ -2x_3 + \frac{\partial \omega}{\partial x_2}(x_1, x_2, \nu) \end{pmatrix}$$

and the two constant vector fields $F_1$ and $F_2$ are

$$F_1(x) = \frac{\partial}{\partial x_3}, F_2(x) = \frac{\partial}{\partial x_4}.$$

We observe that this controlled equation still admits an Hamiltonian formulation. Indeed, using the Legendre transformation

$$q = (q_1, q_2) = (x_1, x_2), p = (p_1, p_2) = (\frac{dx_1}{d\nu} - x_2, \frac{dx_2}{d\nu} + x_1),$$

the Eq. (12) can be written as an Hamiltonian system

$$\frac{dq}{d\nu} = \frac{\partial H_e^c}{\partial p}(q(\nu), p(\nu), u(\nu), \nu), \frac{dp}{d\nu} = -\frac{\partial H_e^c}{\partial q}(q(\nu), p(\nu), u(\nu), \nu)$$

with $H_e^c(q, p, u, \nu) = \frac{1}{2}\|p\|^2 + p_1 q_2 - p_2 q_1 - \omega(q, \nu) - u_1 q_1 - u_2 q_2$. In the rest of this paper, we study the time-minimal trajectories of the elliptic restricted three-body problem between two submanifolds $M_0$ and $M_1$ of $\mathbf{R}^4$, defined as the solutions $x(t)$ the optimal control problem

$$\begin{cases} \dot{x}(\nu) = F_0(\nu, x(\nu)) + u_1(\nu)F_1(x(\nu)) + u_2(\nu)F_2(x(\nu)) \\ \min_{u(.)\in B_{\mathbf{R}^2}(0,\varepsilon))} \int_0^{\nu_f} d\nu \\ x(0) = x_0 \in M_0, x(\nu_f) \in M_1 \end{cases} \tag{14}$$

where $u$ is an admissible control on $[0, \nu_f]$ whose magnitude is bounded by a positive number $\varepsilon$. Notice that what we call time-minimal trajectories of the problem are, in fact, true anomaly-minimal trajectories. However, the true anomaly is a strictly increasing function of the time, since $\dot{\nu}$, given in (7), is strictly positive. Therefore, we can minimize the final time by minimizing the true anomaly. In Sect. 5, the value of the control bound $\varepsilon$, that we will use to simulate time-minimizing space transfers in the Earth-Moon system, corresponds to a 1 N maximum thrust capability for the spacecraft's engine.

## 3 Controllability

In this section we study the controllability of the elliptic restricted three-body problem, our notations follow the ones from [17]. In that paper, the authors establish the controllability of the circular restricted three-body problem, corresponding to the case $e = 0$, over a particular submanifold of $\mathbf{R}^4$ containing the collinear Lagrangian point $L_2$ located between the two primaries, independent of the bound on the control and the value of the mass ratio $\mu$. More precisely, take any $\mu \in (0, 1)$, any positive magnitude $\varepsilon$ on the control and set $Q_\mu = \mathbf{R}^2 \setminus \{(-\mu, 0), (1 - \mu, 0)\}$ to be the region of motion where no collisions with the primaries occur, $X_\mu = TQ_\mu \times \mathbf{R}^2$ as the corresponding phase space, $J_\mu(x)$ to be the Jacobian energy evaluated at a point $x \in X_\mu$ and $j_1(\mu)$ as the Jacobian energy of the Lagrangian point $L_1$. With these, the authors state that the circular restricted three-body problem is controllable on the connected component of the of the subset $\{x \in X_\mu | J_\mu(x) < j_1(\mu)\}$ containing $L_2$. Their proof is based on the classical result of geometric control theory [36] which asserts that any affine control system $\dot{x}(t) = X_0(x(t)) + \sum_{i=1}^{m} u_i(t)X_i(x(t))$ on a connected manifold $M^n$ with $u(t) \in U \subset \mathbf{R}^m$ is controllable provided that the convex hull of $U$ is a neighborhood of the origin, the drift $X_0$ is a recurrent vector field and the family of vector fields $\{X_0, X_1, \ldots, X_m\}$ satisfies the so-called Lie algebra rank condition $\text{Lie}_x\{X_0, X_1, \ldots, X_m\} = T_xM, x \in M^n$.

Our objective is to investigate the generalization of this result for $0 < e < 1$. Notice that, according to the following general Lemma, the Lie algebra rank condition still holds in this case.

**Lemma 1** *A non-autonomous second order controlled system on* $\mathbf{R}^m$

$$\ddot{q}(t) + g(t, q(t), \dot{q}(t)) = u(t)$$

*can be written as a control-affine system on* $\mathbf{R}^{2m}$ *where the distribution* $\mathscr{D}$ *spanned by the vector fields* $X_1, \ldots, X_m$ *is involutive and with a non-autonomous drift* $X_0$ *such that* $\{X_1, \ldots, X_m, [X_0, X_1], \ldots, [X_0, X_m]\}$ *has maximum rank.*

*Proof* The proof is carried out similarly to Lemma 3 in [17], which states the same result for an autonomous second order control system. Indeed, here we have

$$X_0(t, q, \dot{q}) = \sum_{i=1}^{m} \dot{q}_i \frac{\partial}{\partial q_i} - g_i(t, q, \dot{q}) \frac{\partial}{\partial \dot{q}_i}$$

and

$$F_i(q, \dot{q}) = -\frac{\partial}{\partial q_i}$$

so we conclude that $[X_0, X_i] = -\frac{\partial}{\partial q_i} \bmod \mathscr{D}$ for all $1 \leq i \leq m$ which proves the result.

However, due to the explicit dependence of the drift of the elliptic restricted three-body problem on the true anomaly $\nu$, it is no longer possible to define a submanifold of finite volume containing the Lagrangian point $L_2$ on which this vector field is volume preserving, as in the circular restricted problem. As a consequence, Poincaré's recurrence theorem can not be applied and we can no longer claim that the drift of the problem is a recurrent vector field on an adequate submanifold. Let us mention that existing results concerning controllability of nonlinear non-autonomous vector fields are derived by considering the corresponding system augmented with the independent variable and hold for control-affine systems with an autonomous drift, i.e., systems of the form $\dot{x}(t) = F_0(x(t)) + \sum_{i=1}^{m} F_i(t, x(t))u_i(t)$ [9]. Since our system does not fit the hypothesis of these results, we instead examine the properties of local controllability of the problem. First, let us recall the definition of local controllability along a trajectory of a nonlinear control system [22].

**Definition 1** Let $(\bar{x}, \bar{u})$ be a trajectory defined on an interval $[t_0, t_1]$ of the control system $\dot{x} = f(t, x, u)$ where $x \in \mathbf{R}^n$ and $u \in \mathbf{R}^m$. This control system is said to be locally controllable along the trajectory $(\bar{x}, \bar{u})$ if, for every $\varepsilon > 0$, there exists $\eta > 0$ such that, for every $(a, b) \in \mathbf{R}^n \times \mathbf{R}^n$ with $|a - \bar{x}(t_0)| < \eta$ and $|b - \bar{x}(t_1)| < \eta$, there exists a trajectory $(\tilde{x}, \tilde{u})$ defined on $[t_0, t_1]$ such that $\tilde{x}(t_0) = a$, $\tilde{x}(t_1) = b$ and $|\tilde{u}(t) - \bar{u}(t)| \leq \varepsilon$, for all $t \in [t_0, t_1]$.

*Remark 1* It is well-know that any non-linear, non-autonomous control system $\dot{x} = f(t, x, u)$ is locally controllable along a trajectory $(\bar{x}, \bar{u})$ defined on $[t_0, t_1]$ if the linearized control system along $(\bar{x}, \bar{u})$,

$$\dot{x} = \frac{\partial f}{\partial x}(t, \bar{x}(t), \bar{u}(t))x + \frac{\partial f}{\partial u}(t, \bar{x}(t), \bar{u}(t))u, \, t \in [t_0, t_1], \tag{15}$$

is controllable [22]. Notice that, in this reference, the result is given for non-linear autonomous control systems $\dot{x} = f(x, u)$. However, it is easy to verify that the same proof works for non-linear non-autonomous control systems.

Thus, we can state the following.

**Theorem 1** *The elliptic restricted three-body problem is locally controllable along any trajectory $(\bar{x}, \bar{u})$ defined for $\nu \geq 0$ and such that $\bar{x}(.)$ is three times continuously differentiable.*

*Proof* Let $(\bar{x}, \bar{u})$ be a trajectory of the elliptic restricted three-body problem and assume that $\bar{x}$ is 3 times continuously differentiable. We want to show the controllability of the non-autonomous linear control system $\dot{x} = A(\nu)x + B(\nu)u$, where

$$\begin{aligned} A(\nu) &= \frac{\partial}{\partial x}\big(F_0(\nu, \bar{x}(\nu)) + u_1(\nu)F_1(\bar{x}(\nu)) + u_2(\nu)F_2(\bar{x}(\nu))\big) \\ &= \begin{pmatrix} 0 & 0 & 1 & 0 \\ 0 & 0 & 0 & 1 \\ \frac{\partial^2 \omega}{\partial x_1^2}(\nu, \bar{x_1}(\nu), \bar{x_2}(\nu)) & \frac{\partial^2 \omega}{\partial x_1 \partial x_2}(\nu, \bar{x_1}(\nu), \bar{x_2}(\nu)) & 0 & 2 \\ \frac{\partial^2 \omega}{\partial x_1 \partial x_2}(\nu, \bar{x_1}(\nu), \bar{x_2}(\nu)) & \frac{\partial^2 \omega}{\partial x_2^2}(\nu, \bar{x_1}(\nu), \bar{x_2}(\nu)) & -2 & 0 \end{pmatrix} \end{aligned}$$

and

$$B(\nu) = \frac{\partial}{\partial u}\big(F_0(\nu, \bar{x}(\nu)) + u_1(\nu)F_1(\bar{x}(\nu)) + u_2(\nu)F_2(\bar{x}(\nu))\big) = \begin{pmatrix} 0 & 0 \\ 0 & 0 \\ 1 & 0 \\ 0 & 1 \end{pmatrix}.$$

Our assumption on the regularity of $\bar{x}$ asserts that both $A(\nu)$ and $B(\nu)$ are 3 times continuously differentiable. Thus, according to the classical result about the controllability of non-autonomous linear control systems [22, 37, 40], it is sufficient to show that there exist $\nu \geq 0$ satisfying

$$\text{rank}[M_0(\nu)|M_1(\nu)| \cdots |M_3(\nu)] = 4$$

where

$$M_0(\nu) = B(\nu)$$
$$M_{k+1}(\nu) = -A(\nu)M_k(\nu) + \widehat{\dot{M}_k}(\nu), \text{ for } k = 0, \dots, 3.$$

The matrix $B(\nu)$ being constant, computations give, for any $\nu \geq 0$,

$$M_1 = \begin{pmatrix} -1 & 0 \\ 0 & -1 \\ 0 & -2 \\ 2 & 0 \end{pmatrix}$$

so $\det[M_0(\nu)|M_1(\nu)] = 1$ which concludes the proof. □

Then, we can deduce from Theorem 1 the property of small-time local controllability at the Lagrangian points of the problem.

*Remark 2* The term "small-time" is ambiguous here, since the independent variable of the problem is the true anomaly $\nu$ and not the time. However, this terminology is so widespread in the literature that we do not break convention. In the following, we provide the definition of the notion of small-time controllability for a generic control system $\dot{x} = f(t, x, u)$ but the reader should be aware that, in the context of our study, it would be more accurate to talk about "small-true-anomaly" controllability.

**Definition 2** Let $(x_e, u_e)$ be an equilibrium point of the $\dot{x} = f(t, x, u)$. This control system is said to be small-time locally controllable at $(x_e, u_e)$ if, for every $\varepsilon > 0$, there exists $\eta > 0$ such that, for every pair $(x_0, x_1)$ with $|x_0 - x_e| < \eta$ and $|x_1 - x_e| < \eta$, there exists a trajectory $(x, u)$ of the system defined on $[0, \varepsilon]$ satisfying

$$x(0) = x_0, \quad x(\varepsilon) = x_1, \quad |u(t) - u_e| \leq \varepsilon, \text{ for all } t \in [0, \varepsilon]$$

**Corollary 1** *The elliptic restricted 3-body problem is small-time locally controllable at a Lagrangian point $L_i$, $1 \leq i \leq 5$.*

*Proof* This is a consequence of Theorem 1, since any Lagrangian point $L_i$, $1 \leq i \leq 5$, associated with a constant control equal to 0 provide an equilibrium point for the controlled elliptic restricted three-body problem. □

## 4 Structure of the Optimal Control

In this section, we provide an analysis of the system to investigate the structure of the time-minimizing controls and trajectories of the elliptic restricted three-body problem.

## 4.1 Optimality Conditions

Our analysis is based on the application of the Pontryagin maximum principle which provides first-order necessary conditions for optimality [14, 36, 39, 46]. Let an admissible control $u = (u_1, u_2)$ associated with a trajectory $x(.)$, both defined on an interval $[\nu_0, \nu_f]$, be a solution for the time-minimum control of the elliptic restricted three-body problem

$$\begin{cases} \dot{x}(\nu) = F_0(\nu, x(\nu)) + u_1(\nu)F_1(x(\nu)) + u_2(\nu)F_2(x(\nu)) \\ \min_{u \in B_{\mathbf{R}^2}(0,\varepsilon))} \int_{\nu_0}^{\nu_f} d\nu \\ x(\nu_0) = x_0 \in M_0, x(\nu_f) \in M_1 \end{cases} \tag{16}$$

where $F_0, F_1$ and $F_2$ are the vector fields defined in Sect. 2.2 and $M_0$ and $M_1$ are 2 submanifolds of $\mathbf{R}^4$ with tangent spaces at $x_0$ and $x(\nu_f)$. According to the Pontryagin maximum principle, there exist a constant $p^0 \leq 0$ and an adjoint vector function $p : [\nu_0, \nu_f] \to \mathbf{R}^4$ satisfying $(p^0, p(\nu)) \neq 0$ for all $\nu \in [\nu_0, \nu_f]$ such that, for almost every $\nu \in [\nu_0, \nu_f]$

$$\dot{x}(\nu) = \frac{\partial H}{\partial p}(\nu, x(\nu), p(\nu), u(\nu)), \dot{p}(\nu) = -\frac{\partial H}{\partial x}(\nu, x(\nu), p(\nu), u(\nu)) \tag{17}$$

where $H$ is the non-autonomous control Hamiltonian function

$$H(\nu, x, p, u) = p^0 + \langle p, F_0(\nu, x) \rangle + \sum_{i=1}^{2} u_i \langle p, F_i(x) \rangle .$$

Furthermore, the maximization condition

$$H(\nu, x(\nu), p(\nu), u(\nu)) = \max_{v \in U} H(\nu, x(\nu), p(\nu), v) \tag{18}$$

is satisfied for almost every $\nu \in [\nu_0, \nu_f]$. Finally, at $\nu_0$ and $\nu_f$, we have the transversality conditions

$$p(\nu_0) \perp T_{x(\nu_0)}M_0, p(\nu_f) \perp T_{x(\nu_f)}M_1. \tag{19}$$

A 3-tuple $(x, u, p)$ which satisfies the three conditions of the maximum principle is called an extremal. It is said to be normal if $p^0 \neq 0$ and abnormal if $p^0 = 0$. From now on, we will denote $H_0(\nu, x, p) = \langle p, F_0(\nu, x) \rangle$ the non-autonomous Hamiltonian lift of the drift $F_0$ and $H_i(x, p) = \langle p, F_i(x) \rangle$ the autonomous Hamiltonian lift of the vector field $F_i$, for $i = 1, 2$. Thus the Hamiltonian function $H$ is

$$H(\nu, x, p, u) = p^0 + H_0(\nu, x, p) + \sum_{i=1}^{2} u_i H_i(x, p).$$

From the maximization condition (18), we deduce that, for almost every $\nu \in [\nu_0, \nu_f]$ where $(H_1(x(\nu), p(\nu)), H_2(x(\nu), p(\nu))) \neq (0, 0)$, the optimal control is given by

$$u_i(\nu) = \frac{H_i(x(\nu), p(\nu))}{\sqrt{H_1^2(x(\nu), p(\nu)) + H_2^2(x(\nu), p(\nu))}}.$$

Thus, the optimal control is a feedback control and extremals are fully described by the pairs $z = (x, p)$. This observation leads to the definition of the switching function

$$\psi(\nu) = (H_1(x(\nu), p(\nu)), H_2(x(\nu), p(\nu))) \tag{20}$$

and of the switching surface

$$\Sigma = \{z = (x, p) \in \mathbf{R}^4 \times \mathbf{R}^4 | H_1(x, p) = H_2(x, p) = 0\}. \tag{21}$$

Therefore, extremals are divided into two categories. Extremals $z = (x, p)$ that do not lie on $\Sigma$ are called bang extremal and are smooth. Extremals $z = (x, p)$ lies on $\Sigma$ are called singular extremals. Here we call *switching point* a point of contact between a bang arc and a singular arc along a given extremal (although generally switching points may occur in other cases, for example bang-bang, they do not occur in this study). In the following, we study the nature of such contact points to derive the structure of time-minimizing trajectories of the problem.

## 4.2 Singular Flow and Structure of Extremals

**Definition 3** Let $z^s$ be a singular extremal, with corresponding control $u^s$. Then $z^s$ is the flow of the Hamiltonian equation $\dot{z}^s = \vec{H}s(z^s)$ constrained to the set $\Sigma$ (21), called the singular flow of the singular Hamiltonian.

First of all, we recall some useful results, provided in [17] and built upon in [10, 14, 18, 38], from the in-depth study of singularities of the extremal flow of time-minimizing controls of general autonomous, bi-input, control affine systems of the form

$$\dot{x}(t) = F_0(x(t)) + u_1(t)F_1(x(t)) + u_2(t)F_2(x(t)), u_1^2(t) + u_2^2(t) \leq 1 \tag{22}$$

defined on a manifold $M$ of dimension four. Denote $\gamma_b$ a bang extremal arc, $\gamma_s$ a singular extremal arc and, for any $z = (x, p) \in T^*M$, $F_{ij}(x) = [F_i(x), F_j(x)]$, $H_{ij}(z) = \{H_i(z), H_j(z)\}$ for $1 \leq i, j \leq 2$. Make the assumption

(i) $D(x) = \det(F_1(x), F_2(x), F_{01}(x), F_{02}(x)) \neq 0, x \in M$

and consider the stratification $\Sigma = \Sigma_- \cup \Sigma_0 \cup \Sigma_+$ where

$$\begin{aligned} \Sigma_- &= \{z \in \Sigma | H_{12}^2(z) < H_{01}^2(z) + H_{02}^2(z)\} \\ \Sigma_0 &= \{z \in \Sigma | H_{12}^2(z) = H_{01}^2(z) + H_{02}^2(z)\} \\ \Sigma_+ &= \{z \in \Sigma | H_{12}^2(z) > H_{01}^2(z) + H_{02}^2(z)\}. \end{aligned} \tag{23}$$

The following theorem can be stated.

**Theorem 2** *Let $z_0 \in \Sigma_-$; every extremal is locally of the form $\gamma_b \gamma_s \gamma_b$ (where $\gamma_s$ is empty if $H_{12}(z_0) = 0$); every admissible extremal is locally the concatenation of at most two bang arcs. Let $z_0 \in \Sigma_+$; every extremal is locally bang or singular and every optimal extremal is locally bang. Optimal singular extremals are given by the flow of $H_s$ and contained in $\Sigma_0$ (saturating).*

The proof of Theorem 2 is based on the connection between the flow of the specific form of Hamiltonian function $H$ in the singular case and the singular extremals of the problem and a nilpotent approximation [10] around a point $z_0 \in \Sigma \setminus 0$. By defining, for any $x \in M$,

$$\begin{aligned} D_1(x) &= \det(F_1(x), F_2(x), F_{12}(x), F_{02}(x)), \\ D_2(x) &= \det(F_1(x), F_2(x), F_{01}(x), F_{12}(x)), \end{aligned}$$

replacing (i) by

(i') $D_1^2(x) + D_2^2(x) < D^2(x), x \in M$

and assuming

(ii) $\mathscr{D}$ is involutive,

we get the following.

**Theorem 3** *The switching function is continuously differentiable and every extremal is locally bang-bang with switchings of angle $\pi$ ("$\pi$-singularities").*

Finally, assuming

(iii) $F_0 \notin \text{Span}\{F_1, F_2, F_{01}\}$ is involutive,

we have the following.

**Theorem 4** *In the normal case $p^0 \neq 0$, there cannot be consecutive switchings in $\Sigma_1 = \Sigma \cap \{(x, p) \in T^*X | F_0(x) \in Span\{F_1, F_2, F_{01}\}$*

Notice that this analysis does not apply when considering general bi-input control affine systems with a non-autonomous drift

$$\dot{x}(t) = F_0(t, x(t)) + u_1(t)F_1(x(t)) + u_2(t)F_2(x(t)), u_1^2(t) + u_2^2(t) \leq 1 \tag{24}$$

on a manifold of dimension four. Indeed, in this case, the Lie brackets (resp., Poisson brackets) $F_{01}$ and $F_{02}$ (resp., $H_{01}$ and $H_{02}$) are, a priori, non-autonomous. As a consequence, the determinants $D$, $D_1$ and $D_2$ may depend explicitly on $t$ as well and the assumptions (i) and (i') and the stratification (23) are no longer consistent, even though the switching surface $\Sigma$ can still be defined in the exact same way as in (21) since both $H_1$ and $H_2$ remain autonomous. However, in the specific context of the elliptic restricted three-body problem, straightforward computations give

$$F_{01}(\nu, x(\nu)) = \begin{pmatrix} 1 \\ 0 \\ 0 \\ -2 \end{pmatrix}, \quad F_{02}(\nu, x(\nu)) = \begin{pmatrix} 0 \\ 1 \\ 2 \\ 0 \end{pmatrix}.$$

so

$$H_{01}(\nu, x(\nu), p(\nu)) = p_1(\nu) - 2p_4(\nu)$$

and

$$H_{02}(\nu, x(\nu), p(\nu)) = p_2(\nu) + 2p_3(\nu).$$

Thus, even though the drift $F_0$ of the elliptic restricted three-body problem is non-autonomous, the Lie brackets of length 2 (resp., Poisson brackets) $F_{01}$ and $F_{02}$ (resp. $H_{01}$ and $H_{02}$) are autonomous. In fact, they have the exact same values as in the circular restricted three-body problem, and so do the determinants $D, D_1$ and $D_2$ which do not depend explicitly on $\nu$. The first consequence is that the assumption (i), and the stronger one (i'), can be formulated in the context of our study and are satisfied, in accordance with Lemma 1. By examining the expression of the first derivative of the switching function $\psi$, which is necessarily identically zero along a singular arc, we manage to write singular extremal controls as feedback controls. Indeed, exactly as stated in [17], in the neighborhood of a point $z_0 = (x_0, p_0) \in \Sigma$, an extremal control is

$$u_s(x, p) = \Big( -\frac{H_{02}(x, p)}{H_{12}(x, p)}, \frac{H_{01}(x, p)}{H_{12}(x, p)} \Big). \tag{25}$$

Plugging in $H$, we derive the expression of the non-autonomous singular Hamiltonian function

$$H_s(\nu, x, p) = p^0 + H_0(\nu, x, p) - \frac{H_{02}(x, p)}{H_{12}(x, p)} H_1(x, p) + \frac{H_{01}(x, p)}{H_{12}(x, p)} H_2(x, p). \tag{26}$$

In addition, the stratification (23) also makes sense in the conditions of our problem, as well as assumptions (ii) and (iii) which are both clearly verified, once again in accordance with Lemma 1. The rest of the analysis carried out in [17] can then be rigorously reproduced to investigate the structure and regularity of the extremal trajectories of the elliptic restricted three-body problem. As a conclusion, we can state the following.

**Theorem 5** *The elliptic restricted three-body problem has bang-bang time minimizing controls with finitely many $\pi$-singularities.*

# 5 Application to Space Transfers

In this section, we apply our analysis of the time-minimum control of the planar elliptic restricted 3-body problem to simulate time-minimum space transfers between the geostationary orbit and the equilibrium points $L_1$ and $L_2$ in the Earth-Moon system and a rendezvous mission with a near-Earth asteroid.

## *5.1 Numerical Methods*

The numerical simulations presented in this paper are based on locally sufficient second order conditions [1, 11, 14] and indirect methods in optimal control [3, 21]. In this section, we briefly describe these principles which consist of computing solutions to optimal control problems by generating normal extremal curves solutions of the Pontryagin maximum principle whose local optimality is checked using second order conditions. Consider a generic control problem of the form

$$\begin{cases} \dot{x}(t) = f(t, x(t), u(t)) \\ \min_{u(.)\in U} \int_0^{t_f} f^0(t, x(t), u(t))dt \\ x(0) = x_0 \in M_0, x(t_f) \in M_1 \end{cases} \tag{27}$$

where the time $t_f$ is not fixed, $M$ and $U$ are two smooth manifolds of respective dimensions $n$ and $m, f : [0, t_f] \times M \times U \to TM$ and $f^0 : [0, t_f] \times M \times U \to \mathbf{R}$ are smooth, $M_0, M_1$ are two submanifolds of M and $u$ is an admissible control valued in $U$. By applying the Pontryagin maximum principle and using the maximization condition [46], we can, under some generic regularity assumptions [1, 11], write the optimal control $\bar{u}$ solution to (27) as a smooth feedback control $\bar{u}(t, \bar{x}, \bar{p})$, where $(\bar{x}, \bar{p})$ is an extremal trajectory solution to a smooth Hamiltonian system

$$\dot{x}(t) = \frac{\partial H_r}{\partial p}(t, x(t), p(t)), \dot{p}(t) = -\frac{\partial H_r}{\partial x}(t, x(t), p(t)). \tag{28}$$

Define the exponential mapping $\exp_{x_0} : (t, p_0) \to x(t, x_0, p_0)$ as the function which, given a pair $(t, p_0)$, outputs the projection on $M$ of the extremal trajectory $(x, p)$ solution to (28) starting from the initial condition $(x_0, p_0)$ and evaluated at time $t$.

We say that a time $t_c$ is conjugate to 0 along $(x, p)$ if the restriction of the exponential function $p_0 \to x(t_c, x_0, p_0)$ is not an immersion at $p_0$ and we say that $x(t_c, x_0, p_0)$ is a conjugate point. The notion of conjugate time is connected to the property of local optimality through the following sufficient second order condition of optimality: under generic assumptions, we can state that a trajectory $x(\cdot)$ projection of an extremal solution is locally optimal in the $L^\infty$-topology until the first conjugate time along the extremal [1, 11]. Hence, we can develop a process to compute locally optimal numerical solutions to the problem (27). Indeed, the boundary conditions to be satisfied by an extremal trajectory $(x, p)$ can be written in the form

$$R(x(0), p(0), x(t_f, x_0, p_0), p(t_f, x_0, p_0)) = 0_{\mathbf{R}^n}. \tag{29}$$

Furthermore, since $t_f$ is not fixed, the condition $H_r \equiv 0$ holds along any extremal [46]. Thus we can generate an extremal trajectory by solving the shooting equation, i.e., finding a zero to the shooting function

$$S : (t_f, p_0) \to \begin{pmatrix} R(x_0, p_0, x(x_0, p_0, t_f), p(x_0, p_0, t_f)) \\ H_r(x(x_0, p_0, t_f), p(x_0, p_0, t_f)) \end{pmatrix}, \tag{30}$$

and proceeding to a numerical integration of the system (28) with the corresponding initial condition $(x_0, p_0)$. The local optimality of the projection of the extremal is verified by checking that there is no conjugate time along the interval $[0, t_f]$, which amounts to checking a rank condition [11]. The shooting function being smooth, a Newton-type algorithm can be used to determine its zeroes. The most difficult aspect of this approach is to choose an accurate initial guess so that the Newton method converges. This can be achieved by means of a smooth continuation method [13, 14]. The Hamiltonian function $H_r$ is connected to another Hamiltonian function $H_0$ through a family of smooth Hamiltonian functions $(H_\lambda)_{\lambda \in [0,1]}$, associated with a family of exponential mapping $\exp_{x_0}^\lambda$, such that $H_r = H_1$ and the shooting method is easy to solve for $H_0$. Assume that, for every $\lambda \in [0, 1]$, the point $\exp_{x_0}^\lambda(t_f, p_0)$ is not conjugate to $x_0$. Then the solutions of the shooting method form a smooth curve parametrized by $\lambda$ [13]. Thus, the continuation process consists of following this curve to determine a zero of the shooting function (30). This can be managed iteratively: setting up some discretization $0 = \lambda_0 < \lambda_1 < \cdots < \lambda_N = 1$ of the interval $[0, 1]$, we can first solve the shooting method for $H_0$ and then solve the shooting method for each $H_{i+1}$ by using the solution of the shooting method for $H_i$ as an initial guess. As a result, the zero of the shooting function for $\lambda_N$ is a zero of the shooting function (30).

These methods are implemented by using the software *hampath* [15] which allows one to integrate smooth Hamiltonian vector fields, solve shooting equations and evaluate conjugate points along extremal trajectories. Let us mention that this software also allows one to use a differential path-following method, which was not needed to obtain the results presented in this paper.

### 5.2 Numerical Computations

We now present results obtained using the methods above, as applied to three different missions scenarios. We choose to work in the Earth-Moon system, where the mass ratio $\mu = 0.0121536$ and the eccentricity is $e = 0.0549$. This choice is justified by the fact that it is perhaps the most relevant system (arguably, besides the Sun-Earth system) in the design of actual missions, and that there is existing work for the Earth-Moon system in the *circular* restricted three-body problem [16, 17, 19, 45]. For some perspective on how the Earth-Moon system compares to the rest of our Solar System, Table 1 gives the eccentricities of the orbits of several of the other major bodies. As it can be observed, Mercury has the largest eccentricity of all planets orbiting around the Sun but due to its close proximity to the Sun it is not very relevant from spacecraft missions. While still small, the Moon's orbit around the Earth presents an eccentricity that is the largest from other well-known moons orbiting their primary body. This work is centered around the analysis of the impact of a more complete model on the geometry of the transfers and we will consider unrealistic scenarios with higher eccentricities to obtain a broader understanding. For all mission scenarios we assume that the spacecraft has capabilities resembling

**Table 1** Eccentricies of the solar system

| Body | Eccentricity |
|---|---|
| *(a) Planets around Sun* | |
| Mercury | 0.2056 |
| Venus | 0.0068 |
| Earth | 0.0167 |
| Mars | 0.0934 |
| Jupiter | 0.0484 |
| Saturn | 0.0542 |
| Uranus | 0.0472 |
| Neptune | 0.0086 |
| *(b) Moons around planets* | |
| Moon (Earth) | 0.0549 |
| Io (Jupiter) | 0.0041 |
| Europa (Jupiter) | 0.0090 |
| Ganymede (Jupiter) | 0.0013 |
| Callisto (Jupiter) | 0.0074 |
| Mimas (Saturn) | 0.0202 |
| Enceladus (Saturn) | 0.0047 |
| Tethys (Saturn) | 0.0200 |
| Dione (Saturn) | 0.0020 |
| Titan (Saturn) | 0.0288 |
| Iapetus (Saturn) | 0.0286 |

those of an electric propulsion system, with a 1 N maximum thrust capability and a high specific impulse so that the mass variation can reasonably be ignored. This choice was made for simplicity and, again, based on prior work on low-thrust space transfers in the circular problem [16, 17, 19, 45].

We consider three different missions. For each mission, we select a starting departure point for the spacecraft $x_0 \in \mathbf{R}^4$ as well as a final arrival point $x_f \in \mathbf{R}^4$. Both points $x_0$ and $x_f$ provide the desired position and the velocity of the spacecraft at $\nu_0$ and $\nu_f$, where $\nu_0$ is the true anomaly at the mission's start time and $\nu_f$ is at arrival time. The problem is then solved numerically to identify a time-minimal trajectory $x(t)$ so that $x(\nu_0) = x_0$ and $x(\nu_f) = x_f$.

For a chosen eccentricity value, once the initial true anomaly is fixed to a value $\nu_0$ then the initial position and velocity of the spacecraft $x_0$ in the pulsating rotating frame corresponds to a well-defined point in the fixed frame. However, since the final time is free, the final true anomaly $\nu_f$ is free as well which creates some complexity in the problem due to the dependence of the pulsating frame position of the spacecraft with respect to the true anomaly when converting the pulsating rotating coordinates to or from the fixed coordinates. For instance, consider the eccentricity of the Moon's orbit around the Earth, $e = 0.0549$. When the true anomaly $\nu = \frac{\pi}{2}$ the position $(x_1, x_2) = (0, 1)$ in the pulsating rotating frame corresponds to a point one unit from the origin in the fixed frame; however, if the true anomaly is $\nu = \pi$, the position $(x_1, x_2) = (0, 1)$ in the pulsating rotating frame corresponds to a point 1.058 units from the origin in the fixed frame. To overcome this difficulty, if the destination for the spacecraft is a specific point in the dimensional frame, the true anomaly must be fixed at $\nu_f$ rather than at $\nu_0$. In this case, the shooting function integrates time backward instead. Notice however that in the pulsating rotating frame the coordinates of the equilibrium points $L_i$, $i = 1, \ldots, 5$ do not depend on the true anomaly which therefore makes them convenient departing and arrival points to design a mission.

We choose to design missions with the following three scenarios:

Mission 1: The first scenario that we consider is a mission from the Geostationary orbit to the libration point $L_1$. Simulating such a mission is important, as it is a first step to designing optimal Earth-Moon transfers. Indeed, the vicinity of the point $L_1$ is a gateway between the Earth and the Moon gravitational fields. Therefore, Earth-$L_1$ optimal transfers provide good initializations when using a shooting method to compute Earth-Moon optimal transfers [12, 45].

Mission 2: The second scenario is similar to the first one but the destination is a different libration point. The goal is to compute minimal transfers from the Geostationary orbit to $L_2$. This libration point has proved to play an important role as well for transfers to orbit the Moon, see for instance the Artemis mission [4].

Mission 3: Finally, the last scenario we consider is a transfer to a temporarily captured asteroid, namely 2006RH$_{120}$. We choose the starting point to be $L_2$ because it has proved to provide the best transfers in the case of zero eccentricity [44]. Ideally the spacecraft should be considered on a Halo orbit around the $L_2$ point, but for simplicity and as a first step to the analysis of the impact of eccentricity values we assume it unrealistically at the equilibrium point.

### 5.2.1 Transfers to $L_1$ and $L_2$

Existing results from [19, 45] provided time-minimal transfers to $L_1$ and $L_2$ from a geostationary orbit *GEO* in the Earth-Moon circular restricted three-body problem. In this paper, we extend these results by using the elliptic model. The positions and velocities of the libration points in the non-dimensional elliptic frame do not depend on the eccentricity of the system and (the two we consider) are provided in Table 2. The other relevant location, *GEO*, requires more subtlety: a selected geostationary orbit in the *inertial* reference frame, with inertial coordinates (0.0977, 0, 0, 2.9767), corresponds to different elliptic frame coordinates depending on both the eccentricity and the initial true anomaly $\nu_0$. Table 2 also gives the corresponding elliptic frame coordinates for *GEO* for a few different eccentricity and $\nu_0$ ($\nu_f$ for 2006RH$_{120}$) values.

A continuation-based algorithm was used to compute transfers for 91 different eccentricity values $\{e_i\} = \{0.00, .., 0.90\}$ with a step size of 0.01, e.g. $e_1 = 0.01$, $e_{50} = 0.5$, and $e_{90} = 0.9$. The initial true anomaly was assumed $\nu_0 = 0$ for these transfers. A known solution from the CR3BP ($e_0 = 0$) served as the seed for the continuation algorithm, which iterates through the list of eccentricity values in both an increasing and decreasing fashion. The code can be summarized as follows, with some justification afterward:

- for $k = 1, 2, 3, .., 89, 90$
  - if a solution exists for $e_{k-1}$ with transfer time $t_{k-1}$, and either no solution exists for $e_k$ or the best found solution for $e_k$ has a transfer time greater than $t_{k-1}$, initialize the shooting algorithm for $e_k$ with the solution from $e_{k-1}$.

**Table 2** Departure and arrival positions and velocities for the spacecraft in the non-dimensional frame

| Location | $e$ | $\nu$ | $x_1$ | $x_2$ | $x_3$ | $x_4$ |
|---|---|---|---|---|---|---|
| $L_1$ | Any | Any | 0.8369 | 0 | 0 | 0 |
| $L_2$ | Any | Any | 1.1557 | 0 | 0 | 0 |
| *GEO* | 0 | Any | 0.0977 | 0 | 0 | 2.879 |
| *GEO* | 0.1 | $\pi/2$ | 0.0977 | 0 | −0.0098 | 2.879 |
| *GEO* | 0.1 | $\pi$ | 0.0879 | 0 | 0 | 3.2195 |
| *GEO* | 0.5 | $\pi$ | 0.0489 | 0 | 0 | 5.9046 |
| *GEO* | 0.5 | $3\pi/2$ | 0.0977 | 0 | 0.0488 | 2.879 |
| *GEO* | 0.9 | 0 | 0.1856 | 0 | 0 | 1.381 |
| *GEO* | 0.9 | $\pi$ | 0.0098 | 0 | 0 | 29.757 |
| 2006RH$_{120}$ | 0 | 4.019 | 1.1565 | 1.5681 | 1.48 | −1.23479 |
| 2006RH$_{120}$ | 0.0549 | 4.019 | 1.11592 | 1.51308 | 1.4706 | −1.13085 |

- for $k = 89, 88, .., 2, 1, 0$
  - if a solution exists for $e_{k+1}$ with transfer time $t_{k+1}$, and either no solution exists for $e_k$ or the best found solution for $e_k$ has a transfer time greater than $t_{k+1}$, initialize the shooting algorithm for $e_k$ with the solution from $e_{k+1}$.
- If any new solution was found, repeat; otherwise, the algorithm is done.

Continuation methods can be simple yet effective means to compute solutions to a family of problems related by a parameter—in our case, the eccentricity—however, there is no guarantee for convergence at each step of the algorithm. Moreover, it is possible that a locally optimal solution is computed which is far from the global minimum. These points motivate the algorithm described above; looping not only to identify solutions for each eccentricity value, but also retrying calculations that may have converged to much higher (locally optimum) transfer times. Recall that the actual eccentricity of the Earth-Moon system is $e \approx 0.05$, so all other values are strictly hypothetical.

Figure 4 shows the minimal transfer times as a function of eccentricity for *GEO*-to-$L_1$ and -$L_2$ transfers (blue). For comparison, the first conjugate time is also plotted (red) and we see that it is always longer than the transfer time, confirming the local optimality of our solution. For both destinations, higher eccentricity values allow shorter transfer times. Transfer and conjugate times are also given for select eccentricity values in Table 3.

It is interesting to notice the bifurcation that occurs around $e = 0.13$ and $e = 0.34$. At these points, the higher eccentricity of the system seems to enable the spacecraft to make one less revolution of the Earth before heading directly toward its destination. We can consider the spacecraft's trajectory as a closed curve in the plane if we connect the start and end points with a line. Then we can define the *winding number* $w_E$ of the trajectory as the integer representing the total number of times that curve travels

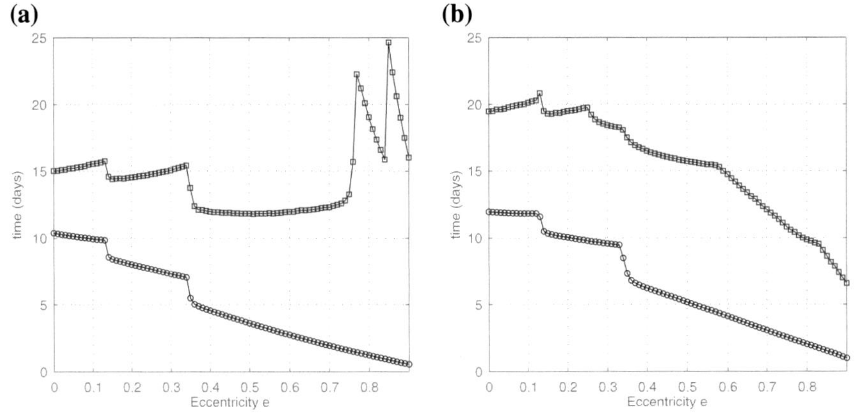


**Fig. 4** Minimum transfer times (*blue circles*) and the corresponding first conjugate times (*red squares*) for $e = 0.0, \ldots, 0.9$. **a** *GEO*-to-$L_1$, **b** *GEO*-to-$L_2$

**Table 3** GEO-to-$L_1$ (*left*) and GEO-to-$L_2$ (*right*) minimum transfer times $t_f^{min}$ and associated first conjugate times $t_{conj}$, in days, and winding number $w_E$, for selected eccentricity values

| $e$ | $t_f^{min}$ | $t_{conj}$ | $w_E$ |
|---|---|---|---|
| 0.0 | 10.38 | 15.02 | 3 |
| 0.1 | 9.95 | 15.55 | 3 |
| 0.2 | 8.04 | 14.55 | 2 |
| 0.3 | 7.32 | 15.08 | 2 |
| 0.4 | 4.57 | 11.99 | 1 |
| 0.5 | 3.63 | 11.86 | 1 |
| 0.6 | 2.75 | 11.97 | 1 |
| 0.7 | 1.95 | 12.32 | 1 |
| 0.8 | 1.23 | 19.05 | 1 |
| 0.9 | 0.54 | 16.03 | 1 |

| $e$ | $t_f^{min}$ | $t_{conj}$ | $w_E$ |
|---|---|---|---|
| 0.0 | 11.96 | 19.48 | 3 |
| 0.1 | 11.82 | 20.12 | 3 |
| 0.2 | 10.04 | 19.51 | 2 |
| 0.3 | 9.59 | 18.45 | 2 |
| 0.4 | 6.23 | 16.50 | 1 |
| 0.5 | 5.18 | 15.74 | 1 |
| 0.6 | 4.12 | 14.76 | 1 |
| 0.7 | 3.07 | 12.13 | 1 |
| 0.8 | 2.05 | 9.87 | 1 |
| 0.9 | 1.00 | 6.58 | 1 |

counterclockwise around the Earth. For example, in Fig. 5a–c, we have $w_E = 3$, 2, and 1, respectively. For the $L_1$ transfers, for $e = 0.00, \ldots, 0.13$ the spacecraft makes three revolutions of the Earth; for $e = 0.14, \ldots, 0.34$ it makes only two revolutions; and finally, for $e = 0.35, \ldots, 0.9$ the craft only makes one revolution of the Earth. Similarly for the $L_2$ transfers, for $e = 0.00, \ldots, 0.12$ the spacecraft makes three revolutions of the Earth; for $e = 0.13, \ldots, 0.33$ it makes only two revolutions; and finally, for $e = 0.34, \ldots, 0.9$ the spacecraft only makes one revolution of the Earth. Table 3 gives $w_E$ for selected eccentricity values.

In Figs. 5 and 6, *GEO*-to-$L_1$ and -$L_2$ transfers are shown, respectively, with $e = \{0, 0.3, 0.8\}$. The images for $e = 0.05 \approx 0.0549$ are indistinguishable from those of $e = 0$—the transfer times and conjugate times are given in Table 3.

### 5.2.2 Transfers to a Near-Earth Asteroid

*Temporarily captured orbiters* (called *minimoons* for short) are a class of near-Earth asteroids gaining recent interest [19, 34]. Informally, minimoons are defined as near-Earth asteroids that are temporarily caught in orbit around the Earth. Although only one minimoon has ever been confirmed, the authors of [34] give rigorous calculations that demonstrate there is a steady state of minimoons in orbit around the Earth. To date, the only confirmed minimoon, known as 2006RH$_{120}$, was discovered in 2006. It is a few meters in diameter and was in orbit around the Earth for about one year. The three dimensional partial trajectory of 2006RH$_{120}$ is shown in Fig. 7 in the inertial geocentric reference frame (ephemeris retrieved from NASA's HORIZONS database). Ongoing research is investigating methods to more regularly detect minimoons.

We now compute a time minimal transfer to rendezvous with 2006RH$_{120}$ starting from the Earth-Moon $L_2$ point. We pre-select a rendezvous point along the trajectory of 2006RH$_{120}$ based on it's vicinity to $L_2$ and zero $z$-coordinate (elliptic frame coordinates are given in Table 2). The rendezvous location is also marked on Fig. 7. It is not in the scope of this paper to optimize the chosen rendezvous location, and

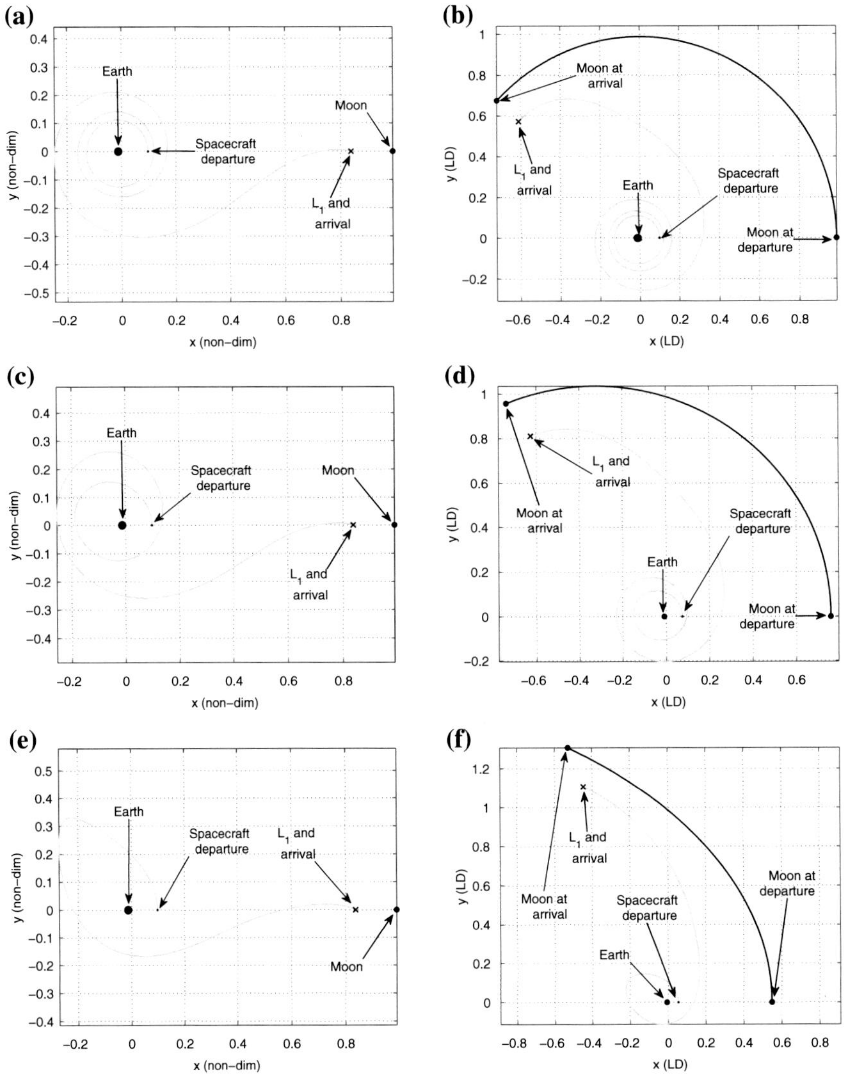


**Fig. 5** Time-minimal transfers from *GEO* to $L_1$, viewed in both the dynamic and fixed frame. **a** $e = 0$, 10.38 days, dynamic frame, **b** $e = 0$, 10.38 days, fixed frame, **c** $e = 0$, 7.32 days, dynamic frame, **d** $e = 0$, 7.32 days, fixed frame, **e** $e = 0$, 1.23 days, dynamic frame, **f** $e = 0$, 1.23 days, fixed frame

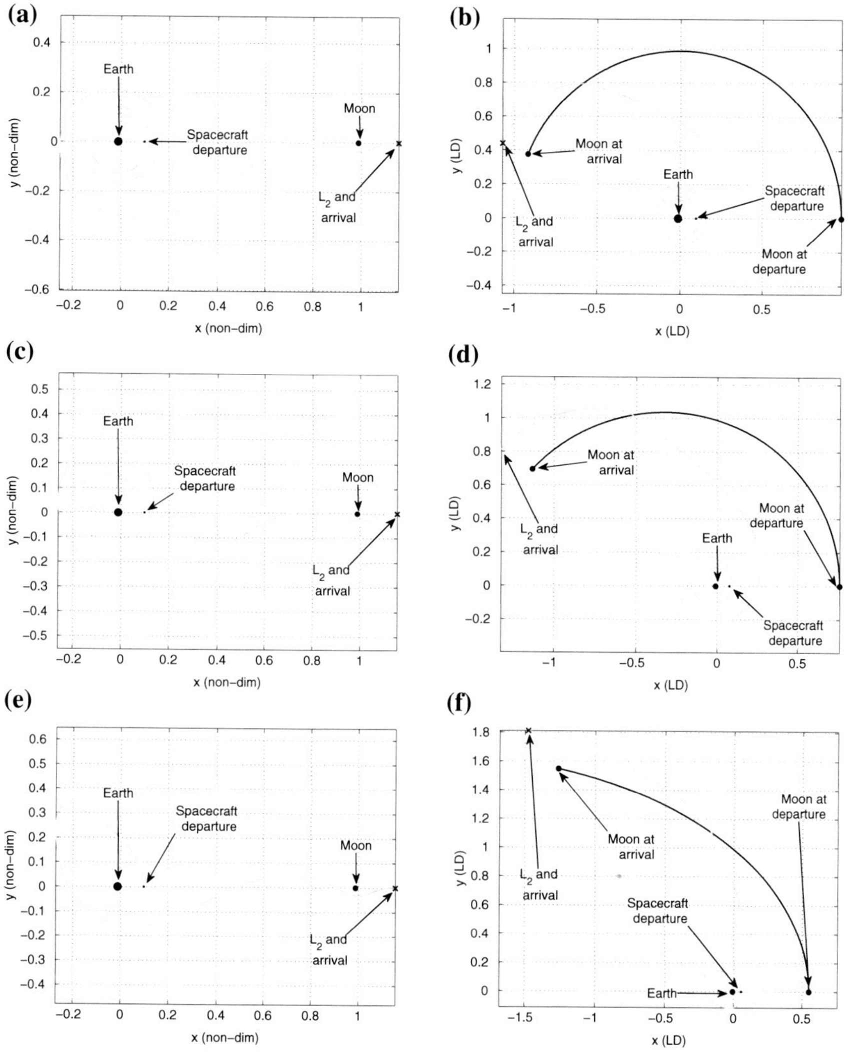


**Fig. 6** Time-minimal transfers from *GEO* to $L_2$, viewed in both the dynamic and fixed frame. **a** $e = 0$, 11.96 days, dynamic frame, **b** $e = 0$, 11.96 days, fixed frame, **c** $e = 0.3$, 9.59 days, dynamic frame, **d** $e = 0.3$, 9.59 days, fixed frame, **e** $e = 0.8$, 2.05 days, dynamic frame, **f** $e = 0.8$, 2.05 days, fixed frame

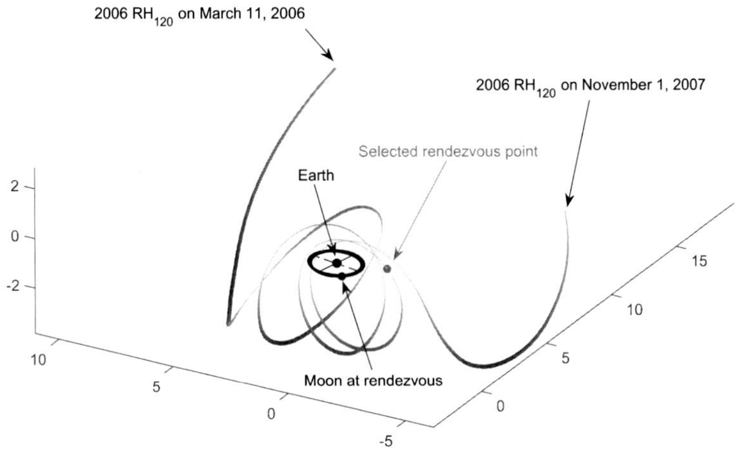


**Fig. 7** Near-Earth asteroid $2006RH_{120}$ (*gray*) viewed in the 3-dimensional fixed frame. Our choice of rendezvous location (*red circle*) was chosen based on low absolute $z$-coordinate and its vicinity to $L_2$. The moon's orbit (*black*) has eccentricity of 0.0549; the major and minor axes of its orbit are plotted as the *solid* and *dashed black lines*, respectively. The true anomaly of the moon at rendezvous is $\nu_f = 4.019$

so this choice is admittedly arbitrary. The true anomaly of the moon at the selected rendezvous point is $\nu_f = 4.019$ radians, also computed from the JPL HORIZONS database.

We use the actual eccentricity and mass ratio of the Earth-Moon system ($\mu = 0.0121536$, $e = 0.0549$), and compare the results to those of the circular problem ($e = 0$). As mentioned, the true anomaly is fixed at rendezvous $\nu_f = 4.019$ radians, and therefore the initial true anomaly $\nu_0$ is free and the shooting method integrates backward in time. Again, existing results in the circular frame are used to initialize the algorithm, and the first conjugate time is calculated to verify the local optimality of our solutions.

Figure 8 shows the time-minimal trajectories for both eccentricity values, in both the non-dimensional and dimensional frames. The trajectories are more or less indistinguishable since the eccentricity of the Earth-Moon system is so low; however, using the actual eccentricity $e = 0.0549$ does provide a transfer time that is 6.4 h faster than with $e = 0$ (10.53 days vs. 10.80 days). The conjugate times for $e = 0$ and $e = 0.0549$ were 17.22 and 16.94 days, respectively. It is likely that missions of longer duration would see larger improvements.

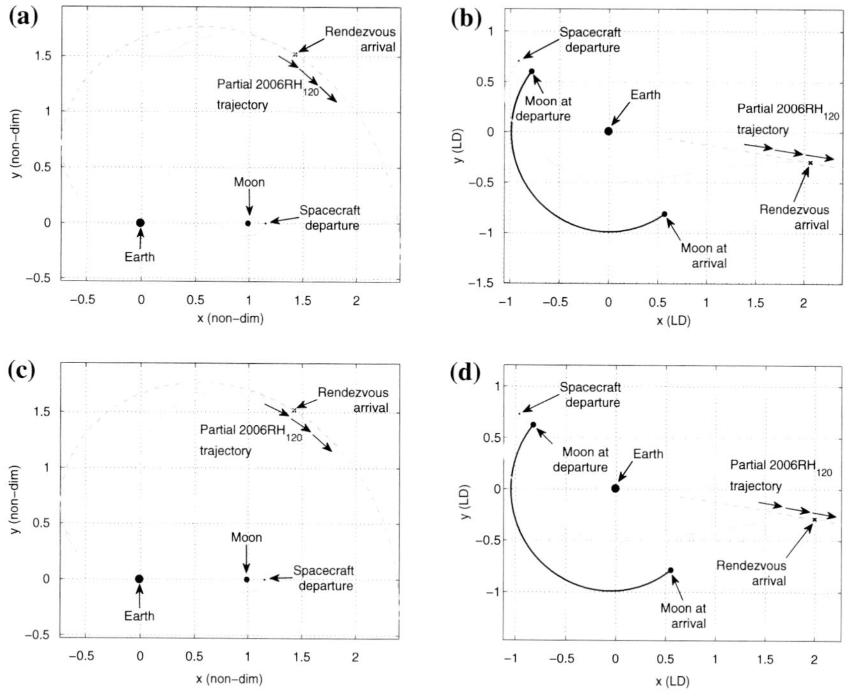


**Fig. 8** Time-minimal transfer from $L_2$ to 2006RH$_{120}$, viewed in both the dynamic and fixed frame. **a** $e = 0$, 10.8 days, dynamic frame, **b** $e = 0$, 10.8 days, fixed frame, **c** $e = 0.0549$, 10.59 days, dynamic frame, **d** $e = 0.0549$, 10.59 days, fixed frame

## 6 Conclusion

In this paper, we generalize some results presented in [17] to the time-minimum control of the planar elliptic restricted three-body problem when the eccentricity $e$ of the Keplerian orbits of the two primaries is strictly positive. The problem is written in the form of a non-autonomous control problem which is shown to be small-time locally controllable in the vicinity of the equilibrium points. We prove that the structure of the time-minimizing controls is preserved, in the sense that the time-minimizing controls are bang-bang with a finite number of $\pi$-singularities. We use this model to compute a collection of time-minimal low-thrust transfers from the geostationary orbit to the equilibrium points $L_1$ and $L_2$ of the Earth-Moon system, for a wide range of eccentricities, by means of a shooting method combined with a continuation method. The local-optimality of these transfers is verified using a second-order optimality condition related to the concept of conjugate points. We observe, numerically a decreasing relation between the minimum time transfer and the eccentricity $e$. Bifurcations occur for $e = 0.13$ and $e = 0.34$, causing the spacecraft to complete less revolutions around the Earth before it reaches its destination. We also simulate a time-minimal rendezvous mission with the near-Earth asteroid 2006RH$_{120}$ in the Earth-Moon system. The initial guesses chosen to initiate our numerical methods are time-minimal transfers in the circular restricted three-body problem computed in [19, 45]. The results show that considering the actual eccentricity of our Moon's orbit around the Earth leads to a slightly shorter rendezvous time with the asteroid 2006RH$_{120}$ than when the eccentricity is neglected. The natural next step of this study will consist of taking into account the significant influence of the Sun on the transfers within the Earth-Moon system. One possibility to achieve this goal would be to derive the equations of a perturbed elliptic three body problem, inspired by the equations of the restricted four-body problem [42] which can be used to model a Sun-perturbed circular restricted three-body problem. The theoretical analysis of the time-minimum control of the perturbed elliptic three body problem will raise an interesting issue from the geometric control point of view. More on the practical side, the main objective will be to compare numerical computations performed with this new model with the ones that are carried out in the present chapter, in order to design even faster low-thrust transfers in the Earth-Moon system. For the sake of realism, another interesting problem would be to consider points on a small halo orbit around the equilibrium points $L_1$ and $L_2$ of the restricted 3-body problem [6] as initial conditions for a rendezvous mission to near-Earth asteroids.

**Acknowledgments** This research is partially supported by the National Science Foundation (NSF) Division of Mathematical Sciences, award #DMS-1109937 and by the NASA, proposal *Institute for the Science of Exploration Targets* from the program Solar System Exploration Research Virtual Institute. Geoff Patterson also received support from the NSF Division of of Graduate Education, award #DGE-0841223.

## References

1. Agrachev AA, Sachkov YL (2004) Control theory from the geometric viewpoint. Springer, New York
2. Alfriend KT, Rand RH (1969) Stability of the triangular points in the elliptic restricted problem of the three bodies. AIAA J 7(6):1024–1028
3. Allgower EL, Georg K (1990) Numerical continuation methods, an introduction. Springer, Berlin
4. Angelopoulos V (2008) The artemis mission. IGPP/ESS UCLA
5. Arnold VI (1989) Mathematical methods of classical mechanics. Springer, New York
6. Bando M, Ichikawa A Formation flying near the libration points in the elliptic restricted three-body problem
7. Baoyin H, McInnes CR (2006) Solar sail equilibria in the elliptical restricted three-body problem. J Guid Control Dyn 29(3):538–543
8. Belbruno E (2007) Fly me to the moon. An insiders guide to the new science of space travel. Princeton University Press
9. Bhat SP (2005) Controllability of nonlinear time-varying systems: applications to spacecraft attitude control using magnetic actuation. IEEE Trans Autom Control 50(11):1725–1735
10. Bonnard B, Caillau J-B, Trélat E (2005) Geometric optimal control of elliptic Keplerian orbits. Discrete Cont Dyn Syst Ser B 4:929–956

11. Bonnard B, Caillau J-B, Trélat E (2007) Second order optimality conditions in the smooth case and applications in optimal control. ESAIM Control Optim Calc Var 13:207–236
12. Bonnard B, Caillau J-B, Picot G (2010) Geometric and numerical techniques in optimal control of the two and three-body problems. Commun Inf Syst 10:239–278
13. Bonnard B, Shcherbakova N, Sugny D (2011) The smooth continuation method in optimal control with an application to quantum systems. ESAIM Control Optim Calc Var 17:267–292
14. Bonnard B, Chyba M (2003) Singular trajectories and their role in control theory. Springer, Berlin
15. Caillau J-B, Cots O, Gergaud J (2012) Differential continuation for regular optimal control problems. Optim Methods Softw 27(2):177–196
16. Caillau J-B, Daoud B, Gergaud J (2012) Minimum fuel control on the planar restricted three-body problem. Celest Mech Dyn Astrom 114(1):137–150
17. Caillau J-B, Daoud B (2012) Minimum time control of the restricted three-body problem. SIAM J Control Optim 50(6):3187–3202
18. Caillau J-B, Noailles J (2001) Coplanar control of a satellite around the Earth. ESAIM Control Optim Calc Var 6:239–258
19. Chyba M, Patterson G, Picot G, Jedicke R, Granvik M, Vaubaillon J (2014) Designing rendezvous missions with mini-moons using geometric optimal control. J Ind Manag Optim **10**(2):477–501
20. Contopoulos G (1967) Integrals of motion in the elliptic restricted three-body problem. Astron J 72:669–673
21. Conway BA (2012) A survey of methods available for the numerical optimization of continuous dynamic systems. J Optim Theory Appl 152:271–306
22. Coron J-M (2007) Control and nonlinearity. mathematical surveys and monographs, vol 136. American mathematical society
23. Danby JMA (1964) Stability of the triangular points in the elliptic restricted problem. Astron J 69(2):165–172
24. Deprit A, Rom A (1970) Characteristic exponents at $L_4$ in the elliptic restricted problem. Astron Astrophys 5:416–425
25. Duffy B (2012) Analytical methods and perturbation theory for the elliptic restricted three-body problem of astrodynamics. Ph.D thesis, The George Washington University
26. Duffy B, Chichka D (2012) Canonical perturbation theory fot the elliptic-restricted three-body problem. Adv Astronaut Sci 143:1267–1286
27. Erdi B, Forgacs-Dajka E, Nagy I, Rajnai R (2009) A parametric study of stability and resonances around $L_4$ in the elliptic restricted three-body problem. Celest Mech Dyn Astronom 104(1–2):145–158
28. Gawlik ES, Marsden JE, Du Toit P, Campagnola S (2009) Lagrangian coherent structures in the planar elliptic restricted three-body problem. Celest Mech Dyn Astronom 103(2):227–249
29. Gergaud J, Haberkorn T (2006) Homotopy method for minimum consumption orbit transfer problem. ESAIM Control Optim Calc Var **12**(2):294–310
30. Gergaud J, Haberkorn T, Martinon P (2004) Low-thrust minimum-fuel orbital transfer: an homotopic approach. J Guidance Control Dyn 27(6):1046–1060
31. Goldstein H, Poole C, Safko J (2002) Classical mechanics. Addison Wesley, San Francisco
32. Gomez G, Koon WS, Lo MW, Marsden JE, Masdemont J, Ross SD (2001) Invariants manifolds, the spatial three-body problem ans space mission design. Adv Astronaut Sci 109:3–22
33. Gomez G, Koon WS, Lo MW, Marsden JE, Masdemont J, Ross SD (2004) Connecting orbits and invariant manifolds in the spatial three-body problem. Nonlinearity 17:1571–1606
34. Granvik M, Vaubaillon J, Jedicke R (2012) The population of natural earth satellites. Icarus 218:262–277
35. Gurfil P, Meltzer D (2006) Stationkeeping on unstable orbits: generalization to the elliptic restricted three-body problem. J Astronaut Sci **54**
36. Jurdjevic V (1997) Geometric control theory. Cambridge University Press
37. Klamka J (1991) Controllability of dynamical systems. Mathematics and Its applications. Kluwer Academic Publishers Group

38. Kupka, I (1987) Generalized Hamiltonians and optimal control: a geometric study of extremals. In: Proceedings of the international congress of mathematicians, Berkeley, CA, pp 1180–1189
39. Ledzewicz U, Schattler H (2012) Geometric optimal control. Theory, methods and examples, interdisciplinary applied mathematics, vol 38. Springer, New York
40. Lee EB, Markus L (1986) Fondations of optimal control theory. Reprint edition, Krieger
41. Lhotka C, Efthymiopoulos C, Dvorak R (2008) Nekhoroshev stability at $L_4$ and $L_5$ in the elliptic restricted three-body problem application to Troyan asteroids. Mon Notice Royal Astron Soc 384(3):1165–1177
42. Mingotti G, Topputo F, Bernelli-Zazzerra F (2007) A method to design sun-perturbed earth-to-moon low-thrust transfers with ballistics capture. XIX congresso nazionale AIDAA
43. Newton I (1966) Principes mathématiques de la philosophie naturelle. Tome I, II (French). Traduction de la marquise du Chastellet, augmentée des commentaires de Clairaut, Librairie scientifique et technique Albert Blanchard, Paris
44. Patterson G (2015) Asteroid rendezvous missions using indirect methods of optimal control. University of Hawaii at Manoa, dissertation
45. Picot G (2012) Shooting and numerical continuation method for computing time-minimal and energy-minimal trajectories in the Earth-Moon system using low-propulsion. Discrete Cont Dyn Syst Ser B 17:245–269
46. Pontryagin LS, Boltyanskii VG, Gamkrelidze RV, Mishchenko EF (1962) The mathematical theory of optimal processes. Wiley, New York
47. Siegel CL, Moser JK (1971) Lectures on celestial mechanics, classics mathematics. Springer, Berlin
48. Szebehely V (1967) Theory of orbits: the restricted problem of three bodies. Academic Press
49. Topputo F, Belbruno E (2015) Earth-Mars transfers with ballistic capture. Celest Mech Dyn Astron 121(4):329–346

# On Local Optima in Minimum Time Control of the Restricted Three-Body Problem

**Jean-Baptiste Caillau and Ariadna Farrés**

**Abstract** The structure of local minima for time minimization in the controlled three-body problem is studied. Several homotopies are systematically used to unfold the structure of these local minimizers, and the resulting singularity of the path associated with the value function is analyzed numerically.

**Keywords** Circular restricted three body problem · Optimal control · Shooting · Homotopy · Swallowtail singularity

## 1 Introduction

There is currently a renewed interest in space missions with electric propulsion. See for instance the BepiColombo [2] or Lisa [9] programs. Very important models for such missions are the two and three-body controlled problems; in particular, the circular restricted three-body problem provides a dynamically relevant and challenging model for missions in the Earth-Moon or Sun-Earth systems. We recall that we take an inertial reference frame such that the line joining the Earth and Moon remains fixed on the $x$-axis, the $z$-axis is parallel to the angular velocity of the couple and the $y$-direction completes a positive triad. We also normalize the units of distance and time such that the distance between the two primaries is one and the period of rotation is $2\pi$, then the equations of motion for the mass-less satellite under the gravitational influence of the two primaries is given by (see, e.g., [5])

J.-B. Caillau (✉)
UBFC & CNRS/INRIA, Math. Institute, Université de Bourgogne,
9 Avenue Alain Savary, 21078 Dijon Cedex, France
e-mail: jean-baptiste.caillau@u-bourgogne.fr

A. Farrés
Dpt. Matematica Aplicada i Analisi, Universitat de Barcelona,
Gran Via de Les Corts Catalanes 585, 08007 Barcelona, Spain
e-mail: ariadna.farres@maia.ub.es



B. Bonnard and M. Chyba (eds.), *Recent Advances in Celestial and Space Mechanics*, Mathematics for Industry 23, DOI 10.1007/978-3-319-27464-5_7

$$\begin{aligned}
\dot{x} &= v_x, \\
\dot{y} &= v_y, \\
\dot{z} &= v_z, \\
\dot{v}_x &= 2v_y + x - (1-\mu)\frac{x+\mu}{r_1^3} - \mu\frac{x+\mu-1}{r_2^3} + \varepsilon u_1, \\
\dot{v}_y &= 2v_x + \left(1 - \frac{(1-\mu)}{r_1^3} - \frac{\mu}{r_2^3}\right) + \varepsilon u_2, \\
\dot{v}_z &= -\left(\frac{(1-\mu)}{r_1^3} - \frac{\mu}{r_2^3}\right) z + \varepsilon u_3,
\end{aligned} \tag{1}$$

where $\mu \in (0, 1/2]$ is the mass parameter of the system ($\mu = m_2/(m_1+m_2)$—for the Earth-Moon case we have $\mu = 0.012153$), $r_1 = \sqrt{(x-\mu)^2 + y^2 + z^2}$, $r_2 = \sqrt{(x-\mu+1)^2 + y^2 + z^2}$ and $u = (u_1, u_2, u_3)$ is the thrust direction on the small satellite (where $|u| \leq 1$) and $\varepsilon$ is the maximal thrust. Our aim is to find a steering law for Earth-Moon transfer orbits minimizing the transfer time. (For the minimum fuel case, see [3]; see [10] as well for a nice numerical study.)

This problem can be formulated as:

$$\begin{aligned}
&\min t_f = \int_0^{t_f} dt, \\
&\dot{x} = F(x,u) = F_0(x) + \varepsilon \sum_{i=0}^{3} F_i(x) u_i, \\
&|u| \leq 1, \\
&x(0) \in \mathscr{X}_0, \\
&x(t_f) \in \mathscr{X}_1.
\end{aligned} \tag{2}$$

where $\mathscr{X}_0$ and $\mathscr{X}_1$ are the initial and final sets; in our case $\mathscr{X}_0$ is a point on a GEO orbit (or the whole GEO orbit) and $\mathscr{X}_1$ will be either $L_2$ or a point on a MO orbit. In this paper, we will focus on the coplanar case and thus we will only consider a two-input control in the orbital plane. More realistic models should, of course, take into account the fact that the initial and final orbits do not need to belong to the orbital plane of the circular motion of the two primaries (see [5] for an example of 3D minimum time computation).

As it is proved in [5], controllability holds for any $\varepsilon > 0$ for a fixed $\mu \in (0, 1)$ (see [8] for the two-body case, $\mu = 0$), provided the Jacobi constant, $J_c$, or *energy*, is not greater that $J_c(L_2)$. (Caveat: In [5], Poincaré terminology is used so what is nowadays termed the $L_1$ Lagrange point was called $L_2$, and conversely. Here, we use the modern standards and by $L_1$ we mean the Lagrange point with lowest energy.) Let us recall that

$$J_c(q, \dot{q}) = \frac{1}{2}|\dot{q}|^2 - \frac{1-\mu}{|q+\mu|} - \frac{\mu}{|q-1+\mu|} - \frac{1}{2}|q|^2.$$

In order to connect orbits around the primaries, it is transparent from the proof that the energy has to be raised beyond $J_c(L_1)$ through the action of the control; it turns out that, for relevant boundary conditions, a time minimizing transfer will actually pass quite close to the $L_1$ point in the phase space. In this respect, for an important class of endpoint values, one can approximately decompose a min time transfer into two two-body problems coupled by an intermediary $L_1$ target. A byproduct of this heuristic interpretation is the role of the rotation numbers (or homology, see [7] for a first numerical study) around each primary. As it is well known, many local minima exist for min time two-body transfers [8], so we propose here a detailed and systematic study of this phenomenon in the restricted three-body setting. The idea is to use a homotopy in the covering angle coordinate of the initial orbit to unfold the structure of these local time minimizers. For a fixed level of thrust $\varepsilon$, we parameterized the solution (computed by single shooting suitably initialized—we build on [5] results) by the angular position on it; performing a homotopy on this angle reveals the connection between the different local minima. Moreover, we investigate the interplay between these local minima and the fact that, when the thrust level is decreased one needs to "make more turns" to depart from the initial orbit (*resp.* reach the final orbit); this analysis is drawn using another homotopy, on $\varepsilon$, to follow the characteristics (aka extremals, that are state and costate solutions of the maximum principle) through some specific singularities.

The paper is organized in two sections and four appendices. Section 2 is devoted to transfers towards the $L_1$ Lagrange points. First, extremals with fixed initial point on the GEO orbit are computed, in combination with homotopies w.r.t either the position angle, $\theta_0$, or the maximum thrust allowed, $\varepsilon$; then extremals with free $\theta_0$ (that is with initial submanifold the whole GEO orbit) are computed, again with a homotopy on $\varepsilon$. The numerical computations reveal the existence of possible cut points, that is of several candidates as global optimizers having the same cost but different structures. The same analysis is performed in Sect. 3 on transfers from the GEO towards an orbit around the Moon, referred to as MO; in this case, a homotopy on the radius of this target circular orbit is also computed. The aim of this paper is to provide a rather extensive numerical study of the problem so that detailed results are archived into several appendices: The first two provide a comprehensive list of tables of shooting initializations allowing to reproduce the results in Sects. 2 and 3, while the last two ones give a precise account of the computations of (what might be) cut points for the GEO to $L_1$ and MO targets, respectively.

## 2 Transfer from a GEO to $L_1$

In this section we will explore the nature of the first phase of an Earth-Moon transfer. Hence, we focus on the minimum-time transfer trajectories from a GEO orbit to the $L_1$ point. Finding a global minima is a hard task as in many cases it will be hard to determine if we have a global minima or just a local one. We will use indirect methods based on the Pontryagin Maximum Principle to determine local minima, using the package `hampath` [4]. First, we use the initial conditions from [5] and refine them so that they meet the constraints of our problem. The solutions in [5] where found doing homotopies from the 2BP to the 3BP, using $\mu$ as the continuation parameter. Then we will perform different homotopies with respect to the position of the initial orbit and with respect to $\varepsilon$, the thrust value. The idea is to understand the global structure, and find all the possible local minimum-time transfer trajectories for a given thrust value, $\varepsilon$, and classify the different type of solutions. In order to analyse this minimum-time transfer we propose two different boundary value problems: (a) a fixed initial condition on the GEO orbit (point-to-point problem), or (b) a free initial condition on the GEO orbit (circle-to-point problem). Both problems will be solved using a shooting method. In order to find a minimum-time transfer trajectory, it is clearly better to leave the initial condition in the departure orbit free. But due to the small radius of convergence of the shooting method, a first exploration fixing the initial position is required. Moreover, it might be possible, that in a concrete mission scenario, the initial position on the GEO orbit is fixed.

## *2.1 Fixed Initial Point on a GEO*

In this section we discuss the results for the minimum-time transfer problem for a fixed point on a GEO orbit to $L_1$. The boundary conditions are:

$$x(t_0) - x_0 = 0, \quad x(t_f) - x_f = 0, \quad h(t_f) = 0, \tag{3}$$

where $x(t)$ represents the position and velocity of the spacecraft at time $t$, $x_0$ is a fixed initial condition on a circular orbit around the Earth of radius $r_0$; $x_f$ is the position in the phase space that we want to reach with minimum time (here $x_f \equiv L_1 = (0.8369, 0, 0, 0)$), and $h(t_f)$ is the Hamiltonian of the PMP that has to be maximised. We parameterise an initial condition on a circular orbit by its radius $r_0$ and angle $\theta_0 \in [-\pi, \pi)$. Accordingly,

$$x_0 = (r_0 \cos\theta_0 - \mu, r_0 \sin\theta_0, -v_0 \sin\theta_0, v_0 \cos\theta_0), \tag{4}$$

where $v_0 = \sqrt{(1-\mu)/r_0}$ (velocity required to have a circular orbit around the Earth using a 2BP approximation). In this section we will use $r_0 = 0.109689855932071$ and $v_0 = 3.000969693845573$. Notice that $r_0 = 0.10968 \approx 42,164$ km which corresponds to a GEO orbit ($\approx 35.786$ km *above the Earth surface*). Later we might want to discuss the effect of taking a smaller $r_0$ but given the nature of the problem, the results should be very similar and we should experience just some more turns around the Earth before getting on an excursion towards $L_1$. In [5], the authors considered $r_0 = 0.109689855932071$, $\tilde{v}_0 = 2.878597058456258$. We have done a homotopy with respect to the initial velocity on the orbit for a fixed point placed at $\theta_0 = \pi$. Tables 1 and 2 summarize these results for different values of $\varepsilon$. (For $\varepsilon = 3$N we had a problem during the continuation process and the local minima found has $t_f < 0$, so we will not use this as reference value.) Caveat: The $T_{\max}$ variable name in graph legends refers to $\varepsilon$.

**Table 1** $\varepsilon$, $t_f$ and $p_0$ of the local minima for $r_0 = 0.109689855932071$, $\tilde{v}_0 = 2.878597058456258$ and $\theta_0 = \pi$ (see [5])

| $\varepsilon$ (N) | $t_f$ (UT) | $p_0$ |
|---|---|---|
| 10.0 | 1.47056664 | (2.57392200 1.58804145 0.06972900 0.07817485) |
| 5.0 | 2.28408175 | (2.54649996 2.42058400 0.10200404 0.11955574) |
| 4.0 | 3.14537620 | (9.72954854 1.31678527 0.05736254 0.42386881) |
| 3.0 | 3.82670885 | (11.03739342 −0.29053787 −0.00260636 0.49490659) |
| 2.5 | 4.39877672 | (11.19394050 −2.02820238 −0.06740069 0.48674283) |
| 1.5 | 6.67073313 | (4.56115506 −0.98210983 −0.02417903 −0.03690623) |
| 1.0 | 8.44011820 | (−22.71568672 9.43440236 0.36634939 −0.86734182) |

**Table 2** $\varepsilon$, $t_f$ and $p_0$ of the local minima for $r_0 = 0.109689855932071$, $v_0 = 3.000969693845573$ and $\theta_0 = \pi$ (results after continuation from Table 1)

| $\varepsilon$ (N) | $t_f$ (UT) | $p_0$ |
|---|---|---|
| 10.0 | 1.4833856 | (3.83493364 1.72669505 0.07642569 0.13229597) |
| 5.0 | 2.3063975 | (5.63281024 2.42739999 0.10473957 0.24553557) |
| 4.0 | 3.3788754 | (14.73978122 0.81593498 0.03725883 0.63113008) |
| 3.0 | −3.6981165 | (−12.78641230 0.12268080 0.01238500 −0.58331391) |
| 2.5 | 4.2681801 | (13.40062243 −1.57809214 −0.05401807 0.60223014) |
| 1.5 | 6.4693309 | (8.27460406 −2.03560608 −0.06799797 0.14339434) |
| 1.0 | 10.4302927 | (41.66289104 −1.40939476 −0.04832354 1.81224566) |

### 2.1.1 Homotopy w.r.t $\theta_0$

As the initial manifold is a whole orbit, the initial position on the GEO shall be left free. Nevertheless, due to the small radius of convergence of the shooting method and the large dimension of the phase space, taking as initial conditions the local minima in Table 2 the shooting method does not converge. This is why we decided to proceed more systematically and solve for a large range of $\theta_0$. We have taken for $\varepsilon = 10$N, 5N and 1N the local minima in Table 2 and perform a homotopy with respect to $\theta_0$ (Eq. 4), i.e. we change the initial position on the GEO orbit. For each $\varepsilon$ we start at $\theta_0 = \pi$ (which corresponds to the values in Table 2) and perform two continuations: one from $\theta = \pi \mapsto 21\pi$ and the other from $\theta = \pi \mapsto -21\pi$. In order to have a good precision along the path, we have divided the homotopies into smaller blocks of length $2\pi$, for example for the homotopy from $\theta = \pi \mapsto 21\pi$ we split it in small homotopies from $\theta = (2k+1)\pi \mapsto (2k+2)\pi$, for $k = -10, \dots, 9$. When we go from one block to another we refine the initial condition taking the last point on the previous block. We proceed in this way not to deprecate precision along the continuation path. The homotopy will be stopped when `hampath` fails to continue for different reasons (e.g., the norm of the shooting function becomes to big, the continuation step-size is to small). Note that the homotopy is actually done on $\theta_0 \in \mathbb{R}$, using implicitly the variable in the covering of the initial orbit diffeomorphic to $\mathbf{S}^1$ (Tables 3, 4, 5, 6, 7, 8, 9, 10 and 11).

Figures 1, 2 and 3 show the homotopy path for $\varepsilon = 10$N, 5N and 1N respectively. On the left-hand side of each Figure we plot the continuation parameter, $\theta_0$, versus the transfer time, $t_f$, and on the right-hand side we plot $\theta_0(\text{mod } 2\pi) \in [-\pi, \pi]$ versus $t_f$ (Tables 12, 13 and 14). The plots on the right-hand side illustrate that for a given initial position on the departure orbit there are different local minima solutions. We recall that each point on the curve is a solution to the minimum-time transfer from GEO to $L_1$ for a fixed initial position on the GEO. The red points correspond to the local minima on the projection of the homotopic path on the $\theta_0$–$t_f$ plane, and they are candidates for being local minima when the initial position on a GEO orbit is left free (Sect. 2.2). The initial conditions for these local minima are summarized in Tables 12, 13 and 14. Notice that the three curves for $\varepsilon = 10$N, 5N and 1N (Figs. 1, 2 and 3 respectively) present a similar behaviour; as we increase $\theta_0$ (i.e. we change the initial position on the GEO following the clockwise direction) the transfer time, $t_f$, decreases until we reach a global minima and then $t_f$ grows drastically. Also notice that for $\varepsilon = 10$N and 5N (Figs. 1 and 2) after $t_f$ has drastically increased the homotopic curve takes a turn and $\theta_0$ starts to decrease and we find different local minima. For both curves we find different local minima and one clear global minima. In the case of $\varepsilon = 1$N (Fig. 3) after reaching the global minima $t_f$ also increases drastically but the continuation scheme is stopped before we can see a similar behaviour to what we obtained previously. We expect that things will continue to grow as for $\varepsilon = 10$N but need a sharper continuation scheme.

As we have mentioned before, for a given initial condition on the GEO orbit and a fixed $\varepsilon$ we have many different local minimum time–transfer orbits from GEO to $L_1$. In Tables 4, 5, 6 and 7 we summarize the different local minima for $\varepsilon = 10$ and $\theta_0(\text{mod } 2\pi) = 0, \pi/2, \pi$ and $-\pi/2$ respectively. In Fig. 4 we plot the transfer trajectory of the solutions in Table 4, corresponding to $\theta_0(\text{mod } 2\pi) = 0$ and $\varepsilon = 10$N and $k = 2$. In Fig. 5 we plot the variation of the Jacobi constant $J_c$ along time for the trajectories that appear in Fig. 4.

As we can see in Fig. 4 there are two types of trajectories. The first type, that we call $\mathscr{T}_1$ and correspond to $k = 1, \ldots, 12$ from Table 4, where the trajectory gives several turns around the Earth and then heads towards $L_1$ directly. The number of turns around the Earth will depend on the initial condition and, as we can see in Fig. 5, the extra turns correspond to a decrease and later increase of $J_c$ before reaching $J_c(L_1)$. For all these orbits the final transfer to $L_1$ is the same. The second type of orbits, that we call $\mathscr{T}_2$ and corresponds to $k = 14$ and 15. Here the trajectories do a large excursion to get to $L_1$. This large excursion corresponds to large increase on $J_c$, high above $J_c(L_1)$. The transition between one kind of trajectories and the other (solution $k = 13$) corresponds to intermediary excursions of the trajectory and are the solutions while $t_f$ drastically increases in Fig. 1. In Figs. 6 and 7 we show the same results for $\varepsilon = 1$N. Here we only find trajectories of type $\mathscr{T}_1$. This is because the continuation method failed at some point, but this does not mean that trajectories of type $\mathscr{T}_2$ do not exist for $\varepsilon = 1$N.

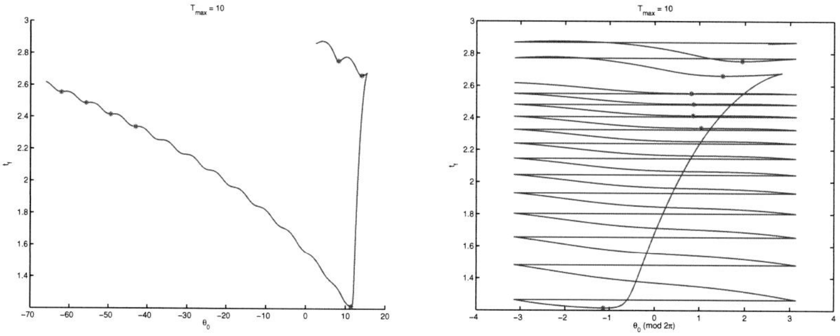


**Fig. 1** For $\varepsilon = 10$N: Projection of the homotopic path $\theta_0$ versus $t_f$ (*left*), $\theta_0 (\text{mod } 2\pi)$ versus $t_f$ (*right*). GEO to $L_1$ transfer, where the initial point on a GEO is fixed to $(r_0 \cos\theta_0 - \mu, r_0 \sin\theta_0, v_0 \cos\theta_0, v_0 \sin\theta_0)$

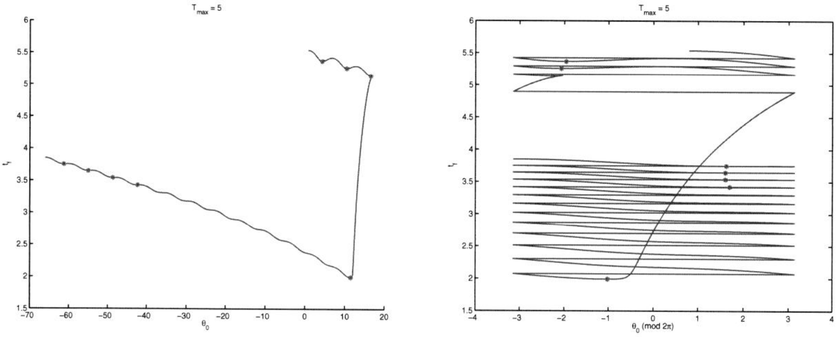


**Fig. 2** For $\varepsilon = 5$N: Projection of the homotopic path $\theta_0$ versus $t_f$ (*left*), $\theta_0 (\text{mod } 2\pi)$ versus $t_f$ (*right*). GEO to $L_1$ transfer, where the initial point on a GEO is fixed to $(r_0 \cos\theta_0 - \mu, r_0 \sin\theta_0, v_0 \cos\theta_0, v_0 \sin\theta_0)$

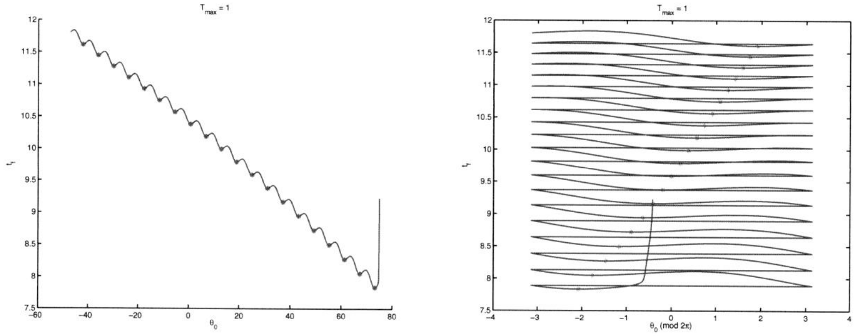


**Fig. 3** For $\varepsilon = 1$N: Projection of the homotopic path $\theta_0$ versus $t_f$ (*left*), $\theta_0 (\text{mod } 2\pi)$ versus $t_f$ (*right*). GEO to $L_1$ transfer, where the initial point on a GEO is fixed to $(r_0 \cos\theta_0 - \mu, r_0 \sin\theta_0, v_0 \cos\theta_0, v_0 \sin\theta_0)$

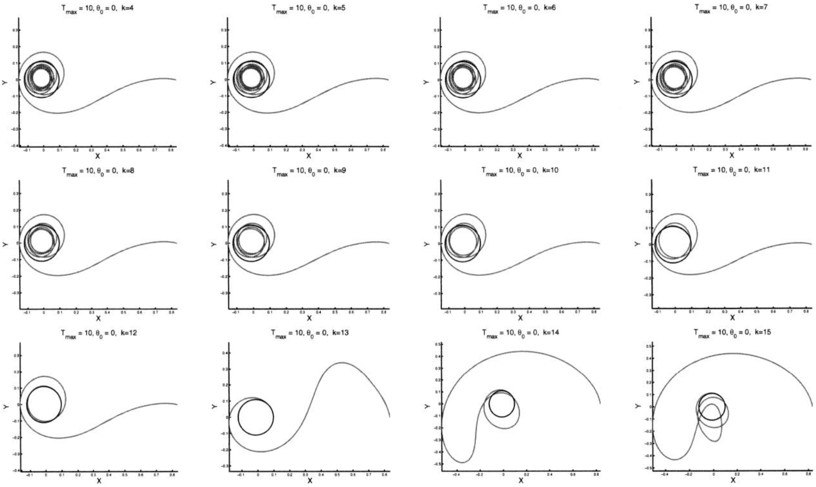


**Fig. 4** Transfer trajectories of the local minima for $\varepsilon = 10\text{N}$ and $\theta_0 = 0$ (Table 4). The initial condition on the GEO orbit is $x_0 = (-r_0 - \mu, 0, 0, v_0)$

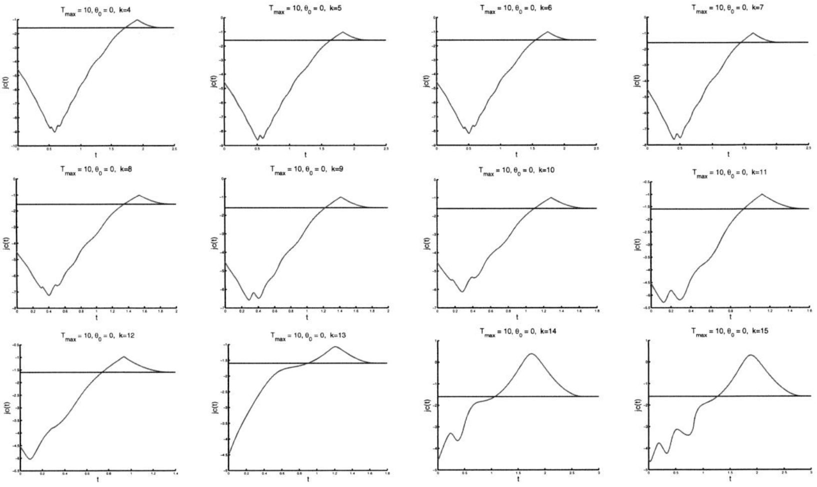


**Fig. 5** Variation of $J_c$ of the local minima for $\varepsilon = 10\text{N}$ and $\theta_0 = 0$ (Table 4). The initial condition on the GEO orbit is $x_0 = (-r_0 - \mu, 0, 0, v_0)$

Finally, we have done a small exploration on the different characteristics of the minimum time–transfer trajectories that appear in Figs. 1, 2 and 3. For each minimum-time transfer orbit, parameterised by $\theta_0$, we have computed the maximal value for $J_c$ along the orbit, $\max_{t\in[t_0:t_f]}(J_c(t))$, the norm of the adjoint vector at $t_0$, $|p(t_0)|$, and the number of turns that the trajectory makes around the Earth, $N_{ET}$, before the trajectory reaches $J_c(L_1)$. These results are summarized in Figs. 8, 9

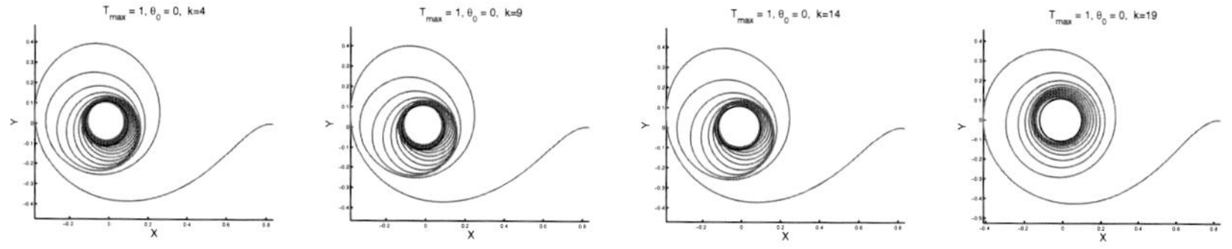


**Fig. 6** Transfer trajectory of the local minima for $\varepsilon = 1\text{N}$ and $\theta_0 = 0$ (Table 8). The initial condition on the GEO orbit is $x_0 = (-r_0 - \mu, 0, 0, v_0)$

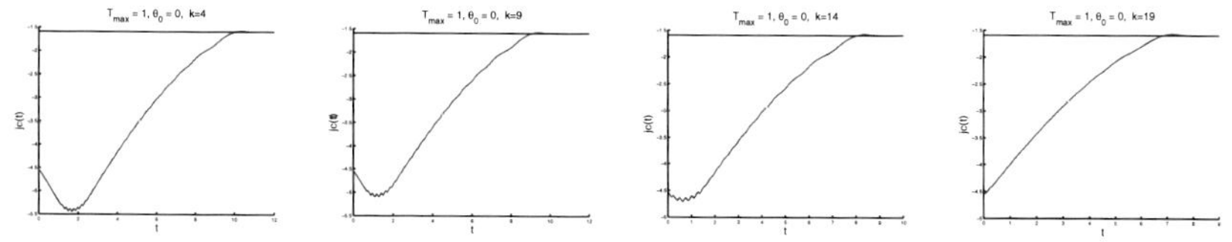


**Fig. 7** Variation of Jacobi constant of the local minima for $\varepsilon = 1\text{N}$ and $\theta_0 = 0$ (Table 8). The initial condition on the GEO orbit is $x_0 = (-r_0 - \mu, 0, 0, v_0)$

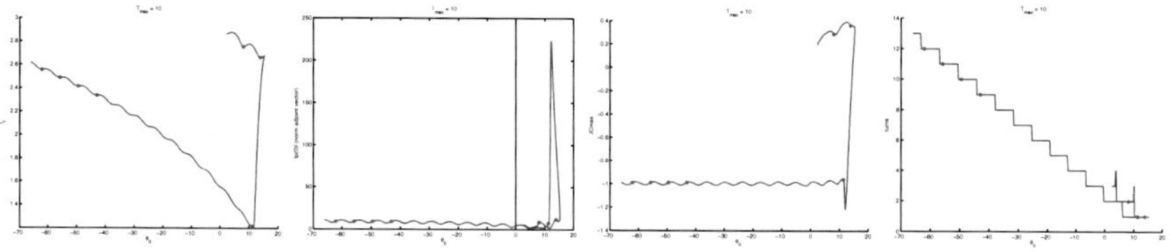

**Fig. 8** For $\varepsilon = 10\text{N}$. From *left* to *right* $\theta_0$ versus $t_f$, $\theta_0$ versus $|p(t_0)|$, $\theta_0$ versus $\max_{t\in[t_0:t_f]}(J_c(t))$, and $\theta_0$ versus $N_{ET}$

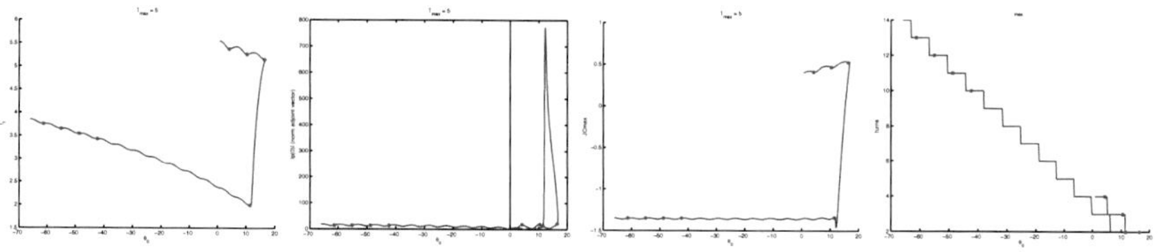

**Fig. 9** For $\varepsilon = 5\text{N}$. From *left* to *right* $\theta_0$ versus $t_f$, $\theta_0$ versus $|p(t_0)|$, $\theta_0$ versus $\max_{t\in[t_0:t_f]}(J_c(t))$, and $\theta_0$ versus $N_{ET}$

and 10 for $\varepsilon = 10\text{N}$, 5N and 1N respectively. In each figure we see: $t_f$ versus $\theta_0$, $|p(t_0)|$ versus $\theta_0$, $\max_{t\in[t_0,t_f]}(J_c(t))$ versus $\theta_0$ and $N_{ET}$ versus $\theta_0$. Notice that the number of turns decreases as we reach the global minima, which will display 1 turn for $\varepsilon = 10\text{N}$, 2 turns for $\varepsilon = 5\text{N}$ and 10 turns for $\varepsilon = 1\text{N}$. We observe that the maximum value reached by the Jacobi constant $J_c$ can be used to filter out solutions far away for what seems to be the global minimum (for this one, the value of $J_c$ is close to $J_c(L_1)$). Finally, looking at $|p(t_0)|$ we see a strong increase when the homotopic path moves from one type of trajectory to the other. We recall that the points in red in Figs. 8, 9 and 10 correspond to the local minima of the projection of the homotopic

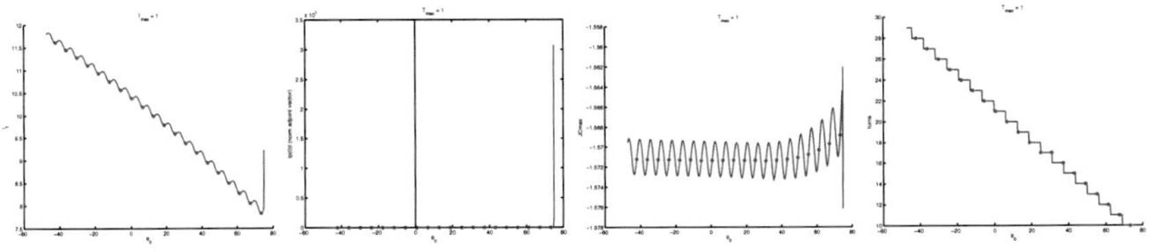

**Fig. 10** For $\varepsilon = 1$N. From *left* to *right* $\theta_0$ versus $t_f$, $\theta_0$ versus $|p(t_0)|$, $\theta_0$ versus $\max_{t\in[t_0:t_f]}(J_c(t))$, and $\theta_0$ versus $N_{ET}$

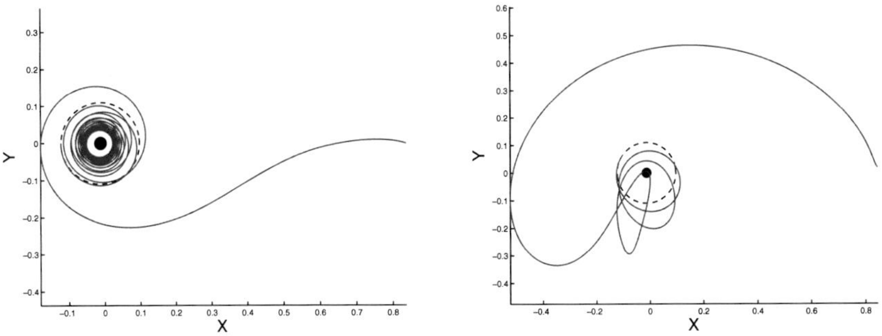


**Fig. 11** Transfer trajectories of the two ends of the homotopic curve $\theta_0$ versus $t_f$

path $\theta_0$ versus $t_f$. All these local minima are summarized in Tables 12, 13 and 14 for $\varepsilon = 10$N, 5N and 1N respectively.

Moreover, for $\varepsilon = 10$N, we have continued the two extremes of the homotopic curve w.r.t $\theta_0$ to see if there are other connections. We have seen that both extremes die when the associated transfer trajectory has a close encounter with the Earth. On one end of the curve we see type $\mathscr{T}_1$ trajectories where the transfer trajectory spirals towards the Earth and then outwards before a direct transfer to $L_1$. We think that if we continue decreasing $\theta_0$ in this direction, the trajectory will collide with the Earth. On the other end of the curve we see type $\mathscr{T}_2$ trajectories, where the transfer trajectory experiences a fast close approach with the Earth. In Fig. 11 we can see the two solutions at the two extremes of the path.

### 2.1.2 Homotopy w.r.t. $\varepsilon$

In this section, we have taken the different solutions for $\varepsilon = 10$N and a fixed initial condition $\theta_0(\text{mod } 2\pi) = 0, \pi/2, \pi$ and $-\pi/2$, summarized in Table 4, 5, 6 and 7. For each initial condition we perform a homotopy with respect to $\varepsilon$ from 10N to 1N. In Fig. 12 we summarize the results for the different values of $\theta_0(\text{mod } 2\pi)$. As we can see, for each fixed $\theta_0(\text{mod } 2\pi)$ the behaviour of the family of homotopic curves is very similar. For most local minima, $t_f$ increases as $\varepsilon$ decreases. In some of the cases, at some point the slope of $\varepsilon(t_f)$ experiences a drastic change and $t_f$

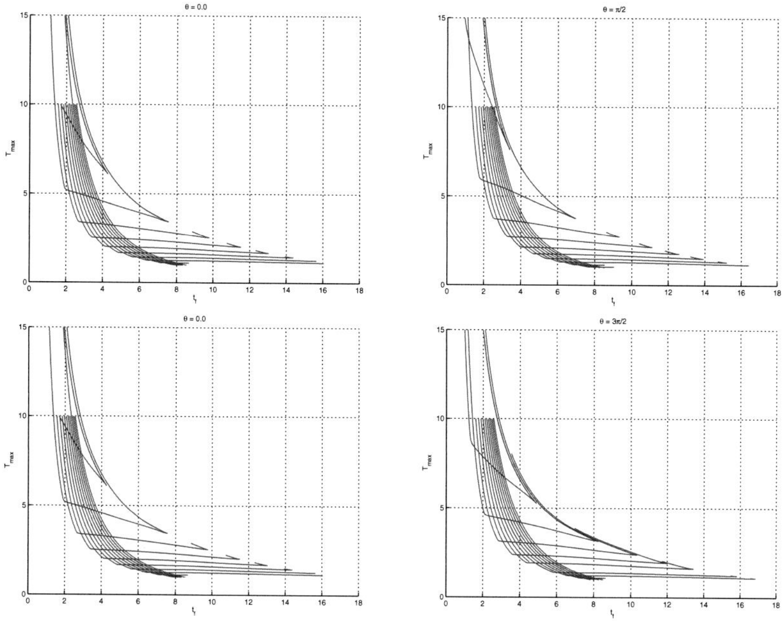


**Fig. 12** Homotopy with respect to $\varepsilon$ for $\theta_0$ fixed: $\theta_0 = 0$ (*top-left*), $\theta_0 = \pi/2$ (*top-right*), $\theta_0 = \pi$ (*bottom-left*), $\theta_0 = -\pi/2$ (*bottom-right*)

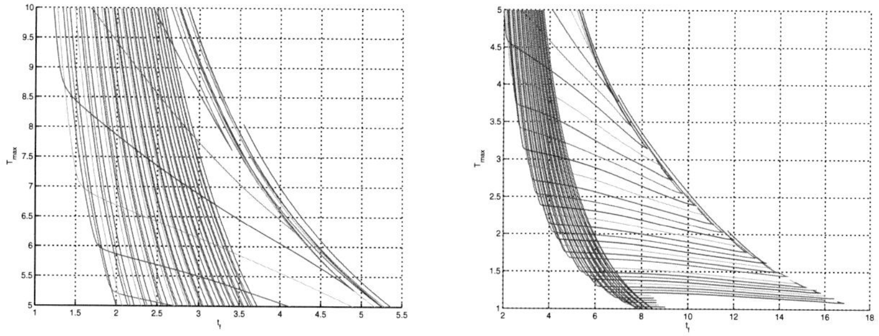


**Fig. 13** Homotopy with respect to $\varepsilon$ for $\theta_0$ fixed: $\theta_0 = 0$ (*magenta*), $\pi/2$ (*red*), $\pi$ (*green*), $-\pi/2$ (*blue*). *Left* zoom for $\varepsilon \in [5:10]$, *Right* zoom for $\varepsilon \in [1:5]$

increases very quickly for small variations of $\varepsilon$. Then at some point the homotopy curve has a turning point and $\varepsilon$ starts to grow, and we are not able to reach lower values for $\varepsilon$. Nevertheless, there are other cases where $\varepsilon$ just decreases and reaches $\varepsilon = 1$N with no drastic changes on the curve. We believe that for these last cases a similar behaviour will be observed for $\varepsilon < 1$N. In Fig. 13 we summarize all the local minima for $\theta_0 (\mathrm{mod}\ 2\pi) = 0$ (magenta), $\pi/2$ (red), $\pi$ (green) and $-\pi/2$ (blue). For $\varepsilon \in [5:10]$ (left) and $\varepsilon \in [1:5]$ (right). In Fig. 14 we show the variation of the type of solutions along time for one of these families ($\theta_0 = 0, k = 2$ in Table 4),

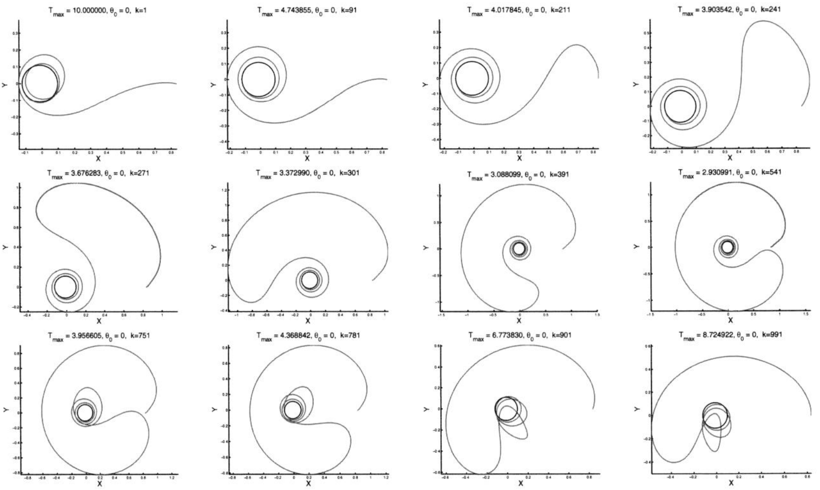


**Fig. 14** Transfer trajectories from GEO to $L_1$ for $\theta_0$ fixed and different $\varepsilon$. Here $\theta_0 = 0$ and all the orbits belong to the homotopic curve generated by $k = 2$ from Table 4 when we use $\varepsilon$ as the homotopy parameter

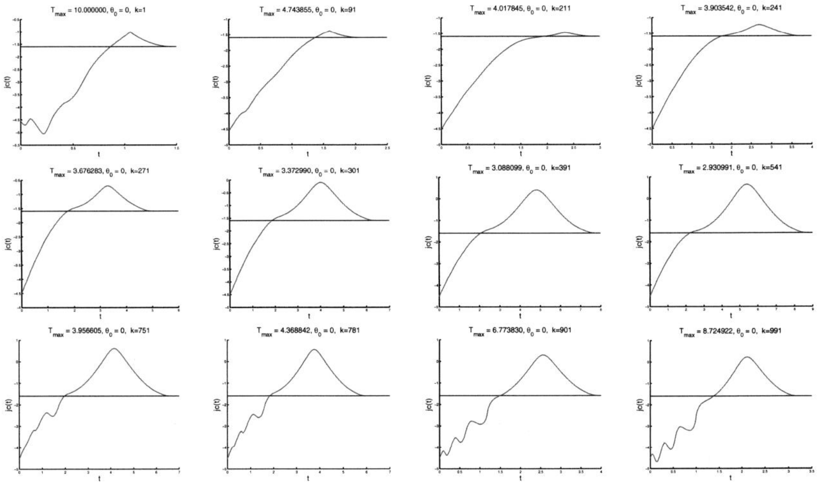


**Fig. 15** Variation of the Jacobi constant along time for a transfer trajectories from GEO to $L_1$ for $\theta_0$ fixed and different $\varepsilon$. Here $\theta_0 = 0$ and all the orbits belong to the homotopic curve generated by $k = 2$ from Table 4 when we use $\varepsilon$ as the homotopy parameter

which corresponds to one of the homotopy paths where $\varepsilon(t_f)$ experiences two drastic changes. As we can appreciate, the orbits on the first part of the homotopic curve are type $\mathscr{T}_1$. When the slope of the path experiences its first drastic change, the transfer trajectories start to do big excursions on the phase space before heading towards $L_1$ and we begin to observe type $\mathscr{T}_2$ trajectories. Eventually, when $\varepsilon$ starts to grow the

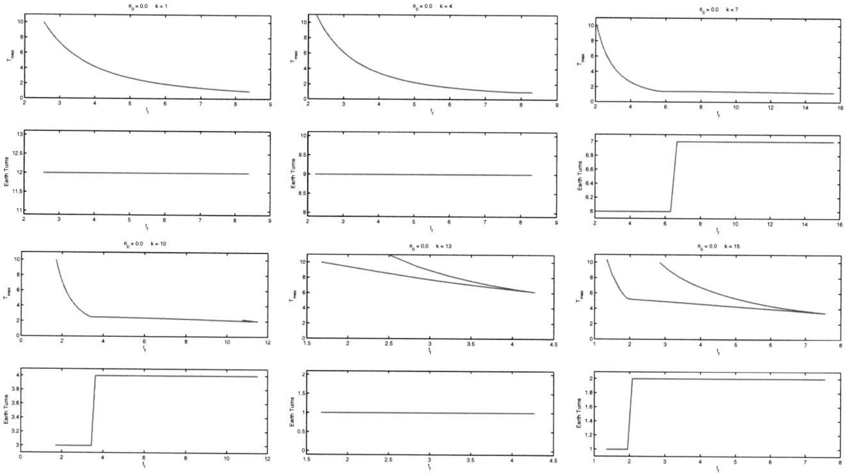


**Fig. 16** For the local minima for $\theta_0 = 0$ and $\varepsilon = 10$ ($k = 1, 4, 7, 10, 13, 15$): homotopic curve $\varepsilon$ versus $t_f$ (subplots on the *top*) and $N_{ET}$ versus $t_f$ (subplots on the *bottom*)

trajectories remain of type $\mathscr{T}_2$ and some of them experience close approaches with the Earth. Hence, these paths connect type $\mathscr{T}_1$ transfer trajectories with type $\mathscr{T}_2$. Finally, in Fig. 15 the variation of the Jacobi constant is displayed with respect to time for the trajectories in Fig. 14.

Finally, we think it is worth saying that for a fixed $\theta_0$, along each of the different homotopic paths that we have generated by varying $\varepsilon$ summarized in Fig. 12, the number of turns the trajectory gives around the Earth, $N_{ET}$, is kept constant before the first drastic change on the homotopic paths slope. There the number of turns increases in 1 and remains constant along that path, even when $\varepsilon$ starts to increase. This phenomena can be seen in Fig. 16, where we plot some of the homotopic curves, the variation of the number of turns around the Earth versus $t_f$. In order to understand better this phenomenon, for each curve we have plotted on top the corresponding homotopic curve ($t_f$ vs. $\varepsilon$) and on the bottom ($t_f$ vs. $N_{ET}$). So if we assume that for a given $\varepsilon^*$ there is a minimum number of turns around the Earth that the trajectory must give before $J_c(x(t)) > J_c(L_1)$, which will allow the trajectory to reach $L_1$, this gives us a criteria to chose one of the solutions for $\varepsilon = 10$N and get to $\varepsilon^*$ by following its homotopic path w.r.t. $\varepsilon$.

## 2.2 *Free Initial Point on a GEO*

In this section we discuss the results for a minimum-time transfer from a GEO orbit to $L_1$. The main difference with respect to the previous section is that here we just impose the initial condition to be on a GEO orbit but we do not fix the position on it. Hence, the boundary conditions are:

$$x(t_0) \in \mathscr{M}_0, \quad p(t_0) \perp T_{x(t_0)}\mathscr{M}_0, \quad x(t_f) - x_f = 0, \quad h(t_f) = 0, \tag{5}$$

where $\mathscr{M}_0$ represents the GEO orbit. The first two boundary conditions are written as

$$\begin{aligned} (x_0 + \mu)^2 + y_0^2 - r_0^2 &= 0, \\ \dot{x}_0^2 + \dot{y}_0^2 - v_0^2 &= 0, \\ (x_0 + \mu)\dot{x}_0 + y_0\dot{y}_0 &= 0, \\ (x_0 + \mu)p_{y0} - y_0 p_{x0} + v x_0 p_{\dot{y}0} - v y_0 p_{\dot{x}0} &= 0, \end{aligned} \tag{6}$$

where $r_0$ is the radius of the GEO orbit and $v_0$ is the corresponding velocity; $x(t_0) = (x_0, y_0, \dot{x}_0, \dot{y}_0)$ is the coordinate vector of the spacecraft and $p(t_0) = (p_{x0}, p_{y0}, p_{\dot{x}0}, p_{\dot{y}0})$ is the adjoint vector, both evaluated at $t_0$. The other boundary conditions are the same as in Sect. 2.1. We recall that all the solutions found in Sect. 2.1, where we fix the initial condition on the GEO orbit, are not necessarily solutions of this more general problem. Here the boundary conditions are more restrictive as we impose the transversality condition on $p(t_0)$. Only the local minima in the projection $(\theta_0, t_f)$ of the homotopic paths in Figs. 1, 2 and 3 will be good initial conditions to be local minima of this problem. The values for $(x_0, p_0)$ are summarized in Tables 12, 13 and 14 for $\varepsilon = 10$N, $5N$ and 1N, respectively. In this section we have taken some of the local minima for $\varepsilon = 10$N and for each one we have performed homotopies with respect to the thrust magnitude $\varepsilon \in [1 : 10]$N. The initial conditions corresponding to the local minima from Figs. 1 and 3 in Sect. 2.1 are summarized in Tables 12 and 14.

On the left-hand side of Fig. 17 we show the continuation curve for the local minima number 4, 5, 6 and 7 for $\varepsilon = 10$N from Table 12. Minima 4 and 5 are $\mathscr{T}_1$ type solutions and the homotopic curves are plotted in magenta and blue respectively. Minima 6 and 7 are $\mathscr{T}_2$ solutions and the homotopic curves are plotted in red and green respectively. Notice that the homotopic paths with respect to $\varepsilon$ do not connect the two type of solutions $\mathscr{T}_1$ and $\mathscr{T}_2$. On the right-hand side of Fig. 17 we compare

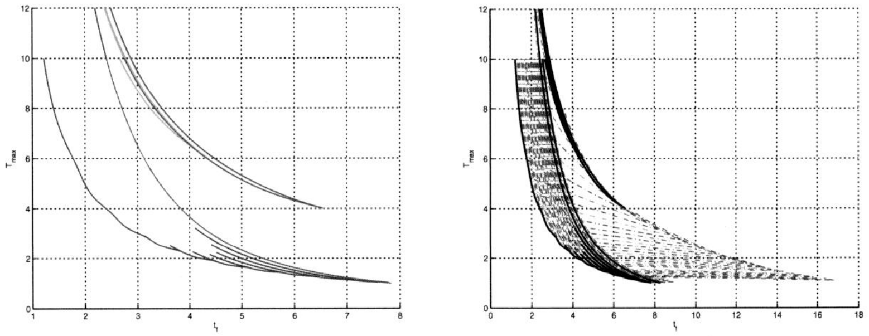

**Fig. 17** Homotopy w.r.t $\varepsilon$ for a transfer orbit from GEO to $L_1$. *Left* the initial condition on the orbit is considered free. *Right* comparison between letting the initial condition free (*black line*) and fixing it to $\theta_0 = 0, \pi/2, \pi, -\pi/2$

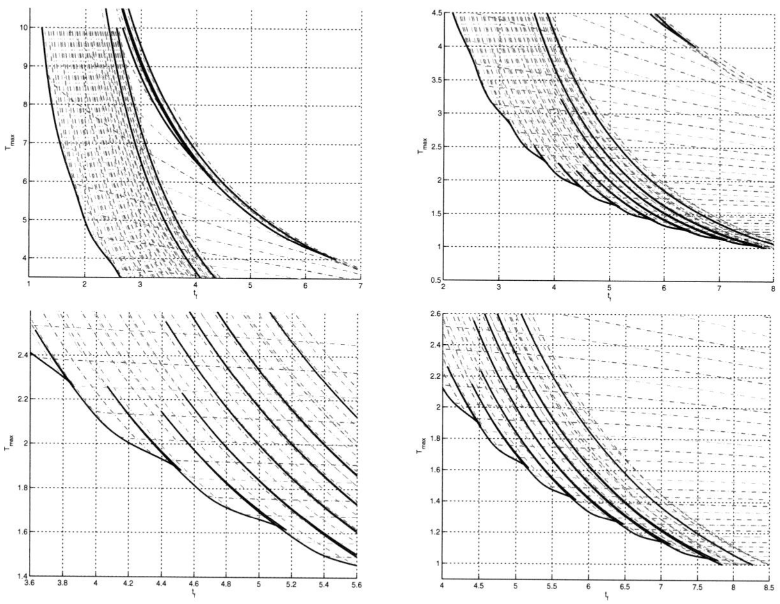


**Fig. 18** Homotopy w.r.t $\varepsilon$: Comparison between letting the initial condition free (*black line*) and fixing it to $\theta_0 = 0, \pi/2, \pi, -\pi/2$

these homotopic curves (now plotted in black) with the homotopic curves for $\theta_0$ fixed. The curves for $\theta_0 = 0, \pi/2, \pi$ and $-\pi/2$ are plotted in blue, magenta, green and red respectively. In Fig. 18 we have plotted different zooms of these last plot. Notice that for $\varepsilon < 3$N the projection of the homotopic path in the $\varepsilon$–$t_f$ plane self-intersects several times. Hence, we have at least two different solutions with the same cost. These are candidates to be cut points. We will describe them in more detail in the next section (Tables 15, 16, 17 and 18).

Finally, in Fig. 19 we show information on the orbital parameters for one of the continuation curves, the one corresponding to the candidate to "global" minima. On the top left hand-side we show $\theta_0$ the argument of the initial condition on the GEO orbit versus $\varepsilon$, and on the top right hand-side we show $\theta_0 (\text{mod } 2\pi)$ verus $\varepsilon$. On the bottom left hand-side we show $\theta_0$ versus the number of turns around the Earth, and on the bottom right hand-side we show $\theta_0$ versus the norm of the adjoint vector for $t_0 = 0$, $|p(t_0)|$. As we can see, $\theta_0$ can be used to parameterize the homotopic curve.

## 2.3 Cut Points

As we can appreciate in Fig. 18, the homotopic path for the GEO to $L_1$ transfer with $\theta_0$ free has several turning points and the path self-intersects several times. These intersections are cut point candidates. We recall that, in optimal control, a *cut point*

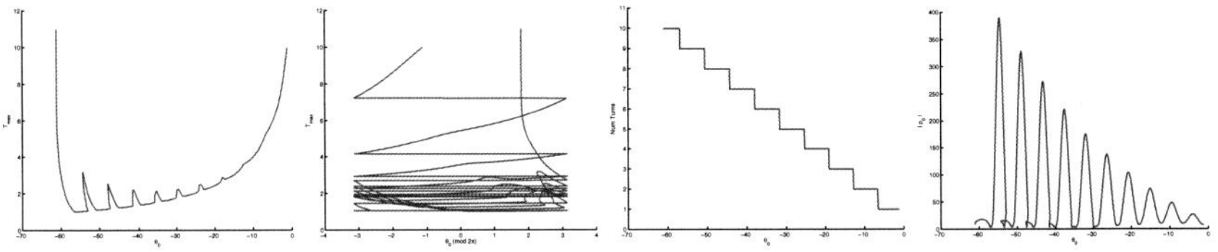

**Fig. 19** Homotopy of the candidate for global minimum with respect to $\varepsilon$. From *left* to *right* $\varepsilon$ versus $\theta_0$ the argument on the GEO orbit, $\varepsilon$ versus $\theta_0 (\text{mod } 2\pi)$, $N_{ET}$ versus $\theta_0$, $|p(t_0)|$ versus $\theta_0$

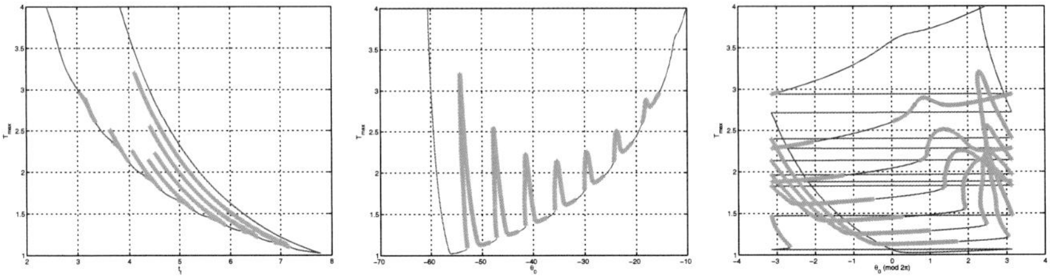

**Fig. 20** Homotopy of the local minima. *Left* $\varepsilon$ versus $t_f$, *Centre* $\varepsilon$ versus $\theta_0$, *Right*: $\varepsilon$ versus $\theta_0$ (mod $2\pi$). The *green* points correspond in both cases the neighbourhood of the cut points ($t_f \in [t_f cut - 0.15, t_f cut + 0.15]$)

is the first point in the extremal where the solution ceases to be globally minimizing. Typically, two different extremals with the same cost are candidates for defining such a point. From now on we call the couple $\{(x_0, p_0), (x_1, p_1)\}$ a cut point if they are two different initial conditions such that for the same $\varepsilon^*$ they generate two local optimal solutions for the GEO to $L_1$ transfer problem with the same transfer time $t_f^*$. We find these initial conditions on the self-intersections of the $\varepsilon$ versus $t_f$ projection of homotopic path in Fig. 17. To fix notation, $(x_0, p_0)$ will be the "first" point on the homotopic path that reaches $(\varepsilon^*, t_f^*)$ and $(x_1, p_1)$ will be the "second" point to reach $(\varepsilon^*, t_f^*)$. In Table 3 we summarize all the cut points that we have found for $\varepsilon \in [1, 10]$. We have computed these points by refining the intersections found in Fig. 18. In Fig. 20 we plot the different projections of the homotopy path that we have already seen, where we have highlighted in green the solutions close to the cut points. On the left hand side of Fig. 20 we have the $t_f$ versus $\varepsilon$ projection, on the centre we have the $\theta_0$ versus $\varepsilon$ projection and on the right hand side we have $\theta_0$ mod $2\pi$ versus $\varepsilon$ projection.

For all seven cut points we have done the same analysis. First of all we have taken both points and computed the transfer trajectory, the variation of $J_c(t)$, the control law $\mathbf{u}(t)$ and $(H_1(t), H_2(t))$. We have also integrated both trajectories backwards and forward in time covering the time range $[-t_f, 2t_f]$, where $t_f$ is the transfer time. As we will see, the main difference between both solutions appears when we look at the integration of the trajectory backwards in time $t \in [-t_f, 0]$ and on the control $u(t)$ at the beginning of the transfer trajectory. Secondly, we have taken a neighbourhood of the cut point and checked the variation of different orbital parameters for the optimal

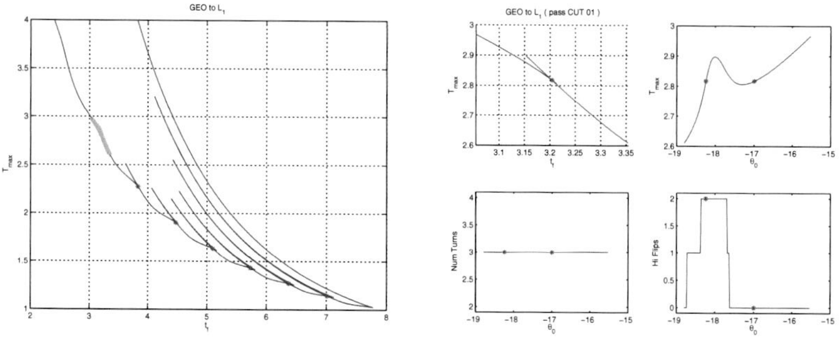


**Fig. 21** For cut point $n^o$ 1: *Left* $t_f$ versus $\varepsilon$ homotopic curve with highlight of the cut passage in *green*; *Right* analysis of the cut passage: (*top-left* subplot) $t_f$ versus $\varepsilon$ zoom, (*top-right* subplot) $\theta_0$ versus $\varepsilon$, (*bottom-left* subplot) $\theta_0$ versus $N_{ET}$, (*bottom-right* subplot) $\theta_0$ versus $N_{ZH}$. *Red* points are values corresponding the each cut point

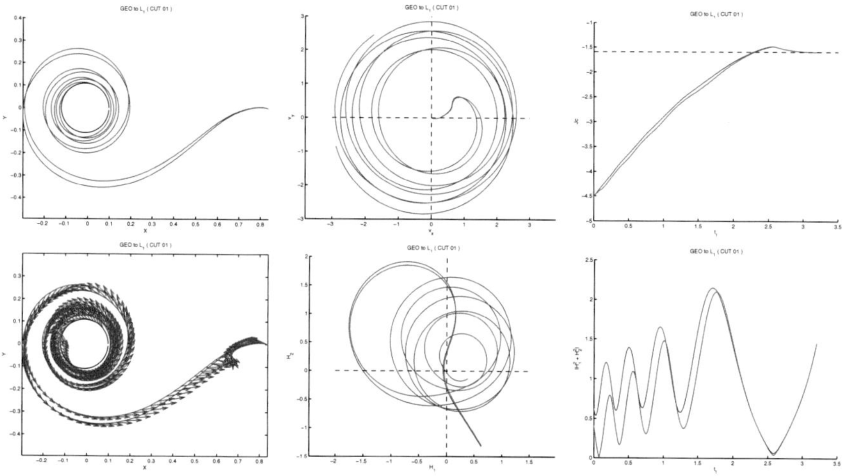


**Fig. 22** For cut point $n^o$ 1: *blue* orbits correspond to the first cut value and *red* orbits to the second cut value. *Top-Left* $\{XY\}$ projection of the transfer trajectory, *Top-Centre* $\{\dot{X}\dot{Y}\}$ projection of the transfer trajectory, *Top-Right* $t$ versus $J_c$ (energy variation along the transfer trajectory), *Bottom-Left* control along the trajectory, *Bottom-Centre* $H_1$ versus $H_2$, *Bottom-Right* $t$ versus $|(H_1, H_2)|$

transfer orbits on the homotopic path. For all cut points, if $\varepsilon^*$, $t_f^*$ is the value of the thrust magnitude and transfer time at the cut point, we analyse the solutions that are close to the cut point, i.e. $t_f \in [t_f^* - 0.15 : t_f^* + 0.15]$. The plots corresponding to these simulations are summarized in Appendix "Summary of the cut points on the GEO to $L_1$ transfer" here we will only describe the results for the cut points number 1 and 5 in Table 3, where Figs. 21, 22 and 23 correspond to the 1st cut point and Figs. 24, 25 and 26 correspond to the 5th cut point. The first cup point occurs at $\varepsilon^* \approx 2.8177314$N and $t_f^* \approx 3.2044271$. On the left-hand side of Fig. 21, we have plotted the whole homotopic curve and highlighted in green the first cut region, i.e.

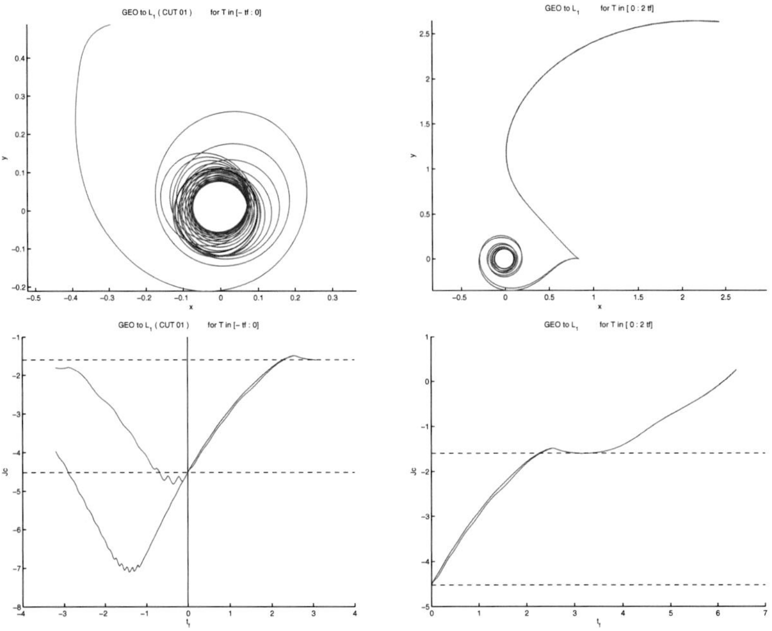


**Fig. 23** For cut point $n^o$ 1: *Left* optimal solutions for $t \in [-t_f, 0]$ ($XY$ projection and $J_c$ variation), *Right* optimal solutions for $t \in [0, 2t_f]$ ($XY$ projection and $J_c$ variation)

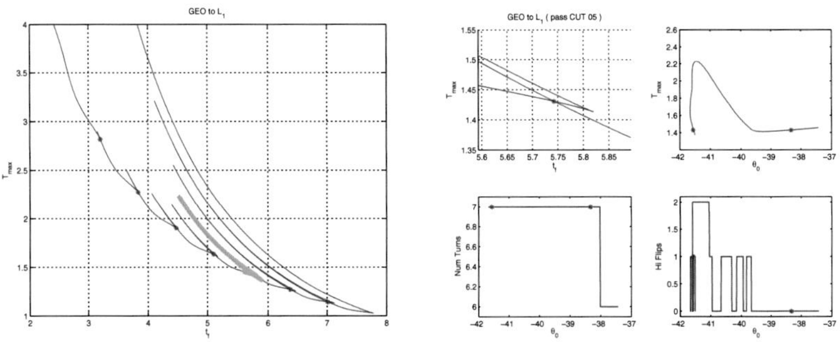


**Fig. 24** For cut point $n^o$ 5: *Left* $t_f$ versus $\varepsilon$ homotopic curve with highlight of the cut passage in *green*; Right analysis of the cut passage: (*top-left* subplot) $t_f$ versus $\varepsilon$ zoom, (*top-right* subplot) $\theta_0$ versus $\varepsilon$, (*bottom-left* subplot) $\theta_0$ versus $N_{ET}$, (*bottom-right* subplot) $\theta_0$ versus $N_{ZH}$. *Red* points are values corresponding the each cut point

solutions with a transfer time between $[t_f - 0.15 : t_f + 0.15]$. On the right-hand side we have 4 subplots, the two subplots on the top are a zoom of the cut region $t_f$ versus $\varepsilon$ (left), and the $\theta_0$ versus $\varepsilon$ projection of the cut region (right). The two subplots on the bottom show $\theta_0$ versus $N_{ET}$, the number of turns around the Earth before going towards $L_1$ (left), and $\theta_0$ versus $N_{ZH}$, the number of times $|(H_1, H_2)|$ get close to

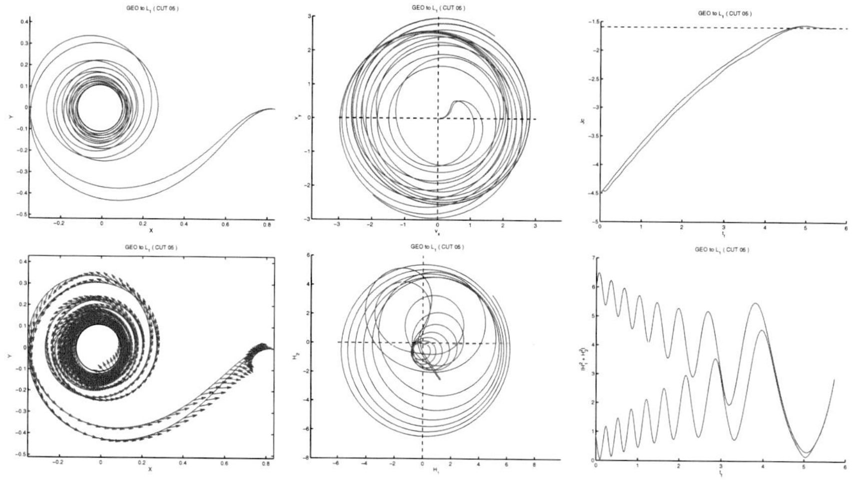

**Fig. 25** For cut point $n^o$ 5: *blue* orbits correspond to the first cut value and *red* orbits to the second cut value. *Top-Left* $\{XY\}$ projection of the transfer trajectory, *Top-Centre* $\{\dot{X}\dot{Y}\}$ projection of the transfer trajectory, *Top-Right* $t$ versus $J_c$ (energy variation along the transfer trajectory), *Bottom-Left* control along the trajectory, *Bottom-Centre* $H_1$ versus $H_2$, *Bottom-Right* $t$ versus $|(H_1, H_2)|$

zero (right). As we can see, the number of turns around the Earth remains constant $N_{ET} = 3$, but $N_{ZH} = 0$ for the first cut point and 2 for the second cut point. In Fig. 22 we show different aspects of both transfer trajectories. The curves in blue correspond to the first cut point, $(x_0, p_0)$, and the curves in red to the second cut point, $(x_1, p_1)$. The three plots on the top, from left to right correspond to the $\{x, y\}$ projection on the transfer trajectory, to the $\{\dot{x}, \dot{y}\}$ projection of the transfer trajectory, and to the evolution of $J_c$ along the transfer trajectory. The three plots on the bottom, from left to right correspond to the $\{x, y\}$ projection of the trajectory and the control law $\mathbf{u}(t)$, to the projection of $(H_1, H_2)$, and to the variation of $|(H_1, H_2)|$ along time. In these plots we can see that the main difference between the two transfer trajectories lies on the control law. The control $\mathbf{u}(t)$ for the second cut point (red curve) has a drastic change of its orientation at the beginning of the transfer trajectory. This translates on $(H_1, H_2)$ passing close to zero—possible discontinuity of the control. This effect is not observed for the first cut point (blue curve). In Fig. 23 we see for both cut points the integration backwards in time ($t \in [-t_f : 0]$) and forward in time ($t \in [0 : 2t_f]$), as well as the corresponding variation of $J_c(t)$. Again, the curves in blue are related to the first cut point and the curves in red to the second cut point. As we can see, there is practically no qualitative difference for the evolution of both transfer trajectories if we integrate forward in time. While we do see a difference between the evolution backwards in time. Notice that the first cut point (blue curve) spirals towards the Earth and $J_c(t)$ decreases drastically, while the second cut point (red curve) spirals outwards and $J_c(t)$ will quickly start to grow.

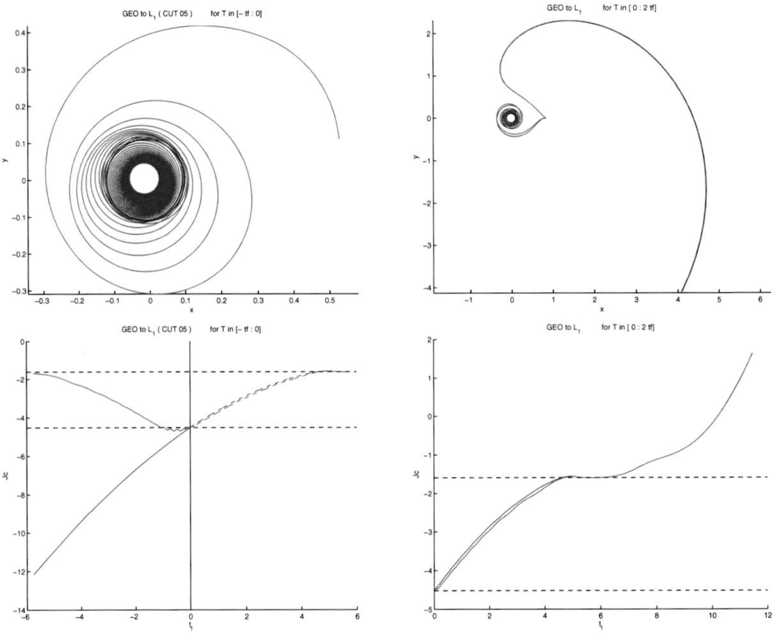


**Fig. 26** For cut point $n^o$ 5: *Left* optimal solutions for $t \in [-t_f, 0]$ ($XY$ projection and $J_c$ variation), *Right* optimal solutions for $t \in [0, 2t_f]$ ($XY$ projection and $J_c$ variation)

Cut point number 5 corresponds to $\varepsilon^* \approx 1.4309308$N and $t_f^* \approx 5.7431841$. On the left hand-side of Fig. 24, we have highlighted in green the cut region. On the right-hand side we also have 4 subplots with a zoom of this cut region and the variation of $N_{ET}$ and $N_{ZH}$ for the different points on the curve. As we can see the main difference between the cut points is the $N_{ZH}$ that is 0 for the first cut point and 1 for the second cut point. In Fig. 25 we compare the transfer orbits of the two cut points $(x_0, p_0)$ (blue curve) and $(x_1, p_1)$ (red curve). Where we show the $\{x, y\}$ and $\{\dot{x}, \dot{y}\}$ projections of the trajectory and $J_c(t)$. We also show the evolution of the control along the trajectory and $(H_1, H_2)$. As it happened for the first cut point, the main difference between both trajectories is on the control at the beginning of the transfer trajectory, where the red curve experiences a drastic change on the direction of $\mathbf{u}(t)$ opposed to a smooth behaviour of $\mathbf{u}(t)$ for the red curve. This can be seen as a close approach of $(H_1, H_2)$ to zero. Finally in Fig. 23 we show the integration backwards and forward in time for the two cut points, and the corresponding variation of $J_c(t)$. As it happened for cut point number 1, there is no qualitative difference for the evolution forward in time. While for the integration backward in time we have the same behaviour as bellow, the first cut point (blue curve) spirals inwards towards the Earth, while the second cut point (red curve) spirals outwards.

As previously indicated, the results for the other cut points are summarized in Appendix "Summary of the cut points on the GEO to $L_1$ transfer". As we can see there is a pattern that repeats for all the cut points that might be useful to detect this phenomena. The main difference between the two solutions appears on the control law at the beginning on the transfer trajectory: where the red curve (corresponding to the second cut point) always experiences a drastic change on its orientation, which is related to $(H_1, H_2)$ passing close to zero, and the blue curve (corresponding to the first cut point) has a smooth behaviour along the first phase of the transfer trajectory. The other difference between both solutions appears when we integrate them backwards in time $t \in [-t_f : 0]$, where one solution spirals towards the Earth and the other outwards.

# 3 Transfer from a GEO to MO

In this section we will focus on the transfer from a GEO to a Moon orbit (MO). Throughout the section we will do a similar analysis to the one done in Sect. 2 for the GEO to $L_1$ transfer. We will also use indirect shooting methods based on Pontryagin maximum principle and the package `hampath` to find different local minima. First we will focus on the two point boundary value problem where the initial condition on the GEO orbit is fixed and perform homotopies with respect to (a) the position on the GEO orbit and (b) the thrust magnitude $\varepsilon$. Second, we will focus on the boundary value problem where the initial condition on the GEO orbit is free. For all these explorations the position on the arrival Moon orbit is free.

## *3.1 Fixed Initial Point on a GEO*

Here we summarize the results for a minimum-time transfer from a GEO to a MO, where the position on the GEO orbit is fixed, hence the Boundary Conditions are:

$$x(t_0) - x_0 = 0, \quad x(t_f) \in \mathcal{M}_1, \quad p(t_f) \perp T_{x(t_f)}\mathcal{M}_1, \quad h(t_f) = 0, \tag{7}$$

where as before, $x(t)$ is the position and velocity of the spacecraft at time $t$; $p(t)$ is the adjoint vector at time $t$; $x_0$ is a fixed initial condition on a GEO orbit; $\mathcal{M}_1$ represents the desired Moon orbit; and $h(t_f)$ is the Hamiltonian of the PMP evaluated at the final point. The two point boundary conditions for the arrival point, $x(t_f)$ and $p(t_f)$, can be written as

$$\begin{aligned} (x_f + \mu - 1)^2 + y_f^2 - r_1^2 &= 0, \\ \dot{x}_f^2 + \dot{y}_f^2 - v_1^2 &= 0, \\ (x_f + \mu - 1)\dot{x}_f + y_f \dot{y}_f &= 0, \\ (x_f + \mu - 1)p_{yf} - y_f p_{xf} + \dot{x}_f p_{\dot{y}f} - \dot{y}_f p_{\dot{x}f} &= 0, \end{aligned} \tag{8}$$

where $r_1$ is the radius of the arrival Moon orbit and $v_1 = \sqrt{\mu/r_1}$ the corresponding velocity for the circular orbit, $x(t_f) = (x_f, y_f, \dot{x}_f, \dot{y}_f)$ are the coordinates of the spacecraft and $p(t_f) = (p_{xf}, p_{yf}, p_{\dot{x}f}, p_{\dot{y}f})$ is the adjoint vector, both evaluated at $t_f$, As in Sect. 2.1, we parameterize the position of the spacecraft on the initial GEO orbit using its radius $r_0$ and the angle $\theta_0 \in [-\pi, \pi]$. Hence, $x_0 = (r_0 \cos\theta_0 - \mu, r_0 \sin\theta_0, -v_0 \sin\theta_0, v_0 \cos\theta_0)$, with $v_0 = \sqrt{(1-\mu)/r_0}$. We will also use: $r_0 = 0.109689855932071$ and the corresponding $v_0 = 3.000969693845573$ for a GEO orbit. For the arrival Moon orbit, we consider $r_1 = 0.034$ and $v_1 = \sqrt{\mu/r_1} = 0.59786$. First of all we need a good initial condition to start exploring type of solutions. We have considered one of the local minima for the GEO to $L_1$ transfer found in Sect. 2.1: $\{t_f, (x_0, p_0)\}$, and use it as initial guess for the GEO to MO problem. For the initial condition to converge we use a slightly larger transfer time as initial guess, i.e. $\hat{t}_f = t_f + h$. We have considered for $\varepsilon = 10$N:

$$\begin{cases} t_f = 1.483385683993085, \\ x_0 = (r_0 - \mu,\ 0.0,\ 0.0,\ -v_0), \\ p_0 = (3.8349336494018,\ 1.7266950508752,\ 0.0764256922941,\ 0.1329597699146), \end{cases}$$

which corresponds to the transfer orbit shown on Fig. 27. Using this as initial guess and taking as transfer time $\hat{t}_f = t_f + h$ for $h = 5 \cdot 10^{-3}, 5 \cdot 10^{-2}$ and $10^{-1}$ we have found three different classes of minimum time transfer solutions from GEO to MO.

1. For $h = 5 \cdot 10^{-3}$, transfer orbit on Fig. 28 left (blue curve):

$$\begin{aligned} \hat{t}_f &= 1.562091470465241, \\ \hat{x}_0 &= (-0.1218428559320,\ 0.0000000000000,\ 0.0000000000000,\ -3.0009696938455), \\ \hat{p}_0 &= (3.9285981330000,\ 1.6544784563269,\ 0.0734194613434,\ 0.1400883421876), \end{aligned}$$

2. For $h = 5 \cdot 10^{-2}$, transfer orbit on Fig. 28 center (red curve):

$$\begin{aligned} \hat{t}_f &= 1.529347472081999, \\ \hat{x}_0 &= (-0.1218428559320,\ 0.0000000000000,\ 0.0000000000000,\ -3.0009696938455), \\ \hat{p}_0 &= (3.9217580218250,\ 1.6961685952114,\ 0.0750545446749,\ 0.1396207935397), \end{aligned}$$

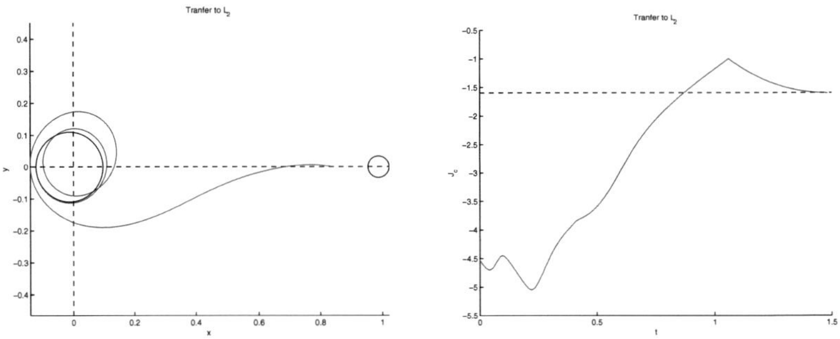


**Fig. 27** Transfer orbit from GEO to $L_1$ that we use as initial condition for the GEO to MO transfer problem ($\varepsilon = 10$N). *Left* $\{XY\}$ projection of the trajectory, *Right* $J_c$ variation

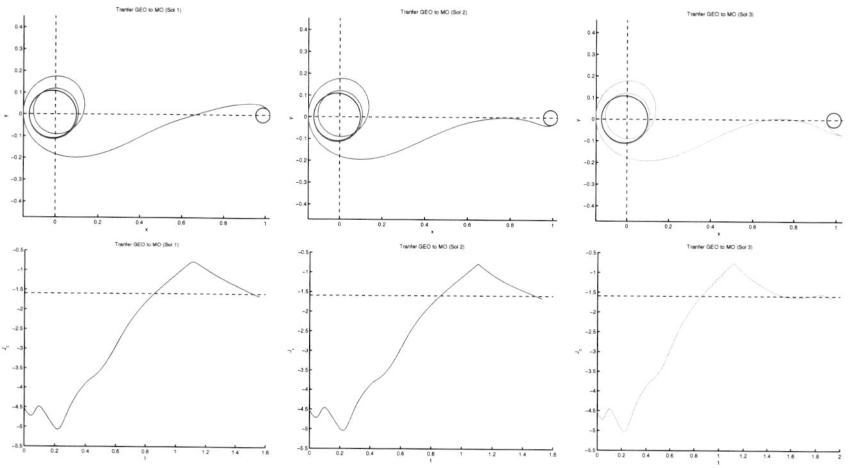

**Fig. 28** Transfer orbits form GEO to MO. Three different solutions found from the initial orbit from Fig. 27. *Top*: $\{XY\}$ projection of the trajectory, *Bottom* $J_c$ variation

3. For $h = 10^{-1}$, transfer orbit on Fig. 28 right (green curve):

$$\begin{aligned}\hat{t}_f &= 1.939506458073425,\\ \hat{x}_0 &= (-0.1218428559320,\ 0.0000000000000,\ 0.0000000000000,\ -3.0009696938455),\\ \hat{p}_0 &= (3.9733475979444,\ 1.7003282465180,\ 0.0751188225176,\ \ 0.1431121866550),\end{aligned}$$

Note that for all transfer orbits in Fig. 28 the first phase of the transfer trajectory (orbiting around the Earth) is the same. The difference is seen on the second part, where we find two orbits that arrives to the Moon following a clockwise sense around the Moon (blue and green orbits) and another orbit following an anticlockwise sense (red orbit). Although we find two orbits that arrive to the Moon in a clockwise sense, there is a big difference in the transfer time, the green orbit taking much more time than the blue one. Moreover, the green orbit starts by approaching the Moon with an anticlockwise orbit, then a cusp occurs (the velocity in the moving frame vanishes) and the end of the trajectory winds again clockwise around the target. Finally, if we look at the control law that produces these three transfer orbits (Fig. 29) we see how these one is very similar to the first part of the transfer while a big difference appears when we approach the $L_1$ neighborhood. There we see how, different ways to decelerate the growth in energy produce different outputs (i.e. transfer orbits). To fix notation, from now on we will call: $\mathscr{C}_1$ the transfer trajectories that arrive to the MO in a clockwise sense, (identified throughout this section by the color blue); $\mathscr{C}_2$ the transfer trajectories that arrive to MO anti-clockwise (identified by the color red); finally $\mathscr{C}_3$ transfer orbits similar to orbit 3 (green), possibly with one cusp before capture by the Moon.

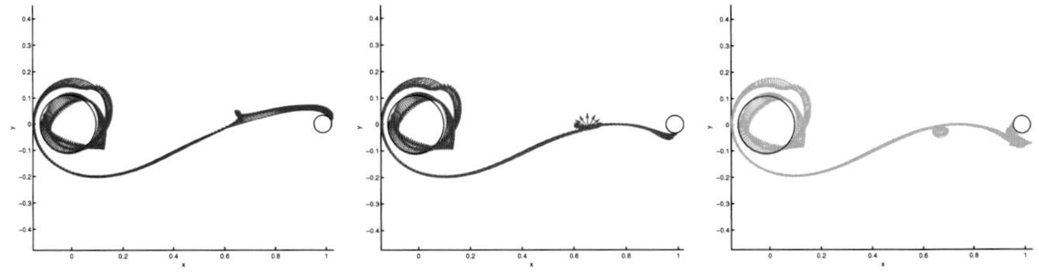

**Fig. 29** Control Law for the transfer orbits form GEO to MO. Three different solutions found from the initial orbit from Fig. 27

### 3.1.1 Homotopy w.r.t. $\theta_0$

Here we have taken the three local minimum time transfer trajectories that appear in Fig. 28 (type $\mathscr{C}_1$, $\mathscr{C}_2$ and $\mathscr{C}_3$), and as in Sect. 2, we have done an homotopy with respect to the initial position on the GEO orbit, $\theta_0$. All three initial orbits are for $\varepsilon = 10$N and $\theta_0 = \pi$, we recall that the initial position on the GEO orbit is given by:

$$x_0 = (r_0 \cos\theta_0 - \mu, r_0 \sin\theta_0, -v_0 \cos\theta_0, v_0 \sin\theta_0).$$

To do this homotopy we proceed as we did in Sect. 2 and compute (if possible) the homotopic path for $\theta = \pi \mapsto 21\pi$ and $\theta = \pi \mapsto -21\pi$, taking small intervals of size $2\pi$ to increase the precision. In Figs. 30, 31 and 32 we see the projection of these curves in the $\theta_0$ versus $t_f$ space. The points in black on the 3 curves are the local minima of these curves, and will be candidates for local minimum time transfer trajectories when the initial condition on the GEO orbit is not fixed (Sect. 3.2). The initial conditions for these local minima are summarized in Tables 27, 28 and 29.

As we can see, the behavior of these three curves has similarities to the results for the GEO to $L_1$ transfer for $\varepsilon = 10$. Notice how as $\theta_0$ increases so does the transfer time, $t_f$, and as $\theta_0$ decreases $t_f$ decreases up to a certain value $\theta_0^*$, there $t_f$ will start to increase drastically for small variations of $\theta_0$ up to some point where the

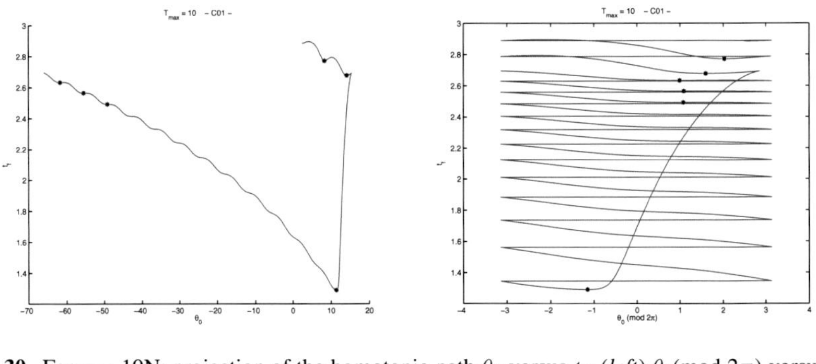

**Fig. 30** For $\varepsilon = 10$N: projection of the homotopic path $\theta_0$ versus $t_f$ (*left*) $\theta_0 (\mathrm{mod}\ 2\pi)$ versus $t_f$ (*right*) for GEO to MO transfer trajectories of type $\mathscr{C}_1$

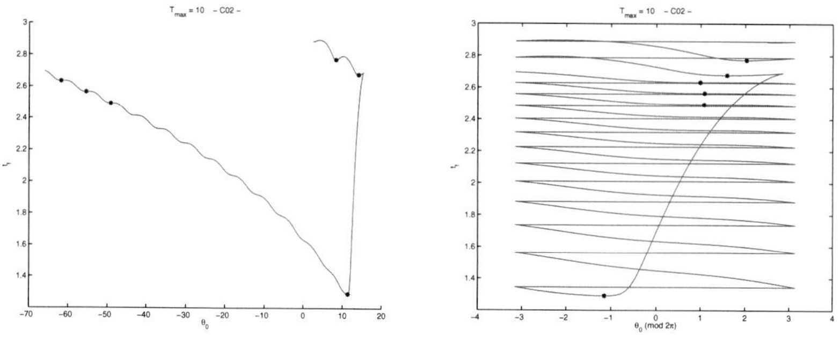


**Fig. 31** For $\varepsilon = 10$N: projection of the homotopic path $\theta_0$ versus $t_f$ (*left*) $\theta_0$(mod $2\pi$) versus $t_f$ (*right*) for GEO to MO transfer trajectories of type $\mathscr{C}_2$

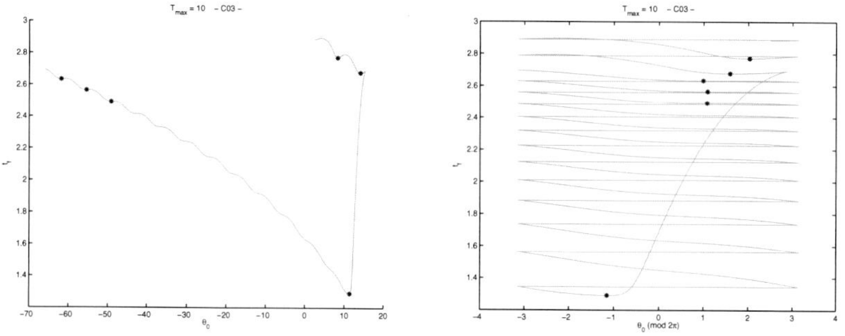


**Fig. 32** For $\varepsilon = 10$: projection of the homotopic path $\theta_0$ versus $t_f$ (*left*) $\theta_0$(mod $2\pi$) versus $t_f$ (*right*) for GEO to MO transfer trajectories of type $\mathscr{C}_3$

homotopic curve has a turning point. The only difference between the three curves is that for a given $\theta_0^*$ the minimum transfer time is different for the three kind of trajectories. In Fig. 39 we can see the three homotopic paths on the same figure, and we can appreciate how in terms of transfer time $\mathscr{C}_1$ is always below $\mathscr{C}_2$, which is always below $\mathscr{C}_3$. For each of the three homotopic curves, in Tables 15, 16, 17, 18, 19, 20, 21, 22 and 23 we have the initial conditions $\{t_f, x_0, p_0\}$ for the local minima for $\theta_0 = \pi, 0, \pi/2$ and $3\pi/2$. In Figs. 33, 35 and 37 we have the transfer trajectories from the three homotopic curves for different initial conditions for $\theta_0(\text{ mod } 2\pi) = 0$. As we can see, for each class, the trajectories along the homotopic path remain of the same class, i.e. the insertion sense on the MO remains always the same for all the orbits on the curve. Moreover, as we can see for each class we also find two type of trajectories, that we can call $\mathscr{T}_1$ and $\mathscr{T}_2$. Type $\mathscr{T}_1$ are trajectories that in the first phase spiral around the Earth to gain $J_c$ and then go directly towards the Moon, passing close to $L_1$. Type $\mathscr{T}_2$ are trajectories that do some turns around the Earth and then do a large excursion before heading towards the Moon. In Figs. 34, 36 and 38 we show the variation of $J_c$ along time for the trajectories that we find in Figs. 33, 35 and 37 respectively. As we can see, for type $\mathscr{T}_1$ trajectories $J_c(t)$ decreases and

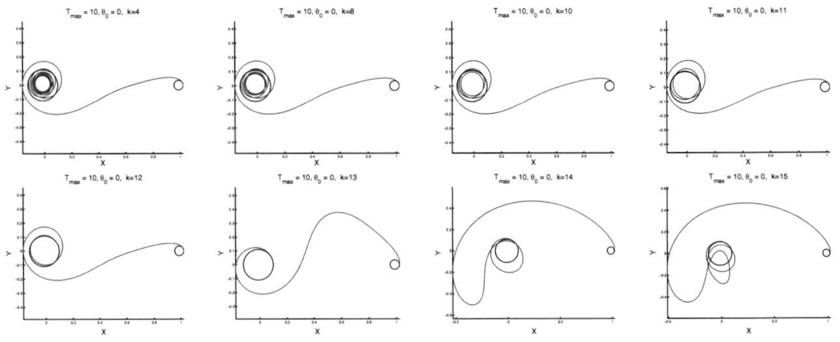


**Fig. 33** For class $\mathscr{C}_1$: Transfer trajectories for $\varepsilon = 10$N and $\theta_0 = 0$ fixed. The initial condition on the GEO orbit is $x_0 = (-r_0 - \mu, 0, 0, v_0)$

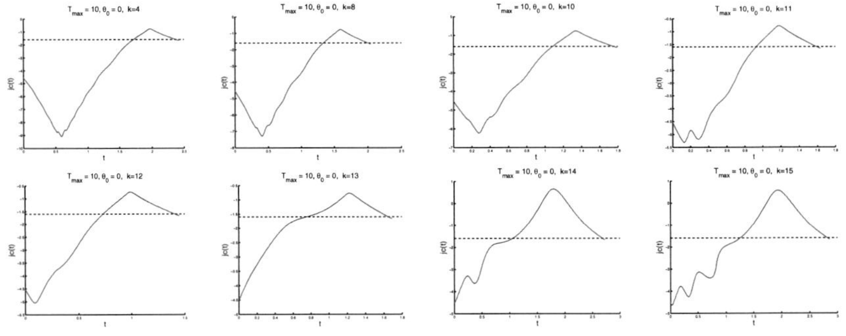


**Fig. 34** For class $\mathscr{C}_1$: Variation of $J_c$ for $\varepsilon = 10$N and $\theta_0 = 0$ fixed. The initial condition on the GEO orbit is $x_0 = (-r_0 - \mu, 0, 0, v_0)$

gains energy while they spiral around the Earth and when $J_c(t)$ is slightly larger than $J_c(L_1)$ this one starts to decreases to meet $J_c$ of the MO. On the other hand, for the type $\mathscr{T}_2$ trajectories, $J_c(t)$ will reach much larger values than $J_c(L_1)$ before decreasing to get to the MO. We recall that the main difference between the behavior of the three classes of transfer trajectories $\mathscr{C}_1$ (blue), $\mathscr{C}_2$ (red) and $\mathscr{C}_3$ (green) is the transfer time. As we can see in Fig. 39 for a fixed $\theta_0$, the transfer time for class $\mathscr{C}_2$ orbits is always less than for class $\mathscr{C}_1$ and class $\mathscr{C}_3$. But there are three cases where these curves intersect each other. Hence, we have trajectories of a different class with the same transfer time (i.e. cost function). It might be interesting to study in more detail these intersections as we have two different classes of strategies with the same cost. In Fig. 40 we have the transfer trajectories and the energy variations of the trajectories corresponding to the 3 intersections that we see in Fig. 39, from left to right $\mathscr{C}_1 \cap \mathscr{C}_2$, $\mathscr{C}_1 \cap \mathscr{C}_3$ and $\mathscr{C}_2 \cap \mathscr{C}_3$. As before the color of the orbit is related to its class ($\mathscr{C}_1$ are in blue, $\mathscr{C}_2$ are in red and $\mathscr{C}_3$ are in green).

Finally, we have done a small exploration on different characteristics for the minimum time–transfer trajectories that appear in Figs. 30, 31 and 32, where for each orbit we have computed the maximal value for $J_c$ along the orbit, the norm of

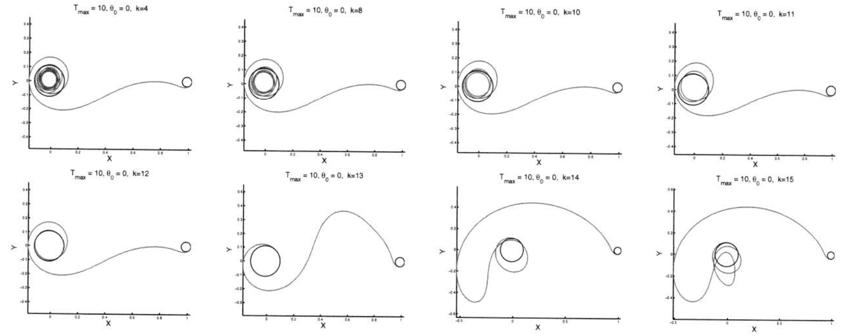


**Fig. 35** For class $\mathscr{C}_2$: Transfer trajectories for $\varepsilon = 10$N and $\theta_0 = 0$ fixed. The initial condition on the GEO orbit is $x_0 = (-r_0 - \mu, 0, 0, v_0)$

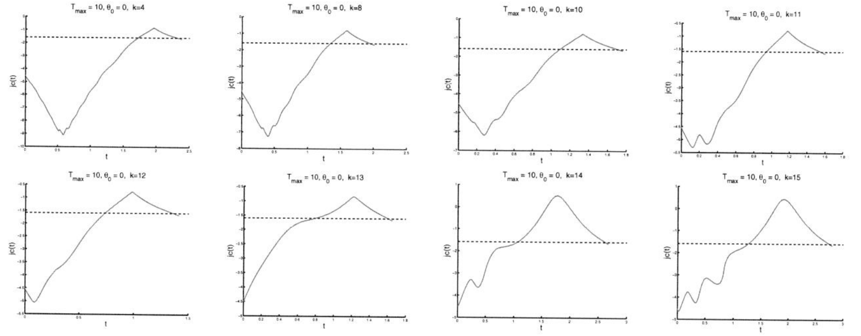


**Fig. 36** For class $\mathscr{C}_2$: Variation of $J_c$ for $\varepsilon = 10$N and $\theta_0 = 0$ fixed. The initial condition on the GEO orbit is $x_0 = (-r_0 - \mu, 0, 0, v_0)$

the adjoint vector at $t_0$, $|p(t_0)|$, and the number of turns the trajectory gives around the Earth before the trajectory goes towards the Moon. These results are summarized in Figs. 41, 42 and 43 for class $\mathscr{C}_1$, $\mathscr{C}_2$ and $\mathscr{C}_3$ respectively, where we can see a similar behavior as the GEO to $L_1$ transfer.

### 3.1.2 Homotopy w.r.t. $\varepsilon$

Given the fact that the transfer time for class $\mathscr{C}_3$ (green) is larger than the transfer time for the other two classes of orbits, from now on we will focus only on the classes $\mathscr{C}_1$ and $\mathscr{C}_2$. We recall that we can distinguish these two classes by the sense of insertion on a Moon orbit (blue = clockwise, red = anticlockwise). In this section we have taken the different solutions for $\varepsilon = 10$N and a fixed initial condition: $\theta_0$ mod $(2\pi) = 0$ and $\pi$. The initial conditions are summarized in Tables 15 and 16 for class $\mathscr{C}_1$ and Tables 19 and 20 for class $\mathscr{C}_2$. For each of the initial conditions we have performed a homotopy with respect to $\varepsilon$ from 10N to 1N. In Fig. 44 and 45 we summarize the results for class $\mathscr{C}_1$ and class $\mathscr{C}_2$ respectively. As we can see, at a first sight, in both cases the behavior is similar to the one found for GEO to $L_1$ transfer

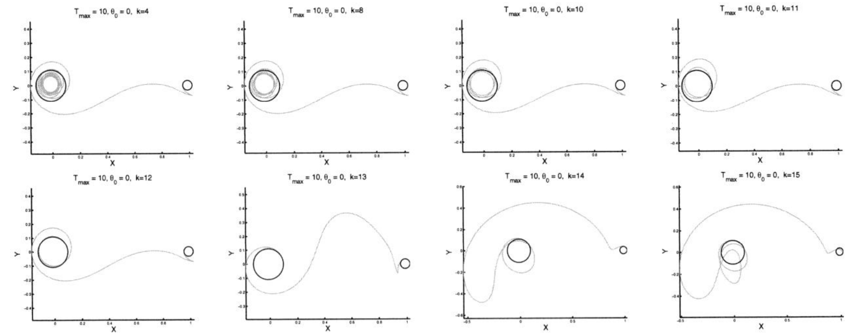


**Fig. 37** For class $\mathscr{C}_3$: Transfer trajectories for $\varepsilon = 10$N and $\theta_0 = 0$ fixed. The initial condition on the GEO orbit is $x_0 = (-r_0 - \mu, 0, 0, v_0)$

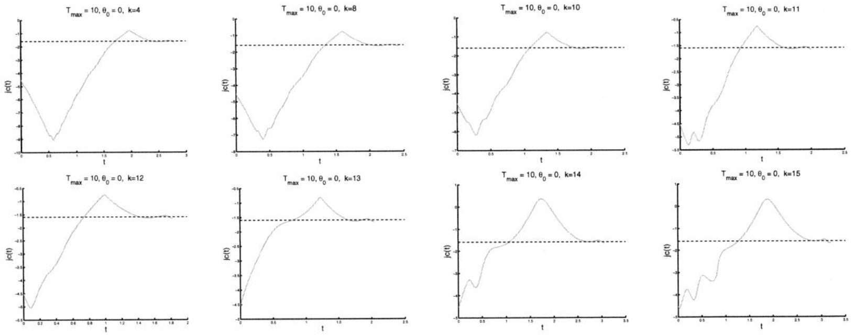


**Fig. 38** For class $\mathscr{C}_3$: Variation of $J_c$ for $\varepsilon = 10$N and $\theta_0 = 0$ fixed. The initial condition on the GEO orbit is $x_0 = (-r_0 - \mu, 0, 0, v_0)$

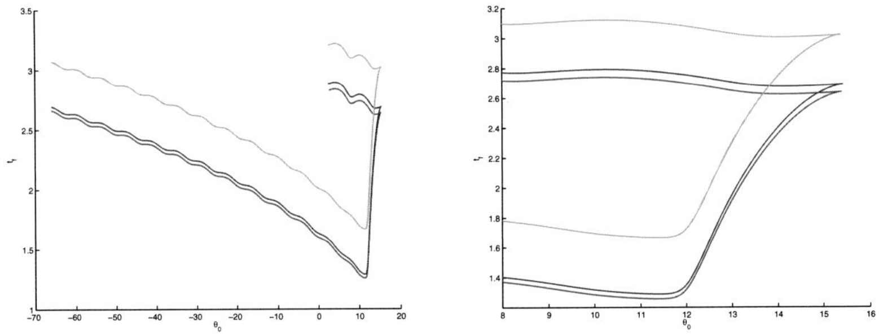


**Fig. 39** For $\varepsilon = 10$N: homotopic path $\theta_0$ versus $t_f$ for $\mathscr{C}_1$ (*blue*), $\mathscr{C}_2$ (*red*) and $\mathscr{C}_3$ (*green*)

trajectories (Sect. 2). So in both cases, as $\varepsilon$ decreases the transfer time, $t_f$, increases. In some cases, at some point the slope of $\varepsilon(t_f)$ experiences a drastic change and $t_f$ increases very quickly for small variations of $\varepsilon$. Then at some point the homotopic curve has a turning point and $\varepsilon$ grows, not being able to find solutions for lower values of $\varepsilon$. Nevertheless, there are other curves where $\varepsilon$ decreases with no problem reaching $\varepsilon = 1$N.

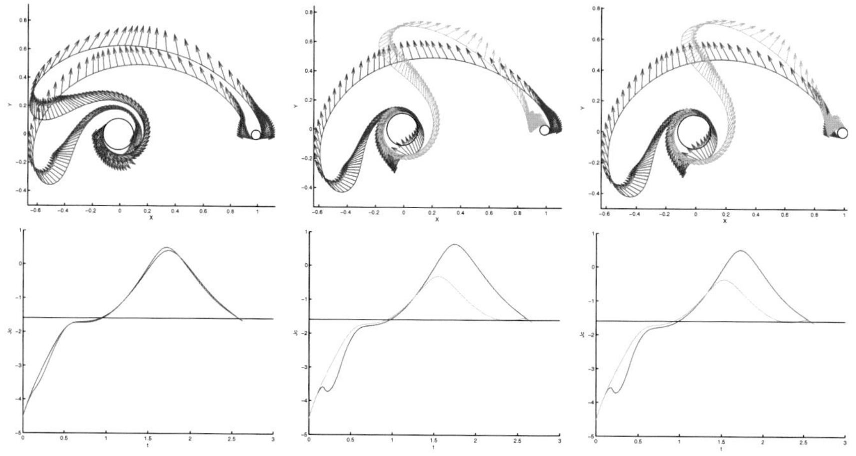

**Fig. 40** Transfer trajectories of the intersections between the homotopic path in Fig. 39. *Top plots* transfer trajectories and the associated control law; *Bottom plots* variation of $J_c$ w.r.t. time. From *left* to *right* $\mathscr{C}_1 \cap \mathscr{C}_2$, $\mathscr{C}_1 \cap \mathscr{C}_3$ and $\mathscr{C}_2 \cap \mathscr{C}_3$. The color of the curves is associated to the class, *blue* for class $\mathscr{C}_1$, *red* for class $\mathscr{C}_2$ and *green* for class $\mathscr{C}_3$

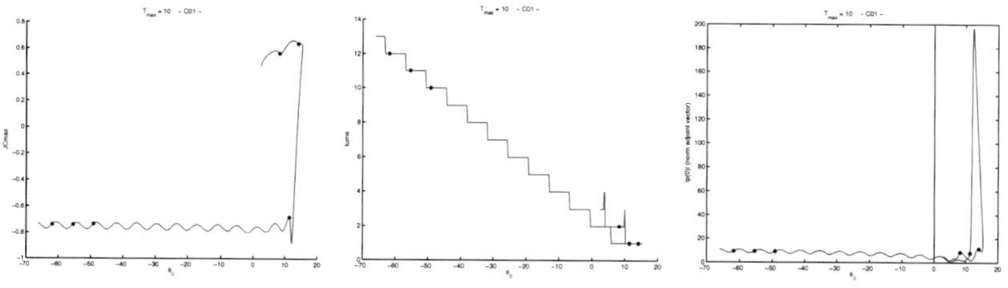

**Fig. 41** For class $\mathscr{C}_1$ and $\varepsilon = 10$N from *left* to *right* $\theta_0$ versus $\max_{t\in[t_0:t_f]}(J_c(t))$, $\theta_0$ versus $N_{ET}$ and $\theta_0$ versus $|p(t_0)|$

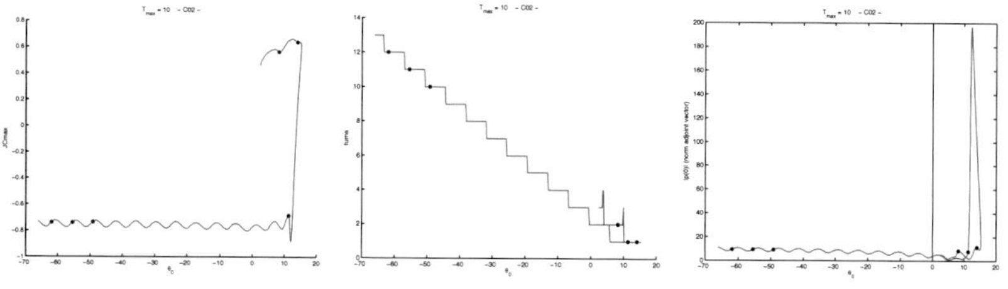

**Fig. 42** For class $\mathscr{C}_2$ and $\varepsilon = 10$N from *left* to *right* $\theta_0$ versus $\max_{t\in[t_0:t_f]}(J_c(t))$, $\theta_0$ versus $N_{ET}$ and $\theta_0$ versus $|p(t_0)|$

The main difference between the two class of orbits, appears in the region where $\varepsilon \in [1:2]$. While for class $\mathscr{C}_1$ the behavior is as we have mentioned, for class $\mathscr{C}_2$ the homotopic path experiences turning points and self-intersections, finding for some of these curves cut points for a fixed $\theta_0$. In Fig. 46 we have zoomed these area for both class of orbits and $\theta_0 = 0$, but the same phenomena is observed for $\theta_0 = \pi$. In Figs. 47

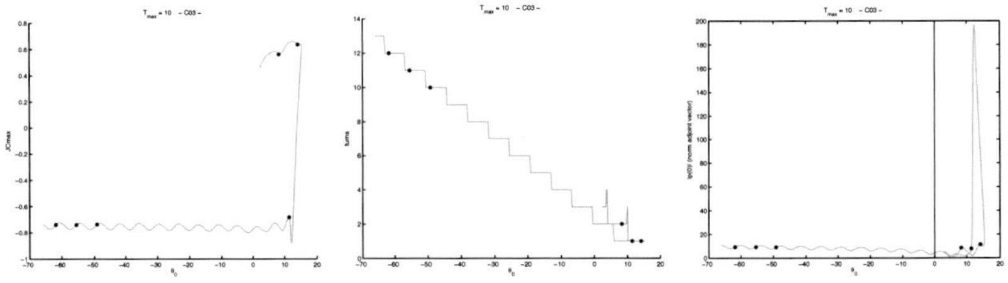


**Fig. 43** For class $\mathscr{C}_3$ and $\varepsilon = 10$N from *left* to *right* $\theta_0$ versus $\max_{t\in[t_0:t_f]}(J_c(t))$, $\theta_0$ versus $N_{ET}$ and $\theta_0$ versus $|p(t_0)|$

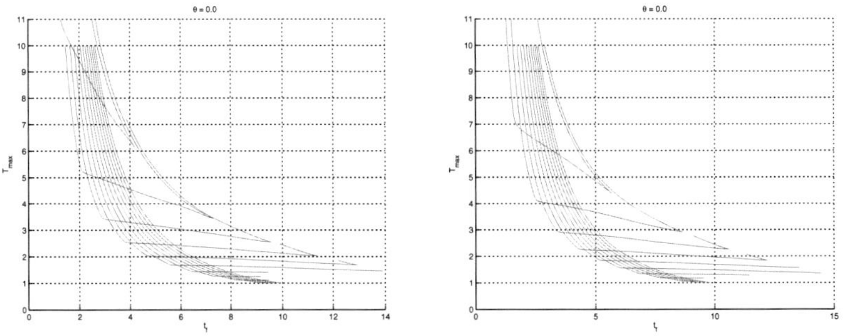


**Fig. 44** For class $\mathscr{C}_1$: Homotopy with respect to $\varepsilon$ for $\theta_0$ fixed, $\theta_0 = 0$ (*left*), $\theta_0 = \pi$ (*right*)

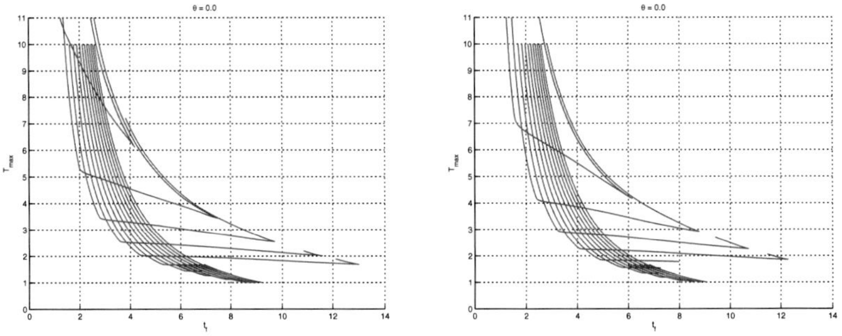


**Fig. 45** For class $\mathscr{C}_2$: Homotopy with respect to $\varepsilon$ for $\theta_0$ fixed, $\theta_0 = 0$ (*left*), $\theta_0 = \pi$ (*right*)

and 49 we show the type of solutions that we find along one of the homotopic curves of Figs. 44 and 45 respectively, for $\theta_0 = 0$ fixed, and varying $\varepsilon$. The plots correspond to the homotopy path starting by: $\theta_0 = 0$ and $k = 12$ from Table 16 for the class $\mathscr{C}_1$, and $\theta_0 = 0$ and $k = 12$ from Table 20 for the class $\mathscr{C}_2$. Both cases correspond to homotopic paths where $\varepsilon(t_f)$ experiences two drastic changes. As we can see the trajectories remain within their class along the homotopic path. We also notice that on the first part of the path, the trajectories are of type $\mathscr{T}_1$. While when the slope of the homotopic curve changes, the trajectories start to do big excursions before going towards the Moon, i.e. type $\mathscr{T}_2$ trajectories appear. In Figs. 48 and 50, we show the variation of $J_c$ along time for the trajectories plotted in Figs. 47 and 49 respectively (Tables 24, 25 and 26).

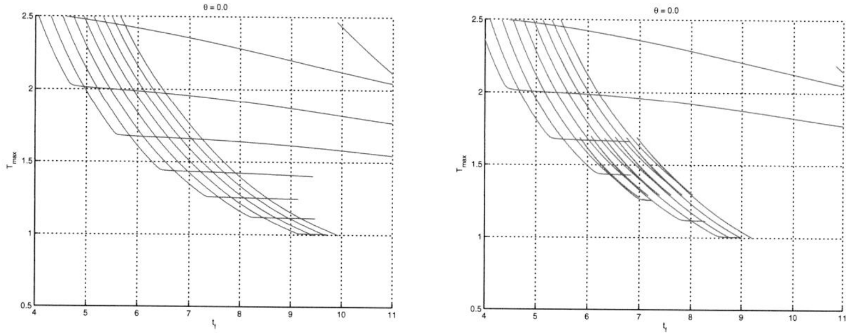


**Fig. 46** For class $\mathscr{C}_1$ (*left*) and $\mathscr{C}_2$ (*right*): homotopy with respect to $\varepsilon$ for $\theta_0 = 0$ fixed. Zoom for $\varepsilon \in [1 : 2]$

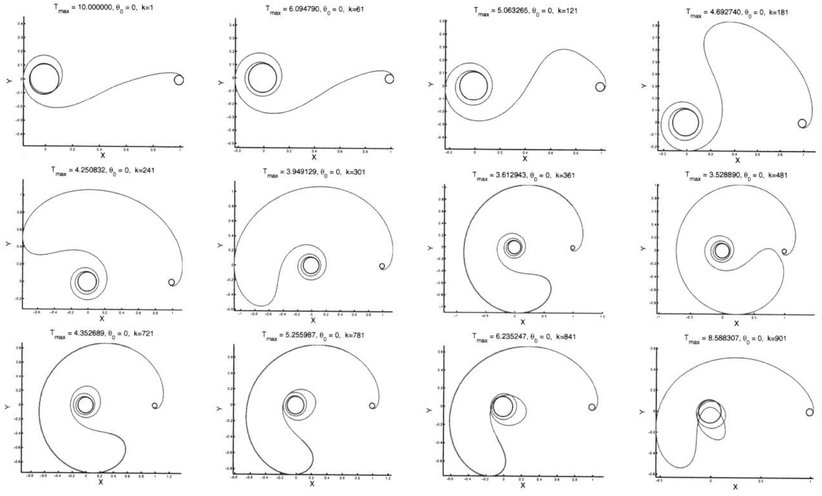


**Fig. 47** For class $\mathscr{C}_1$: Transfer trajectories from GEO to MO for $\theta_0$ fixed and varying $\varepsilon$. Here $\theta_0 = 0$ and all the orbits belong to the homotopic curve generated by $k = 12$ from Table 16 when we use $\varepsilon$ as the homotopy parameter

## 3.2 Free Initial Point on a GEO

In this section we summarize the results for a minimum-time transfer from a GEO to a MO, where the position on the GEO orbit is left free. Hence, the Boundary Conditions are:

$$x(t_0) \in \mathscr{M}_0, \quad p(t_0) \perp T_{x(t_0)}\mathscr{M}_0, \quad x(t_f) \in \mathscr{M}_1, \quad p(t_f) \perp T_{x(t_f)}\mathscr{M}_1, \quad h(t_f) = 0, \tag{9}$$

where $\mathscr{M}_0$ represents the GEO orbit, and $\mathscr{M}_1$ the MO. The first two Boundary Conditions are written as,

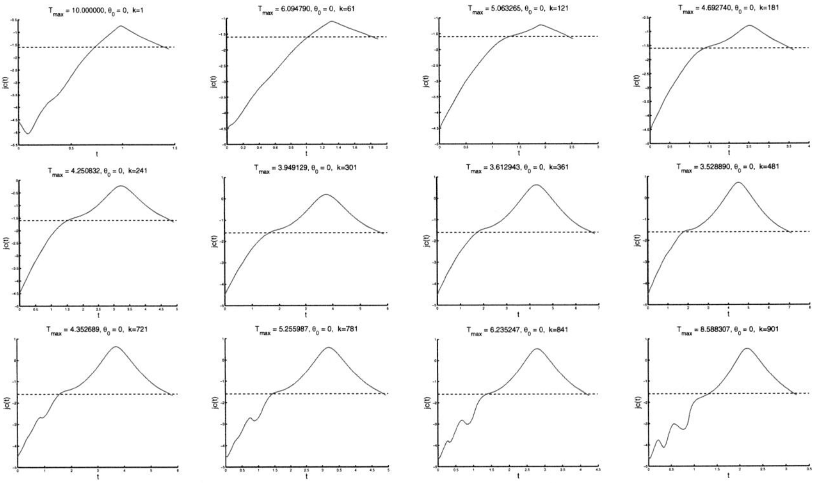


**Fig. 48** For class $\mathscr{C}_1$: Variation of the Jacobi constant along time for a transfer trajectories from GEO to MO for $\theta_0$ fixed and varying $\varepsilon$. Here $\theta_0 = 0$ and all the orbits belong to the homotopic curve generated by $k = 12$ from Table 16 when we use $\varepsilon$ as the homotopy parameter

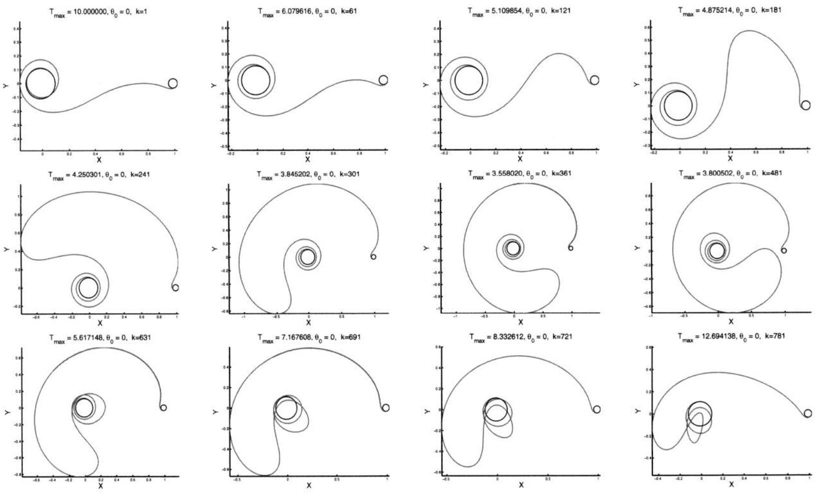


**Fig. 49** For class $\mathscr{C}_2$: Transfer trajectories from GEO to MO for $\theta_0$ fixed and varying $\varepsilon$. Here $\theta_0 = 0$ and all the orbits belong to the homotopic curve generated by $k = 12$ from Table 20 when we use $\varepsilon$ as the homotopy parameter

$$\begin{aligned} (x_0+\mu)^2 + y_0^2 - r_0^2 &= 0, \\ \dot{x}_0^2 + \dot{y}_0^2 - v_0^2 &= 0, \\ (x_0+\mu)\dot{x}_0 + y_0\dot{y}_0 &= 0, \\ (x_0+\mu)p_{y0} - y_0 p_{x0} + \dot{x}_0 p_{\dot{y}0} - \dot{y}_0 p_{\dot{x}0} &= 0, \end{aligned} \tag{10}$$

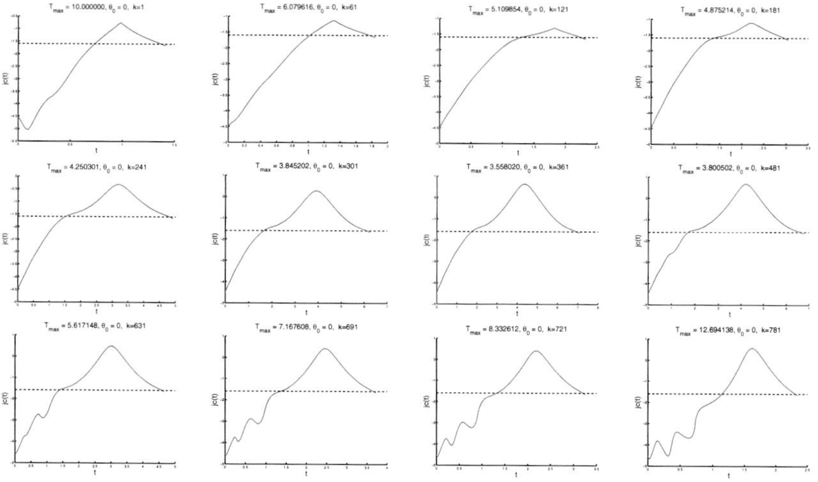


**Fig. 50** For class $\mathscr{C}_2$: Variation of the Jacobi constant along time for a transfer trajectories from GEO to MO for $\theta_0$ fixed and varying $\varepsilon$. Here $\theta_0 = 0$ and all the orbits belong to the homotopic curve generated by $k = 12$ from Table 20 when we use $\varepsilon$ as the homotopy parameter

where $x(t_0) = (x_0, y_0, \dot{x}_0, \dot{y}_0)$ are the coordinates of the spacecraft and $p(t_0) = (p_{x0}, p_{y0}, p_{\dot{x}0}, p_{\dot{y}0})$ is the adjoint vector, both evaluated at $t = t_0$. While the second two:

$$\begin{aligned} (x_f + \mu - 1)^2 + y_f^2 - r_1^2 &= 0, \\ \dot{x}_f^2 + \dot{y}_f^2 - v_1^2 &= 0, \\ (x_f + \mu - 1)\dot{x}_f + y_f \dot{y}_f &= 0, \\ (x_f + \mu - 1) p_{yf} - y_f p_{xf} + \dot{x}_f p_{\dot{y}_f} - \dot{y}_f p_{\dot{x}_f} &= 0, \end{aligned} \tag{11}$$

where $x(t_f) = (x_f, y_f, \dot{x}_f, \dot{y}_f)$ are the coordinates of the spacecraft and $p(t_f) = (p_{xf}, p_{yf}, p_{\dot{x}f}, p_{\dot{y}f})$ is the adjoint vector, both evaluated at $t = t_f$. Moreover, $r_0$ is the radius of the GEO orbit, $r_1$ is the radius of the arrival Moon orbit and $v_0 = \sqrt{(1-\mu)/r_0}$, $v_1 = \sqrt{\mu/r_1}$ the corresponding velocities on the GEO and the arrival MO so that the orbits are circular at first order. We recall that all the solutions found in the previous section, for a fixed initial condition on the GEO orbit, are not necessarily solutions of this problem. Only the local minima in the homotopic paths in Figs. 30, 31 and 32 are good initial guesses to find the local minima of this new problem. The values for $x(t_0)$, $p(t_0)$ are summarized in Tables 27, 28 and 29 for classes $\mathscr{C}_1$, $\mathscr{C}_2$ and $\mathscr{C}_3$ respectively and $\varepsilon = 10$N. In this section we have taken some of the local minima of the homotopic curve for $\varepsilon = 10$N and class $\mathscr{C}_1$ and $\mathscr{C}_2$, and performed homotopies with respect to the thrust magnitude $\varepsilon \in [1 : 10]$. For both class of transfer orbits we have taken as initial condition the local minima numbers 5, 6 and 7 in Tables 27 and 28. On the left hand side of Fig. 51 we can these homotopic paths. As usual, the curve in blue represents the solution for class $\mathscr{C}_1$ transfer orbits and the curve in red the solutions for class $\mathscr{C}_2$ transfer orbits. Notice that the red curve is always

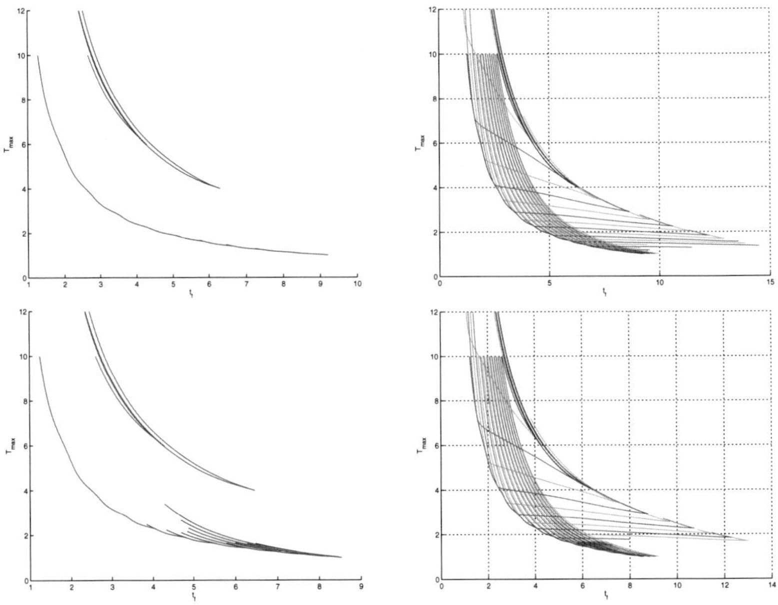

**Fig. 51** Homotopy w.r.t. $\varepsilon$ for transfers orbit from GEO to MO. *Left* Local minimum for $\theta_0$ free. *Right* Comparison between the local minima for $\theta_0$ free and $\theta_0$ fixed. *Top* results for class $\mathscr{C}_1$, $\theta_0$ free (*blue line*) and $\theta_0$ fixed (*green* and *magenta lines*). *Bottom* results for class $\mathscr{C}_2$, $\theta_0$ free (*red line*) and $\theta_0$ fixed (*green* and *magenta lines*)

bellow the blue curve, hence, the class $\mathscr{C}_2$ transfer trajectories are always better than the class $\mathscr{C}_1$ ones. Also notice that the red curve presents a more complex structure for $\varepsilon \in [1:3]$N. It can be seen that, as it happened for the GEO to $L_1$ transfer, all the solutions generated by the local minima number 5 are $\mathscr{T}_1$ type transfer orbits, while solutions generated by local minima number 6 and 7 are type $\mathscr{T}_2$ transfer orbits. Moreover, when we let $\theta_0$ free these two type of solutions do not connect if we compute the homotopic paths varying $\varepsilon$. On the right hand side of Fig. 51 we compare for both class of trajectories, $\mathscr{C}_1$ (top) and $\mathscr{C}_2$ (bottom), the solutions for $\theta_0$ free and $\theta_0$ fixed. As we can see, in both cases, the curve for $\theta_0$ free is always below the curves for $\theta_0$ fixed. In Figs. 52 and 53 we have zoomed different areas of these two curves for comparisons, class $\mathscr{C}_1$ and $\mathscr{C}_2$ respectively. In both cases, the curves for $\theta_0$ free present self-intersections, i.e. cut point candidates. Although the structure for $\mathscr{C}_2$ is much more complex presenting different kind of self-intersections. In the next section we will study in more detail these phenomena.

## 3.3 Cut Points

As we can appreciate in Figs. 52 and 53, for both classes ($\mathscr{C}_1$ and $\mathscr{C}_2$) the homotopic paths for the GEO to MO transfer with $\theta_0$ free have several turning points and the

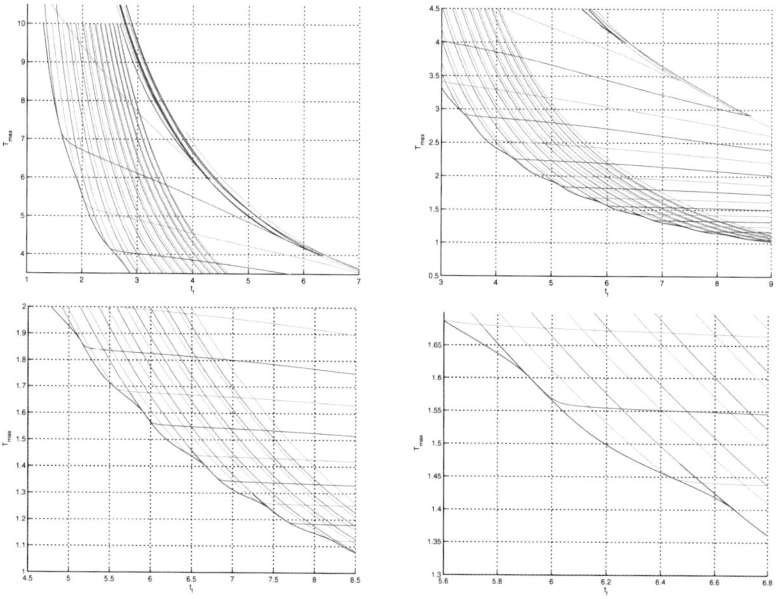

**Fig. 52** For class $\mathscr{C}_1$: zoomed regions of the comparison between the local minima for $\theta_0$ free (*blue line*) versus $\theta_0$ fixed (*green* and *magenta lines*)

path self-intersects several times. These intersections are cut point candidates. As in Sect. 2.3, we call the couple $\{(x_0, p_0), (x_1, p_1)\}$ a cut point, if they are two different initial conditions such that for the same $\varepsilon$ they generate two different local optimal transfer orbits from GEO to MO with the same transfer time $t_f^*$. We find these initial conditions on the self-intersections on the projection of homotopic path $\varepsilon$ versus $t_f$, Figs. 52 and 53. To fix notation, $(x_0, p_0)$ is the first point on the homotopic path that reaches $(\varepsilon, t_f)$ and $(x_1, p_1)$ is the second point on the same curve that reaches $(\varepsilon, t_f)$. In Tables 30 and 31 we summarize all the cut points that we have found for $\varepsilon \in [1, 10]$N for class $\mathscr{C}_1$ and $\mathscr{C}_2$ respectively. All these points have been computed by refining the intersections found in Figs. 52 and 53. Finally in Figs. 54 and 55 we plot different projections of the homotopic path for the two class of orbits, and we have highlighted in green the regions close to the different cut points.

We do the same analysis as in Sect. 2.3. First we have taken both solutions $(x_0, p_0)$ and $(x_1, p_1)$ and compute the transfer trajectory, the variation of $J_c(t)$, the control law $\mathbf{u}(t)$ and $(H_1(t), H_2(t))$. Second we have integrated both trajectories backward and forward in time on $[-t_f, 2t_f]$, where $t_f$ is the transfer time. We have also taken the solutions in the neighborhood of the cut point and checked the variation of different parameters. We recall, that as we did in Sect. 2.3, if $\varepsilon^*$, $t_f^*$ are the thrust magnitude and the minimum transfer time for the cut point, we consider a solution to be in the cut neighborhood if $t_f \in [t_f^* - 0.15 : t_f^* + 0.15]$. In the Appendix "Summary of the cut points on the GEO to MO transfer" we have the plots summarizing this analysis for all the cut points. In this section we will only plot some of them and discuss

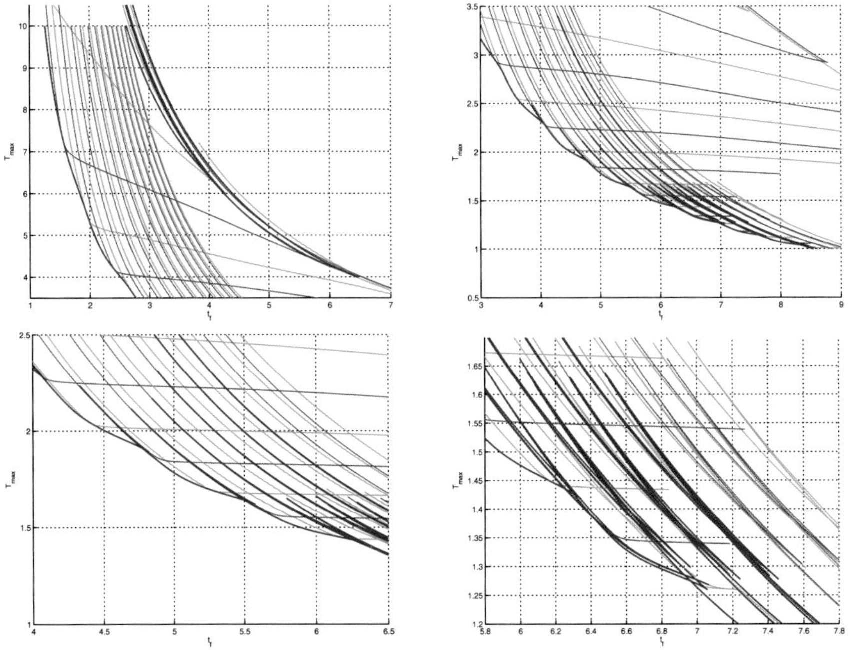

**Fig. 53** For class $\mathscr{C}_2$: zoomed regions of the comparison between the local minima for $\theta_0$ free (*red line*) versus $\theta_0$ fixed (*green* and *magenta lines*)

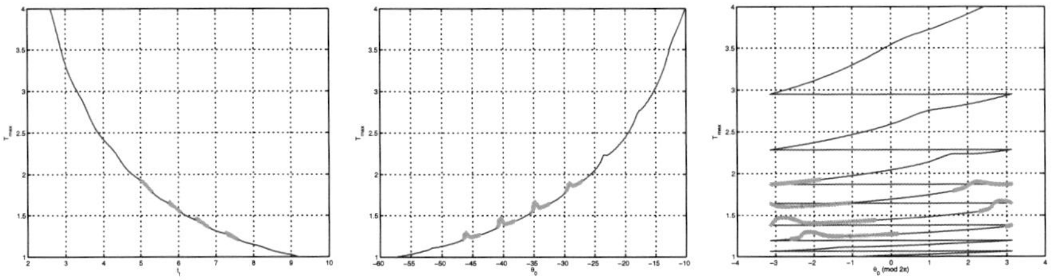

**Fig. 54** Homotopy of the local minima of class $\mathscr{C}_1$, for the GEO to MO control problem. *Left* $\varepsilon$ versus $t_f$, *Center* $\varepsilon$ versus $\theta_0$, *Right* $\varepsilon$ versus $\theta_0$ mod $2\pi$

the most relevant results. In the case of class $\mathscr{C}_1$ orbits (Table 30) all the cut points are of the same kind and present, qualitatively, a similar behavior. This is why here we only show the results for the cut point number 3. This cut point corresponds to $\varepsilon^* \approx 1.6073723$ and $t_f^* \approx 5.9023179$ (see Table 30). On the left hand side of Fig. 56 we have the homotopic curve and highlighted in green the region that we want to study. On the right hand side of the Fig. we have 4 subplots, one is a zoom of the cut point region showing $t_f$ versus $\varepsilon$, the other is the same zoom but plotting $\theta_0$ versus $\varepsilon$. The two subplots on the bottom show $\theta_0$ versus $N_{ET}$ and $\theta_0$ versus $N_{ZH}$, being $N_{ET}$ the number of turns around the Earth before $J_c(t) > J_c(L_1)$ and $N_{ZH}$ the number of times $|(H_1(t), H_2(t))|$ is close to zero.

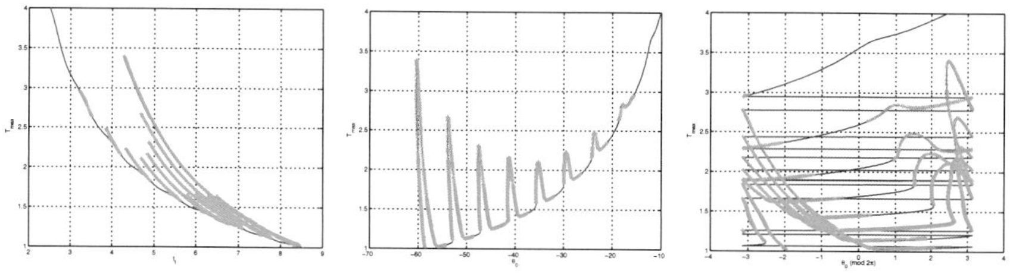

**Fig. 55** Homotopy of the local minima of class $\mathscr{C}_2$, for the GEO to MO control problem. *Left* $\varepsilon$ versus $t_f$, *Center* $\varepsilon$ versus $\theta_0$, *Right* $\varepsilon$ versus $\theta_0$ mod $2\pi$

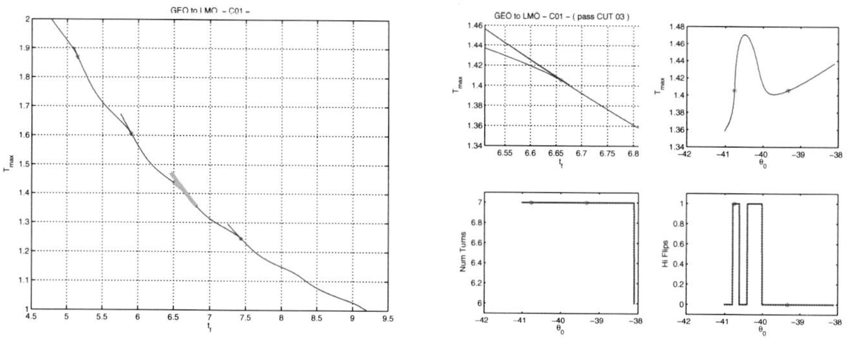


**Fig. 56** $\mathscr{C}_1$ cut point $n^o$ 3: *Left* $t_f$ versus $\varepsilon$ homotopic curve with highlight of the cut passage in *green*; *Right* versus analysis of the cut passage: (*top-left* subplot) $t_f$ versus $\varepsilon$ zoom, (*top-right* subplot) $\theta_0$ versus $\varepsilon$, (*bottom-left* subplot) $\theta_0$ versus $N_{ET}$, (*bottom-right* subplot) $\theta_0$ versus $N_{ZH}$. *Red* points are values corresponding the each cut point

In Fig. 57 we plot different aspects of both cut transfer trajectories. In all the plots, the curves in blue correspond to the first cup point $(x_0, p_0)$ and the curves in red correspond to the second cut point $(x_1, p_1)$. The three plots on the top show, from left to right: the $\{x, y\}$ projection of the transfer trajectory, the $\{\dot{x}, \dot{y}\}$ projection of the transfer trajectory and $J_c(t)$ along the transfer trajectory. The three plots on the bottom show, from left to right: the $\{x, y\}$ projection of the transfer trajectory and the control law $\mathbf{u}(t)$, the projection $(H_1, H_2)$ and the variation of $|(H_1, H_2)|$ along the transfer trajectory. As we can see, the main difference between both trajectories appears on the control law at the beginning of the transfer, where the second cut point (red curve) experiences a drastic change.

In Fig. 58 we plot the integration backward in time ($t \in [-t_f : 0]$) and forward in time ($t \in [0 : 2t_f]$) for both transfer trajectories and the variation of $J_c$ for each of the trajectories. As we can see, when we integrate backwards in time, we have a similar behaviors to the one experienced by the different cut points in the GEO to $L_1$ transfer problem. We have that the first cut points spirals away form the Earth (red curve) while the second cut points spirals towards the Earth (blue curve). This is also reflected on the behavior of $J_c(t)$ where in the first case will start to grow, while in the second case this one will decrease. Moreover, this behavior is repeated for all the cut points of class $\mathscr{C}_1$. If we look at the behavior of the two trajectories for $t \in [0 : 2t_f]$, it is true that there is a difference between the two trajectories. But as we can see in

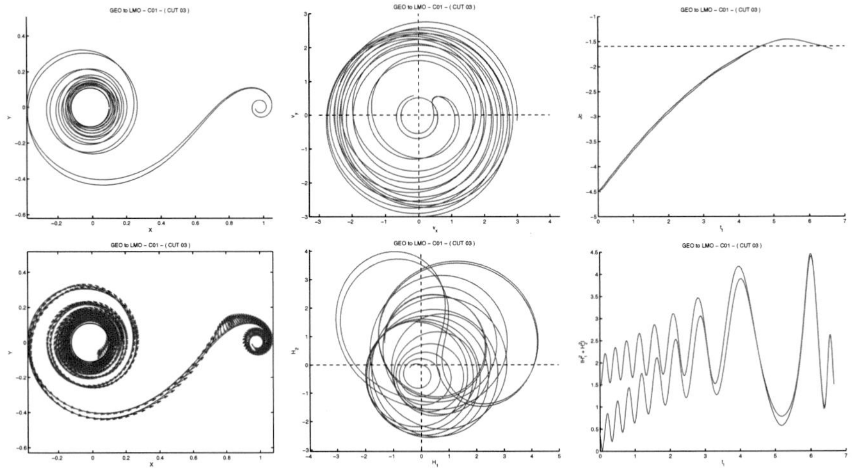


**Fig. 57** $\mathscr{C}_1$ cut point $n^o$ 3: *blue* orbits correspond to the first cut value and *red* orbits to the second cut value. *Top-Left* $\{XY\}$ projection of the transfer trajectory, *Top-Center* $\{\dot{X}\dot{Y}\}$ projection of the transfer trajectory, *Top-Right* $t$ versus $J_c$ (energy variation along the transfer trajectory), *Bottom-Left* control along the trajectory, *Bottom-Center* $H_1$ versus $H_2$, *Bottom-Right* $t$ versus $|(H_1, H_2)|$

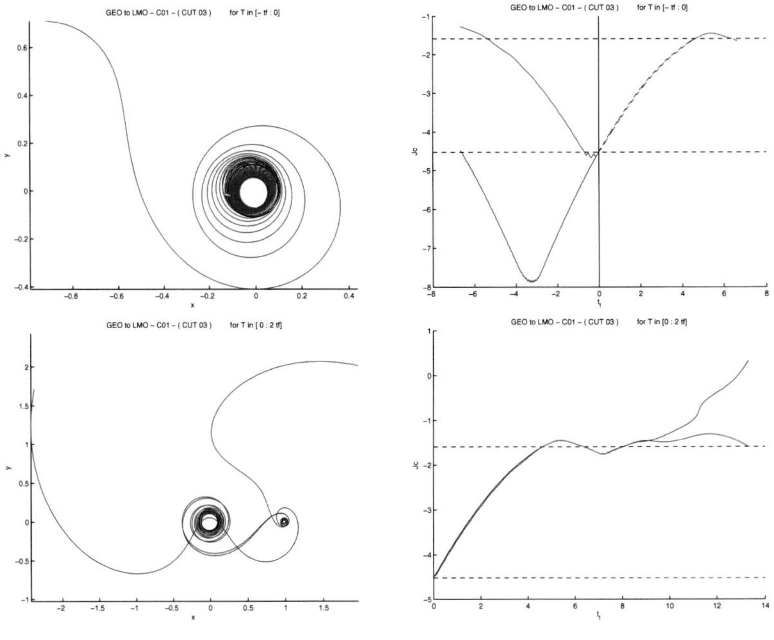


**Fig. 58** $\mathscr{C}_1$ cut point $n^o$ 3: *Left* optimal solutions for $t \in [-t_f, 0]$ ($\{XY\}$ projection and $J_c$ variation), *Right* optimal solutions for $t \in [0, 2t_f]$ ($\{XY\}$ projection and $J_c$ variation)

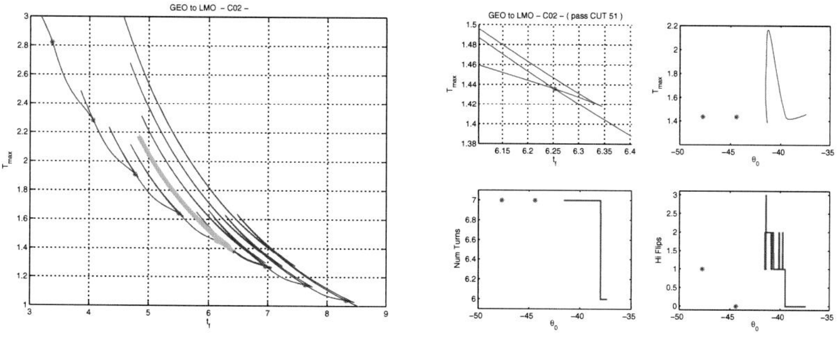


**Fig. 59** $\mathscr{C}_2$ cut point $n^o$ 5: *Left* $t_f$ versus $\varepsilon$ homotopic curve with highlight of the cut passage in *green*; *Right* analysis of the cut passage: (*top-left* subplot) $t_f$ versus $\varepsilon$ zoom, (*top-right* subplot) $\theta_0$ versus $\varepsilon$, (*bottom-left* subplot) $\theta_0$ versus $N_{ET}$, (*bottom-right* subplot) $\theta_0$ versus $N_{ZH}$. *Red* points are values corresponding the each cut point

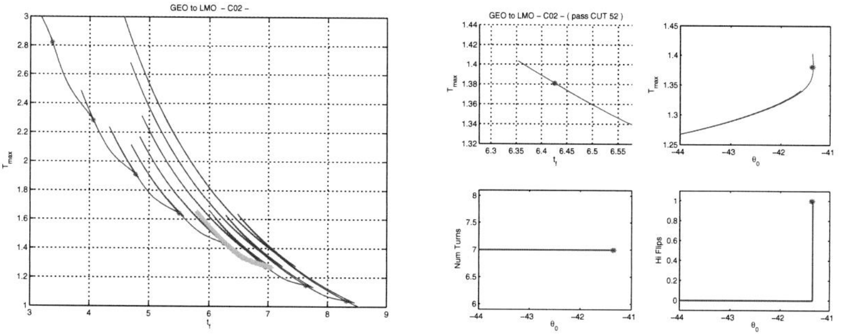


**Fig. 60** $\mathscr{C}_2$ cut point $n^o$ 6: *Left* $t_f$ versus $\varepsilon$ homotopic curve with highlight of the cut passage in *green*; *Right* analysis of the cut passage: (*top-left* subplot) $t_f$ versus $\varepsilon$ zoom, (*top-right* subplot) $\theta_0$ versus $\varepsilon$, (*bottom-left* subplot) $\theta_0$ versus $N_{ET}$, (*bottom-right* subplot) $\theta_0$ versus $N_{ZH}$. *Red* points are values corresponding the each cut point

Appendix "Summary of the cut points on the GEO to MO transfer", this behavior does not show a distinctive pattern between the four cut points of class $\mathscr{C}_1$. On the other hand, not all the cut points of class $\mathscr{C}_2$ (Table 31) experience a similar behavior. As the plots in Appendix "Summary of the cut points on the GEO to MO transfer" show, we have cut point number 1, 2, 3, 4, 5, 8 and 9 that present a similar behavior and cut points 6 and 7 that show another. Here we show the results for cut point 5 and cut point number 6 and we will briefly comment on their main differences. A more extensive study on these two kinds of cut points should be done in detail. In Figs. 59 and 60 we plot the behavior of the trajectories close to the cut point for cut point number 5 and 6 respectively. Where we have the on the right hand side the variation of $N_{ET}$ and $N_{ZH}$ for the different solutions.

In Figs. 61 and 62 we see the behavior of the two transfer trajectories for cut point number 5 and 6 respectively. As before, red is assigned to the first cut point $(x_0, p_0)$ and blue to the second one $(x_1, p_1)$. On the top we have, from left to right, the $\{x, y\}$

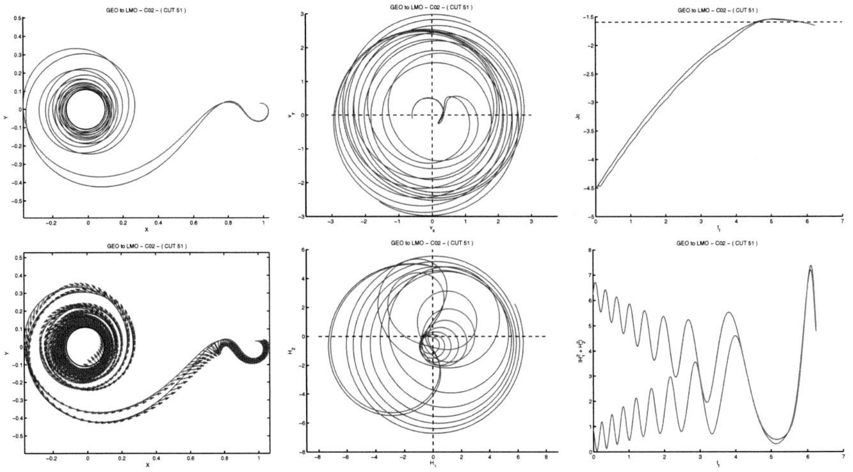


**Fig. 61** $\mathscr{C}_2$ cut point $n^o$ 5: *blue* orbits correspond to the first cut value and *red* orbits to the second cut value. *Top-Left* $\{XY\}$ projection of the transfer trajectory, *Top-Center* $\{\dot{X}\dot{Y}\}$ projection of the transfer trajectory, *Top-Right* $t$ versus $J_c$ (energy variation along the transfer trajectory), *Bottom-Left* control along the trajectory, *Bottom-Center* $H_1$ versus $H_2$, *Bottom-Right* $t$ versus $|(H_1, H_2)|$

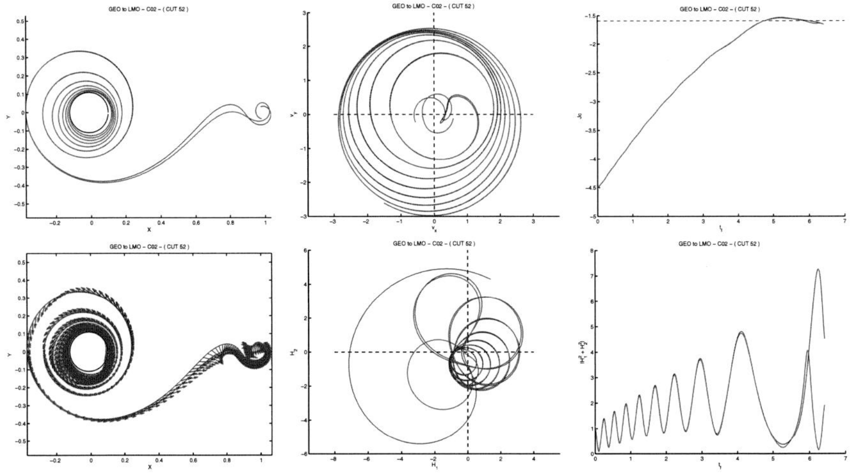


**Fig. 62** $\mathscr{C}_2$ cut point $n^o$ 6: *blue* orbits correspond to the first cut value and *red* orbits to the second cut value. *Top-Left* $\{XY\}$ projection of the transfer trajectory, *Top-Center* $\{\dot{X}\dot{Y}\}$ projection of the transfer trajectory, *Top-Right* $t$ versus $J_c$ (energy variation along the transfer trajectory), *Bottom-Left* control along the trajectory, *Bottom-Center* $H_1$ versus $H_2$, *Bottom-Right* $t$ versus $|(H_1, H_2)|$

projection, the $\{\dot{x}, \dot{y}\}$ projection and the variation of $J_c(t)$. On the bottom we have, from left to right, the $\{x, y\}$ projection and the control law $\mathbf{u}(t)$, $(H_1, H_2)$ projection and the variation of $|(H_1, H_2)|$ along time. As we can see, for cut point number 5, the main difference between the two cut trajectories is seen at the beginning of the orbit, where again we see a drastic change on the orientation of $\mathbf{u}$. Which can be related to $|(H_1, H_2)|$ passing close to zero for one of the two trajectories. On the other hand, for cut point number 6, the main difference between the two cut trajectories appears at the end of the transfer, during the insertion to the Moon orbit. Where we can see how the arrival point on the Moon orbit is very different for the two trajectories (i.e. this is not the case in cut point number 5). Moreover, the control law is very different at the end of the transfer, and the second cut point (red curve) experiences a drastic change on its orientation.

Finally, in Figs. 63 and 64 we show for cut point number 5 and 6 respectively, the behavior of the transfer trajectories when we integrate backward in time $t \in [-t_f : 0]$ and forward in time $t \in [0 : 2t_f]$. As we can see, in the case of cut point number 5, the difference between the two cut point appears when we look at the behavior of the trajectories backward in time. Where we find similar results to the ones we have already observed. While if we look at the behavior of the transfer trajectories forward in time for cut point number 5 both are qualitatively the same. On the contrary, if we look at the behavior of cut point number 6 backward in time, both trajectories have a similar behavior, they both spiral away from the Earth and $J_c(t)$ increases. But if we look at their behavior forward in time we do see different behaviors between them.

To summarize, we can say that for class $\mathscr{C}_1$ transfer orbits, the cut points present a similar behavior to the cut points that we found when we studied the GEO to $L_1$ minimum-time transfer problem. Where the main difference between the two cut points is found at the beginning of the transfer trajectory. On the other hand, for the

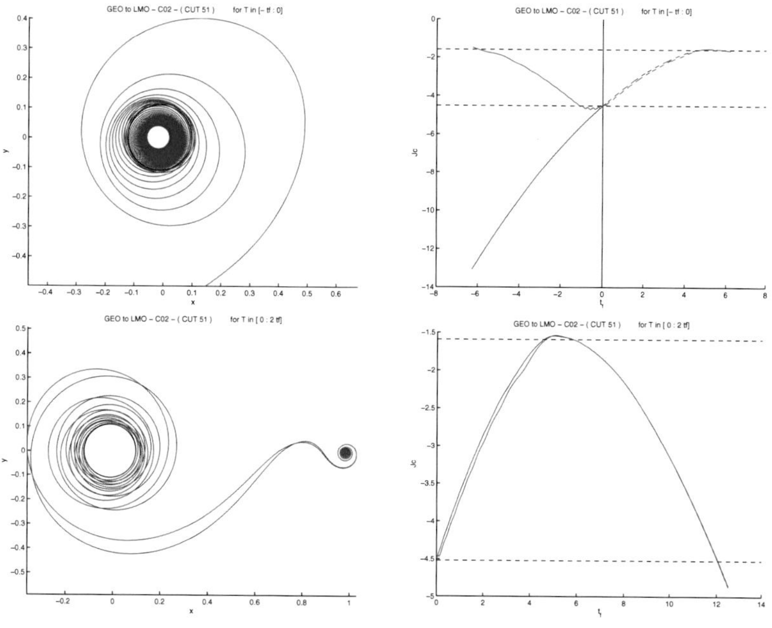


**Fig. 63** $\mathscr{C}_2$ cut point $n^o$ 5: *Top* optimal solutions for $t \in [-t_f, 0]$ ($XY$ projection and $J_c$ variation), *Bottom* optimal solutions for $t \in [0, 2t_f]$ ($XY$ projection and $J_c$ variation)

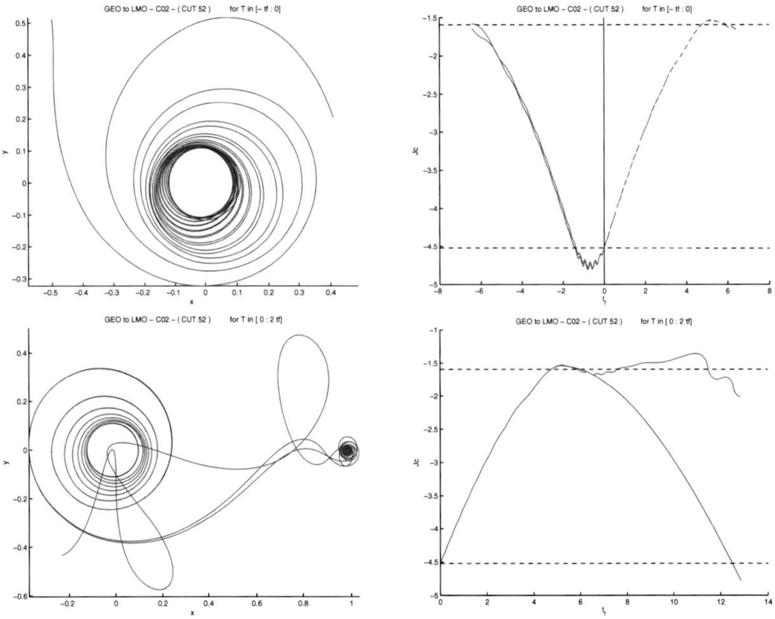


**Fig. 64** $\mathscr{C}_2$ cut point $n^o$ 6: *Top* optimal solutions for $t \in [-t_f, 0]$ ($XY$ projection and $J_c$ variation), *Bottom* optimal solutions for $t \in [0, 2t_f]$ ($XY$ projection and $J_c$ variation)

class $\mathscr{C}_2$ transfer orbits, the cut point structure is more complex. There are many type of self-intersections and type of cut behavior. We find many cut points where the behavior shows similarities with the cut points for class $\mathscr{C}_1$. But we also find two cut points where the difference between the two transfer trajectories appears in the second phase of the transfer, i.e. when we get to the Moon orbit. This is probably because we enter this orbit in an anti-clock wise sense and the structure of the $L_1$ to Moon orbit has a similar behavior. Further studies in this direction should be done in order to draw further conclusions.

## 3.4 Homotopy w.r.t $r_1$

Here we have considered the minimum time solutions found for $\varepsilon = 1\text{N}$ and $\theta_0$ free found in Sect. 3.2. The trajectories of the transfer orbit for both classes are in Fig. 65. We recall that the blue orbit corresponds to class $\mathscr{C}_1$ and the red orbit corresponds to class $\mathscr{C}_2$. In this section we will perform homotopies of these solutions with respect to $r_1$ the size of the arrival orbit, to find transfer trajectories to a circular orbit closer to the Moon. To be more specific, we have considered a transfer from GEO to circular MO where the position on the departure and arrival orbits is not fixed. We recall that the boundary conditions of these problem are written as:

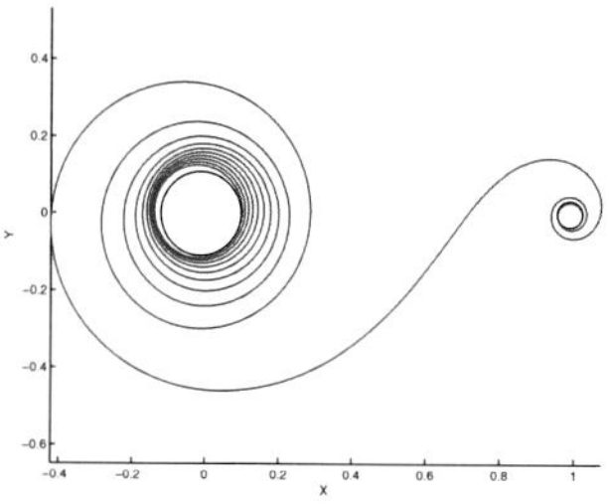


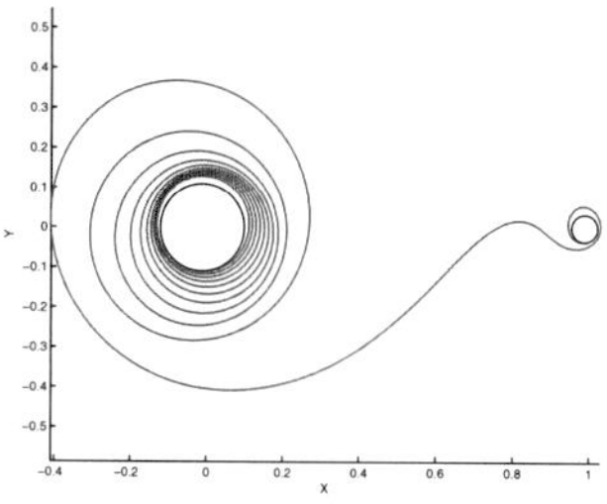


**Fig. 65** Minimum-time transfer trajectories of class $\mathscr{C}_1$ (*left*) and class $\mathscr{C}_2$ (*right*), for $\varepsilon = 1N$, and $\theta_0$ free

$$
\begin{aligned}
(x_0+\mu)^2 + y_0^2 - r_0^2 &= 0,\\
\dot{x}_0^2 + \dot{y}_0^2 - v_0^2 &= 0,\\
(x_0+\mu)\dot{x}_0 + y_0\dot{y}_0 &= 0,\\
(x_0+\mu)p_{y0} - y_0 p_{x0} + \dot{x}_0 p_{\dot{y}0} - \dot{y}_0 p_{\dot{x}0} &= 0,\\
(x_f+\mu-1)^2 + y_f^2 - r_1^2 &= 0,\\
\dot{x}_f^2 + \dot{y}_f^2 - v_1^2 &= 0,\\
(x_f+\mu-1)\dot{x}_f + y_f\dot{y}_f &= 0,\\
(x_f+\mu-1)p_{yf} - y_f p_{xf} + \dot{x}_f p_{\dot{y}_f} - \dot{y}_f p_{\dot{x}_f} &= 0.
\end{aligned}
\tag{12}
$$

We recall that $x(t_0) = (x_0, y_0, \dot{x}_0, \dot{y}_0)$ and $x(t_f) = (x_f, y_f, \dot{x}_f, \dot{y}_f)$ are the coordinates of the spacecraft at $t = t_0$ and $t_f$ respectively; $p(t_0) = (p_{x0}, p_{y0}, p_{\dot{x}0}, p_{\dot{y}0})$ and $p(t_f) = (p_{xf}, p_{yf}, p_{\dot{x}f}, p_{\dot{y}f})$ are the coordinates of the adjoint vector at $t = t_0$ and $t_f$ respectively; $r_0$ is the radius of the GEO, $r_1$ the radius of the MO, and $v_0 = \sqrt{(1-\mu)/r_0}$, $v_1 = \sqrt{\mu/r_1}$ the corresponding velocities such that these orbits are circular using the two-body problem approximation. In this work we have used $r_0 = 0.109689855932071$ for the GEO orbit and $r_1 = 0.034$ for the MO, which corresponds to $r_0 \approx 42,164$ km and $r_1 \approx 13,069.6$ km (we recall that $r_M = 1,737.10$ km). We have performed an homotopy with respect to $r_1$, from 0.034 to 0.015 ($r_1 = 0.015$ corresponds to a MO of radius $\approx 5,766$ km). To perform this homotopy we have also used the package `hampath`.

In Fig. 66 we show the homotopic curve found by varying $r_1$, showing the projection $t_f$ versus $r_f$. The blue curve corresponds to the path found for class $\mathscr{C}_1$ transfer orbits and the red curve to the path for class $\mathscr{C}_2$. Notice that for class $\mathscr{C}_1$ the curve decreases slowly having larger transfer times for smaller $r_1$ as expected. On the other hand, for class $\mathscr{C}_2$ orbits the curve presents a more complex structure. It is still true that the transfer time increases as $r_1$ gets smaller, but we find several self-intersections. Having cut points and different local minimum solutions for the same $r_1$. A more detailed study on the structure of these "cut" points should be done. Moreover, notice that the transfer time for class $\mathscr{C}_2$ is always smaller that the one for class $\mathscr{C}_1$ orbits. In Fig. 67 we plot the two minimum-time transfer trajectories found for $\varepsilon = 1$N and $r_1 = 0.015$, each orbit corresponds to one class. For both orbits we have also plotted the variation of $J_c$ with respect to time. Notice that now $J_c(t_f)$ is

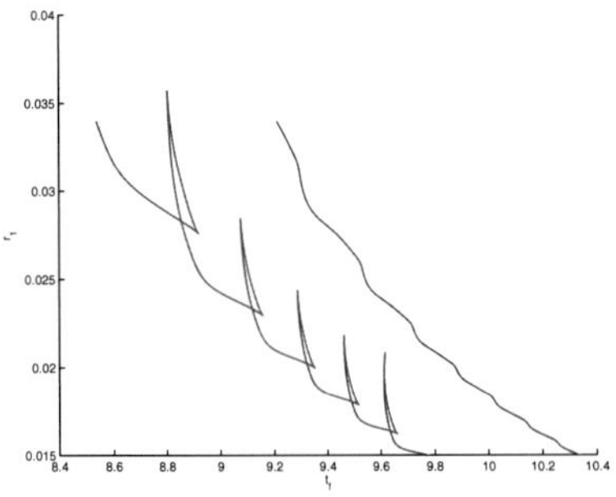


**Fig. 66** Homotopic path w.r.t. $r_1 \in [0.034 : 0.015]$, for $\varepsilon = 1N$ ($t_f$ vs. $r_1$). In *blue* solutions for class $\mathscr{C}_1$ and in *red* solutions for class $\mathscr{C}_2$

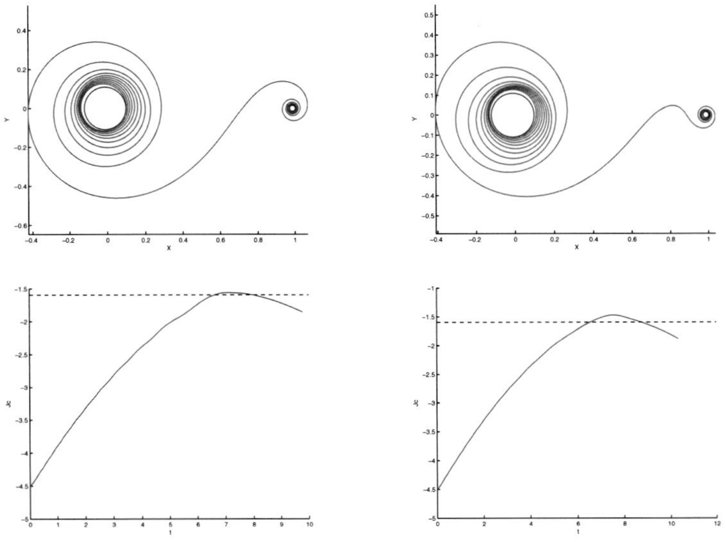

**Fig. 67** *Top* $\{XY\}$ projection of minimum-time transfer trajectories for $\varepsilon = 1$N and $r_1 = 0.015$. *Bottom* variation of $J_c$ along the transfer trajectories. *Left* (and *blue curves*) class $\mathscr{C}_1$ trajectories, *Right* (and *red curves*) class $\mathscr{C}_2$ trajectories

much smaller than $J_c(L_1)$. We can also see that in both cases, the transfer trajectory is split in three phases. A first phase where the orbit spirals around the Earth and gains $J_c$, a second phase where the orbit goes from the vicinity of the Earth to the vicinity of the Moon, and a third phase where the orbit spirals towards the Moon and $J_c$ decreases. The difference between the two class of orbits would happen in the second phase where the orbit chooses different kind of paths and control laws to reach the Moon orbit.

Finally, in Fig. 68 we show the $X$, $Y$, $J_c$ projection of different transfer trajectories for $\varepsilon = 1$N. On the left hand side we have the two transfer orbits for $r_1 = 0.034$ and on the right hand side the two transfer orbits for $r_1 = 0.015$. Here we can see clearly

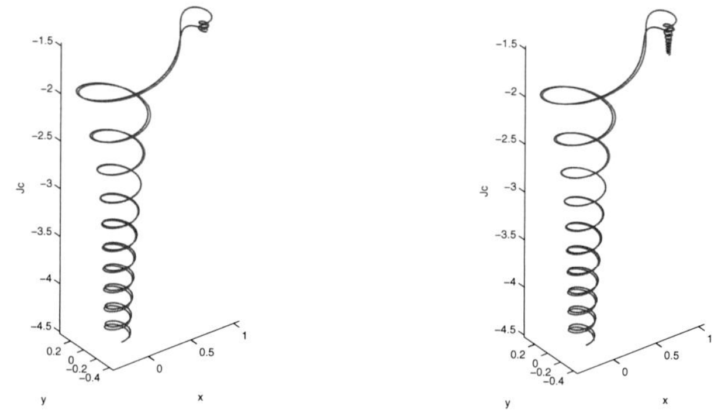


**Fig. 68** $\{XYJ_c\}$ projection of minimum-time transfer trajectories for $\varepsilon = 1\text{N}$, for trajectories of class $\mathscr{C}_1$ (*blue*) and class $\mathscr{C}_2$ (*red*). *Left* $r_1 = 0.034$ and *Right* $r_1 = 0.015$

the structure of the transfer orbit, how $J_c$ increases and decreases spiraling around one of the primaries. The difference between the two plots can be seen in $J_c(t_f)$ that will vary from one problem to the other. This plot suggests that one might be able to describe the strategies in terms of $J_c(t_0)$ and $J_c(t_f)$.

# 4 Conclusion

The detailed numerical study conducted in Sects. 2 and 3 illustrates two features of the three−body problem that are obviously due to the particular topology of the two-body problem (as explained in the introduction, for typical boundary conditions the controllability analysis of [5] entails that one can view the problem as two 2BP coupled by an $L_1$ target): (i) For a given level of thrust, a homotopy in the covering of the angle defining the initial position on the initial orbit allows to unfold and connect local minima associated to different rotation numbers; in particular, local minima of different types (some with many revolutions around the primary, some with large excursions—both clearly not globally minimizing) are indeed connected. (ii) The systematic study of these local minima for fixed thrust level allows to confirm numerically that, when leaving the position free on the initial orbit (here the geostationary one for all tests) and using a homotopy on the level of thrust, one actually follows a path of (at least seemingly) global optimizers for some time. When the thrust is decreased sufficiently, one has to add an extra turn around the initial primary at some point (and possibly around the target one as well), which could result in a bifurcation of the path, or even lead to a discontinuity, that is to the requirement to jump to another branch. It turns that the relevant phenomenon, at least for what is observable in the current computations, is a classic swallowtail singularity [1]: No

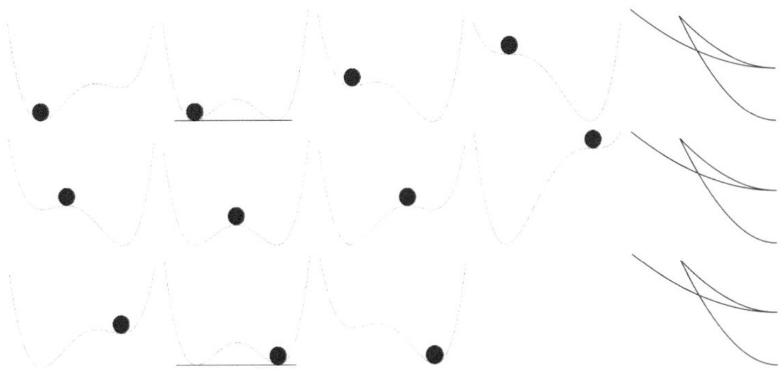

**Fig. 69** Swallowtail singularity. On the first line, the global minimum is changed into a local one (passing through a configuration with two equal global minima, corresponding to the first crossing of the self intersection on the rightmost graph), then into a critical point which is neither a local minimum or maximum (corresponding to the first cusp or turning point on the path). On the second line, a branch of local maxima is described, up to another critical point (second cusp or turning point). On the last line of subplots, local then global maxima are retrieved (passing now through a configuration with two equal global minima corresponding to the second crossing of the self intersection on the rightmost graph). All three rightmost subplots are schematic views of these three connected branches of the $(t_f, \varepsilon)$-path. In this optimal control setting, turning points are associated with conjugate points while self-intersections correspond to cut points.

discontinuity is encountered, and the path connects the (apparently) global solutions for, say, $\varepsilon_1$ and $\varepsilon_2$ ($< \varepsilon_1$) by going through a branch of global then local minima (change at a first cut point), a first turning point (which is also a conjugate point—see the analysis in [5]—and is neither a local minimum or maximum), a branch of local maxima, a second turning point (again a conjugate point), a branch of local then global maxima (change at a second cut point with same cost as the previous one). See Fig. 69 for a schematic picture. The rest of the picture consists of connecting in the $(t_f, \varepsilon)$-space such swallowtail singularities to form the global path (see, e.g., Fig. 21). The typical cut-like point encountered in these situations correspond to very similar though different control strategies; in particular, the two extremals may have the same rotation numbers (see Fig. 22). But there is actually of wealth of extremals corresponding to various structures of the control, and we are far from understanding the global picture at this stage. (See for instance Fig. 40 illustrating three different strategies for the same problem, possibly living on different branches of the $(t_f, \varepsilon)$-homotopy.)

# 5 Tables for GEO to $L_1$ Transfer Problem

**Table 3** Cut Points for GEO to $L_1$ minimum-time transfer problem

| $N^o$ | $t_f$ | $\varepsilon$ (N) | $(x_0, p_0)$ |
|---|---|---|---|
| 1 | 3.2044271 | 2.8177314 | x0 = (−4.3004036e−02 1.0526195e−01 −2.8798280e+00 −8.4404363e−01)<br>p0 = (−8.0686913e+00 1.8446666e+01 −6.1769698e−01 −8.3732696e−02) |
| | 3.2044218 | 2.8177314 | x1 = (7.7386132e−02 6.3360936e−02 −1.7334716e+00 2.4496725e+00)<br>p1 = (5.7983831e+00 8.0411480e+00 −1.3156893e−01 3.8933850e−01) |
| 2 | 3.8375203 | 2.2761716 | x0 = (−1.2063951e−01 1.6203154e−02 −4.4329690e−01 −2.9680477e+00)<br>p0 = (−4.4671254e+01 5.8135647e+00 −2.4592639e−01 −1.4365076e+00) |
| | 3.8375175 | 2.2761716 | x1 = (5.2514551e−02 8.8600069e−02 −2.4239810e+00 1.7692189e+00)<br>p1 = (4.3721245e+00 1.1331595e+01 −3.5478899e−01 4.0145337e−01) |
| 3 | 4.4722368 | 1.9042420 | x0 = (−8.7103192e−02 −8.0089532e−02 2.1911439e+00 −2.0505383e+00)<br>p0 = (−5.5133744e+01 −5.7358673e+01 1.9780637e+00 −1.7979208e+00) |
| | 4.4722337 | 1.9042420 | x1 = (6.9733651e−03 1.0800947e−01 −2.9549967e+00 5.2327211e−01)<br>p1 = (−6.5935699e−01 1.4254660e+01 −6.0595336e−01 2.2366696e−01) |
| 4 | 5.1076234 | 1.6346218 | x0 = (1.8465131e−03 −1.0879282e−01 2.9764280e+00 3.8300820e−01)<br>p0 = (1.2560860e+01 −1.2124635e+02 4.1425055e+00 6.4421919e−01) |
| | 5.1076249 | 1.6346218 | x1 = (−4.7448985e−02 1.0385595e−01 −2.8413617e+00 −9.6565156e−01)<br>p1 = (−8.1438007e+00 1.4247984e+01 −7.4247922e−01 −1.3165896e−01) |
| 5 | 5.7431841 | 1.4309308 | x0 = (7.7334294e−02 −6.3434128e−02 1.7354740e+00 2.4482543e+00)<br>p0 = (1.3695137e+02 −1.0313743e+02 3.3137469e+00 4.9871067e+00) |
| | 5.7431642 | 1.4309308 | x1 = (−9.5138164e−02 7.1730935e−02 −1.9624637e+00 −2.2703646e+00)<br>p1 = (−1.5939313e+01 9.6554337e+00 −6.4632581e−01 −5.7341753e−01) |
| 6 | 6.3789344 | 1.2718365 | x0 = (9.5773675e−02 1.9588193e−02 −5.3590712e−01 2.9527314e+00)<br>p0 = (2.2901189e+02 3.4599146e+01 −1.7065479e+00 7.9999291e+00) |
| | 6.3789252 | 1.2718365 | x1 = (−1.2039301e−01 1.7775384e−02 −4.8631103e−01 −2.9613039e+00)<br>p1 = (−2.0965394e+01 2.8593691e−01 −2.6964121e−01 −9.3925785e−01) |
| 7 | 7.0149950 | 1.1442218 | x0 = (5.1158197e−02 8.9574309e−02 −2.4506349e+00 1.7321108e+00)<br>p0 = (1.8011622e+02 2.3963981e+02 −8.8959222e+00 5.8951334e+00) |
| | 7.0149569 | 1.1442218 | x1 = (−1.1314330e−01 −4.2811500e−02 1.1712661e+00 −2.7629612e+00)<br>p1 = (−2.0185746e+01 −1.1936458e+01 3.2074658e−01 −1.0480066e+00) |

Results obtained from Fig. 18

**Table 4** Initial conditions for minimum-time transfer orbits for $\varepsilon = 10$N and $\theta_0 = 0$ (fixed)

| $k$ | $t_f$ | $p_0$ |
|---|---|---|
| 1 | 2.5599110579 | (−8.10302600967e+00, 3.02060685402e−01, 7.51746305081e−03, −2.72282719406e−01) |
| 2 | 2.4917687101 | (−7.99512361116e+00, 3.62833827849e−01, 9.87071526042e−03, −2.70747978337e−01) |
| 3 | 2.4189259511 | (−7.77134183976e+00, 4.70243193749e−01, 1.38944882768e−02, −2.63770704241e−01) |
| 4 | 2.3422687228 | (−7.61830506790e+00, 5.25624650131e−01, 1.59953031798e−02, −2.60958013110e−01) |
| 5 | 2.2591797538 | (−7.35595976228e+00, 6.30264231036e−01, 1.98961901745e−02, −2.53456263812e−01) |
| 6 | 2.1709334950 | (−7.11929765212e+00, 6.79535789150e−01, 2.16728858950e−02, −2.48023436490e−01) |
| 7 | 2.0733748800 | (−6.78411359981e+00, 7.70584361967e−01, 2.49883101080e−02, −2.38936843908e−01) |
| 8 | 1.9682235824 | (−6.37730302696e+00, 8.14920691231e−01, 2.63907417363e−02, −2.27864801770e−01) |

(continued)

**Table 4** (continued)

| $k$ | $t_f$ | $p_0$ |
|---|---|---|
| 9 | 1.8483950863 | (−5.88075426291e+00, 8.70998413743e−01, 2.81742195599e−02, −2.14116786836e−01) |
| 10 | 1.7156981531 | (−5.06426694761e+00, 9.19001172403e−01, 2.92798458710e−02, −1.89107342495e−01) |
| 11 | 1.5565271364 | (−4.11188635118e+00, 8.96978816242e−01, 2.74753371581e−02, −1.59911895051e−01) |
| 12 | 1.3726059062 | (−1.28979835829e+00, 1.13483961580e+00, 3.40752669425e−02, −6.71095824899e−02) |
| 13 | 1.6797923278 | (2.06357664128e+02, 4.43098626559e+00, 3.96574729885e−01, 7.29822417118e+00) |
| 14 | 2.7126920399 | (3.46017661782e+00, −2.74792593807e+00, −1.17714841093e−01, 3.75634764455e−02) |
| 15 | 2.8397955185 | (−2.03473423827e+00, −2.51445819123e+00, −1.05824599138e−01, −1.77547444466e−01) |

Here, $x_0 = (9.7536855, 0.00, 0.00, 3.0009696)$

**Table 5** Initial conditions for minimum-time transfer orbits for $\varepsilon = 10$N and $\theta_0 = \pi/2$ (fixed)

| $k$ | $t_f$ | $p_0$ |
|---|---|---|
| 1 | 2.5572956763 | (−2.44641634194e−01, −1.01506238330e+01, 4.12858647068e−01, 9.76331017638e−03) |
| 2 | 2.4886728553 | (−1.89607461766e−01, −1.00207778997e+01, 4.08352174314e−01, 7.44373886471e−03) |
| 3 | 2.4160055626 | (−1.58685283710e−01, −9.88623604487e+00, 4.05766385543e−01, 6.13888229610e−03) |
| 4 | 2.3384047229 | (−9.13707083469e−02, −9.68400986037e+00, 3.98201189251e−01, 3.23659167759e−03) |
| 5 | 2.2550784768 | (−4.14972187431e−02, −9.48955468735e+00, 3.93446236657e−01, 1.11399808019e−03) |
| 6 | 2.1651688933 | (3.84033740410e−02, −9.16568566700e+00, 3.80989948634e−01, −2.44111301378e−03) |
| 7 | 2.0665590453 | (1.14827639931e−01, −8.85889814521e+00, 3.71852942749e−01, −5.75137698271e−03) |
| 8 | 1.9585116066 | (2.08588953344e−01, −8.30708874791e+00, 3.50763039836e−01, −1.01280325044e−02) |
| 9 | 1.8359151993 | (3.21190272472e−01, −7.76041413477e+00, 3.31853813310e−01, −1.51754873201e−02) |
| 10 | 1.6977423611 | (4.38111145210e−01, −6.68069683079e+00, 2.92026306333e−01, −2.10584817444e−02) |
| 11 | 1.5315914535 | (5.95238470964e−01, −5.50225170504e+00, 2.47157039202e−01, −2.86754751281e−02) |
| 12 | 1.3337000522 | (8.41013821159e−01, −2.36998678661e+00, 1.40666039253e−01, −4.16602092444e−02) |
| 13 | 2.4460134134 | (−3.41756924766e+00, 9.30607903262e+01, −3.30975784324e+00, 2.28357418848e−01) |
| 14 | 2.6648851780 | (1.04603325391e+00, 1.15237952755e+01, −5.20036306956e−01, −3.77212192660e−02) |
| 15 | 2.7603423501 | (2.59359121331e+00, 6.50388228992e+00, −3.36307995906e−01, −1.04272573878e−01) |

Here, $x_0 = (-0.0121530, 0.10968985, -3.000969693, 0.00)$

**Table 6** Initial conditions for minimum-time transfer orbits for $\varepsilon = 10$N and $\theta_0 = \pi$ (fixed)

| $k$ | $t_f$ | $p_0$ |
|---|---|---|
| 1 | 2.5527521947 | (1.06375513461e+01, 1.47617758655e+00, 5.76423437473e−02, 4.26278852503e−01) |
| 2 | 2.4822720399 | (1.04148331743e+01, 1.51294121857e+00, 5.94360744304e−02, 4.15282668813e−01) |
| 3 | 2.4076155799 | (1.01096843737e+01, 1.56726510132e+00, 6.19943495282e−02, 4.01904923185e−01) |
| 4 | 2.3275603558 | (9.79781572723e+00, 1.59332432531e+00, 6.34927961111e−02, 3.86869947665e−01) |
| 5 | 2.2416129890 | (9.40007362986e+00, 1.65348347180e+00, 6.64030056517e−02, 3.69487185012e−01) |
| 6 | 2.1483459356 | (8.93393057118e+00, 1.66987158824e+00, 6.77164579804e−02, 3.48137299904e−01) |
| 7 | 2.0461306463 | (8.40135607228e+00, 1.72740096089e+00, 7.07069439315e−02, 3.24679976788e−01) |
| 8 | 1.9331012440 | (7.61364755861e+00, 1.74635539983e+00, 7.24930725671e−02, 2.91092295102e−01) |
| 9 | 1.8052056817 | (6.83154051800e+00, 1.76666004320e+00, 7.43502285645e−02, 2.56569966663e−01) |
| 10 | 1.6583042568 | (5.29715773055e+00, 1.81736951791e+00, 7.81883129707e−02, 1.94875672822e−01) |
| 11 | 1.4833856840 | (3.83493364971e+00, 1.72669505097e+00, 7.64256922974e−02, 1.32959769935e−01) |
| 12 | 1.2663896517 | (−6.72434494581e−01, 1.82222989708e+00, 8.21403775003e−02, −4.33956185640e−02) |
| 13 | 2.7731212505 | (−6.53927977825e+00, −3.06988032726e+00, −1.17632989071e−01, −2.73303542519e−01) |
| 14 | 2.8708627397 | (−4.21419866917e+00, −3.15031550790e+00, −1.19145737733e−01, −2.14919176388e−01) |

Here, $x_0 = (-0.1218428559, 0.00, 0.00, -3.00096969)$

**Table 7** Initial conditions for minimum-time transfer orbits for $\varepsilon = 10\text{N}$ and $\theta_0 = -\pi/2$ (fixed)

| $k$ | $t_f$ | $p_0$ |
|---|---|---|
| 1 | 2.5907703198 | (−1.61091775746e+00, 8.96037592684e+00, −3.04417077032e−01, 6.67655131141e−02) |
| 2 | 2.5224151691 | (−1.55066132206e+00, 8.65746432330e+00, −2.90472826121e−01, 6.48366258398e−02) |
| 3 | 2.4501991195 | (−1.47912650562e+00, 8.34215239832e+00, −2.75824570744e−01, 6.24872471641e−02) |
| 4 | 2.3732312047 | (−1.38995583017e+00, 7.99971550025e+00, −2.60564073245e−01, 5.94653100445e−02) |
| 5 | 2.2909754645 | (−1.31509971463e+00, 7.55737613122e+00, −2.40862906381e−01, 5.71424148173e−02) |
| 6 | 2.2024783905 | (−1.18473581289e+00, 7.17707239902e+00, −2.24753134059e−01, 5.25318628076e−02) |
| 7 | 2.1060684965 | (−1.09371385568e+00, 6.54755398598e+00, −1.97864122169e−01, 4.97973648749e−02) |
| 8 | 2.0011242565 | (−9.22196318408e−01, 6.07973280392e+00, −1.79300208002e−01, 4.36502684663e−02) |
| 9 | 1.8831711949 | (−7.94589175887e−01, 5.15470225726e+00, −1.41779328238e−01, 3.98082008222e−02) |
| 10 | 1.7521393123 | (−5.80577218930e−01, 4.42683184881e+00, −1.14541558497e−01, 3.22645112679e−02) |
| 11 | 1.5971627494 | (−3.91234096149e−01, 2.89940049164e+00, −5.62359857709e−02, 2.66110707927e−02) |
| 12 | 1.4178043955 | (−7.38577985007e−02, 1.04802410397e+00, 1.11883639221e−02, 1.54629906135e−02) |
| 13 | 1.2200224705 | (6.62105764118e−01, −4.51174644544e+00, 2.10821364039e−01, −1.97873327616e−02) |
| 14 | 2.7726988590 | (1.56502812470e+00, −2.70258668378e+00, −2.67064625097e−03, −5.02608779760e−02) |
| 15 | 2.8760265384 | (1.98895030587e+00, 6.05889722576e−01, −1.29117271405e−01, −7.01041330191e−02) |

Here, $x_0 = (-0.012153, -0.10968985, 3.0009696, 0.00)$

**Table 8** Initial conditions for minimum-time transfer orbits for $\varepsilon = 1$N and $\theta_0 = 0$ (fixed)

| $k$ | $t_f$ | $p_0$ |
|---|---|---|
| 1 | 1.1734093510e+01 | (−5.00293794181e+01, −8.92810787059e+00, −3.59642774825e−01, −1.86911263217e+00) |
| 2 | 1.1552149062e+01 | (−4.64542148014e+01, −8.85541434942e+00, −3.57354266774e−01, −1.68458131918e+00) |
| 3 | 1.1366823527e+01 | (−4.27162042580e+01, −8.50085033910e+00, −3.43959041446e−01, −1.49382338146e+00) |
| 4 | 1.1178437340e+01 | (−3.90281882445e+01, −7.78774378223e+00, −3.16398638411e−01, −1.30767766999e+00) |
| 5 | 1.0987120064e+01 | (−3.51231910879e+01, −6.81437528378e+00, −2.78577923093e−01, −1.11807707778e+00) |
| 6 | 1.0793552884e+01 | (−3.14356984450e+01, −5.56528120994e+00, −2.29828878667e−01, −9.43717166154e−01) |
| 7 | 1.0597583348e+01 | (−2.79786185901e+01, −4.07511367761e+00, −1.71505714926e−01, −7.88549733091e−01) |
| 8 | 1.0399786390e+01 | (−2.46548436680e+01, −2.46000281169e+00, −1.08203666943e−01, −6.47556492294e−01) |
| 9 | 1.0200113173e+01 | (−2.19410610646e+01, −7.51713433563e−01, −4.10953162593e−02, −5.43897525038e−01) |
| 10 | 9.9985467473e+00 | (−1.93444226096e+01, 9.50690537301e−01, 2.58472323136e−02, −4.54373346619e−01) |
| 11 | 9.7951297247e+00 | (−1.73746355786e+01, 2.54951122509e+00, 8.88514567696e−02, −4.02551176887e−01) |
| 12 | 9.5893495406e+00 | (−1.55102319471e+01, 4.03093493029e+00, 1.47292632024e−01, −3.63484827095e−01) |
| 13 | 9.3811110629e+00 | (−1.41307893780e+01, 5.24777692114e+00, 1.95387270812e−01, −3.53693100684e−01) |
| 14 | 9.1697915366e+00 | (−1.24787976530e+01, 6.30173897074e+00, 2.37044906641e−01, −3.40124616465e−01) |
| 15 | 8.9564212414e+00 | (−9.96070518965e+00, 7.14686809501e+00, 2.70390736552e−01, −3.01555539885e−01) |
| 16 | 8.7418485088e+00 | (−6.83558281878e+00, 7.74226722168e+00, 2.93836955496e−01, −2.47118606219e−01) |
| 17 | 8.5274300169e+00 | (−2.87799323127e+00, 8.02137698122e+00, 3.04743143954e−01, −1.67260037676e−01) |
| 18 | 8.3147860737e+00 | (2.35469713546e+00, 7.93462860048e+00, 3.01178035452e−01, −4.34661909985e−02) |
| 19 | 8.1065871814e+00 | (9.08133786760e+00, 7.45204156306e+00, 2.82147510584e−01, 1.35684066835e−01) |

Here, $x_0 = (9.7536855, 0.00, 0.00, 3.0009696)$

**Table 9** Initial conditions for minimum-time transfer orbits for $\varepsilon = 1\text{N}$ and $\theta_0 = \pi/2$ (fixed)

| $k$ | $t_f$ | $p_0$ |
|---|---|---|
| 1 | 1.1617836482e+01 | (−5.86265374666e−01, −3.55563514170e+01, 1.05337756761e+00, 1.33730970732e−02) |
| 2 | 1.1448706422e+01 | (−1.95081480500e+00, −3.38390778413e+01, 1.00258679684e+00, 6.71037566339e−02) |
| 3 | 1.1278103045e+01 | (−3.20864825547e+00, −3.26693195566e+01, 9.81204357826e−01, 1.16722579010e−01) |
| 4 | 1.1106156553e+01 | (−4.35464384443e+00, −3.14871164620e+01, 9.65464703194e−01, 1.61949064207e−01) |
| 5 | 1.0932241536e+01 | (−5.31187787666e+00, −3.08348948035e+01, 9.76384410476e−01, 1.99807373235e−01) |
| 6 | 1.0756369590e+01 | (−6.10327738038e+00, −3.00934747065e+01, 9.88636669326e−01, 2.31120029517e−01) |
| 7 | 1.0577724397e+01 | (−6.66192909273e+00, −2.97815306058e+01, 1.02109064306e+00, 2.53299344487e−01) |
| 8 | 1.0396252516e+01 | (−7.00372755398e+00, −2.93378374216e+01, 1.05145175237e+00, 2.66902151015e−01) |
| 9 | 1.0210946491e+01 | (−7.11002130272e+00, −2.91066374962e+01, 1.09142290415e+00, 2.71227569949e−01) |
| 10 | 1.0021678797e+01 | (−6.95232923626e+00, −2.87290157156e+01, 1.12625828544e+00, 2.65104252329e−01) |
| 11 | 9.8273162387e+00 | (−6.57280516945e+00, −2.82964650663e+01, 1.15892547212e+00, 2.50208359620e−01) |
| 12 | 9.6276234212e+00 | (−5.90387876777e+00, −2.76421558052e+01, 1.18053824596e+00, 2.23843275857e−01) |
| 13 | 9.4214770376e+00 | (−5.02818856177e+00, −2.65752199328e+01, 1.18564798030e+00, 1.89258893999e−01) |
| 14 | 9.2093222798e+00 | (−3.92185928576e+00, −2.45693574627e+01, 1.15400456124e+00, 1.45449008253e−01) |
| 15 | 8.9910694916e+00 | (−2.55235178987e+00, −2.15715237005e+01, 1.08180604209e+00, 9.11178102638e−02) |
| 16 | 8.7671882439e+00 | (−9.03453037648e−01, −1.75530736265e+01, 9.65375348447e−01, 2.56172123980e−02) |
| 17 | 8.5383024028e+00 | (1.01520847489e+00, −1.20346776264e+01, 7.83887782000e−01, −5.07298023085e−02) |
| 18 | 8.3056624144e+00 | (3.15812939064e+00, −4.39749794305e+00, 5.11247116113e−01, −1.36177598815e−01) |
| 19 | 8.0723986254e+00 | (5.42827481965e+00, 6.63751133717e+00, 9.73408800576e−02, −2.26732255341e−01) |

Here, $x_0 = (-0.0121530, 0.10968985, -3.000969693, 0.00)$

**Table 10** Initial conditions for minimum-time transfer orbits for $\varepsilon = 1$N and $\theta_0 = \pi$ (fixed)

| $k$ | $t_f$ | $p_0$ |
|---|---|---|
| 1 | 1.1646032796e+01 | (4.35820317970e+01, −6.97203875251e+00, −2.67592133177e−01, 1.66301217141e+00) |
| 2 | 1.1485236628e+01 | (4.37585430926e+01, −6.69039944005e+00, −2.56610149092e−01, 1.70903376102e+00) |
| 3 | 1.1321046566e+01 | (4.37240293957e+01, −6.25620042964e+00, −2.39574105051e−01, 1.74655569983e+00) |
| 4 | 1.1152733288e+01 | (4.37675745318e+01, −5.62229590496e+00, −2.14649926510e−01, 1.78436089077e+00) |
| 5 | 1.0980154177e+01 | (4.35089211994e+01, −4.82799184320e+00, −1.83347465212e−01, 1.80865451918e+00) |
| 6 | 1.0802454544e+01 | (4.32226530120e+01, −3.84571340396e+00, −1.44603533348e−01, 1.82732965694e+00) |
| 7 | 1.0619484295e+01 | (4.25243806843e+01, −2.70997766117e+00, −9.97386371548e−02, 1.82657605409e+00) |
| 8 | 1.0430292744e+01 | (4.16628910456e+01, −1.40939476929e+00, −4.83235458265e−02, 1.81245666478e+00) |
| 9 | 1.0234702320e+01 | (4.02817932505e+01, 1.81467268683e−02, 8.18204640558e−03, 1.77302515080e+00) |
| 10 | 1.0031771052e+01 | (3.84999672540e+01, 1.54946772480e+00, 6.88626431563e−02, 1.70778591237e+00) |
| 11 | 9.8212253563e+00 | (3.60965807585e+01, 3.15294489415e+00, 1.32484589900e−01, 1.61129767954e+00) |
| 12 | 9.6024262387e+00 | (3.30037280021e+01, 4.73129817551e+00, 1.95230574044e−01, 1.47523999498e+00) |
| 13 | 9.3749180129e+00 | (2.91609229803e+01, 6.27129559125e+00, 2.56569834778e−01, 1.30076781743e+00) |
| 14 | 9.1390248144e+00 | (2.39414709043e+01, 7.67238191823e+00, 3.12601135363e−01, 1.06314010901e+00) |
| 15 | 8.8950906687e+00 | (1.75803390429e+01, 8.84229708176e+00, 3.59632148056e−01, 7.70429590121e−01) |
| 16 | 8.6444531496e+00 | (9.88532815690e+00, 9.64112370784e+00, 3.92074238174e−01, 4.16075141018e−01) |
| 17 | 8.3893607950e+00 | (2.91016617542e−01, 9.91527824166e+00, 4.03440280820e−01, −1.84749248199e−02) |
| 18 | 8.1349942117e+00 | (−1.21682629238e+01, 9.49406823282e+00, 3.85908930770e−01, −5.63637830626e−01) |
| 19 | 7.8908221594e+00 | (−2.81958350894e+01, 8.32334870019e+00, 3.37763204171e−01, −1.23276529113e+00) |

Here, $x_0 = (-0.1218428559, 0.00, 0.00, -3.00096969)$

**Table 11** Initial conditions for minimum-time transfer orbits for $\varepsilon = 1$N and $\theta_0 = -\pi/2$ (fixed)

| $k$ | $t_f$ | $p_0$ |
|---|---|---|
| 1 | 1.1836259008e+01 | (1.23953384370e−01, 5.58014204190e+01, −2.34355121544e+00, 7.31217010374e−04) |
| 2 | 1.1668173067e+01 | (−1.12637871873e+00, 5.49368632729e+01, −2.31314733956e+00, 5.02034025451e−02) |
| 3 | 1.1494977738e+01 | (−2.43013759564e+00, 5.37800558199e+01, −2.26636836799e+00, 1.01834227745e−01) |
| 4 | 1.1316444909e+01 | (−3.73867482357e+00, 5.21481628093e+01, −2.19301350363e+00, 1.53734809391e−01) |
| 5 | 1.1132147759e+01 | (−5.03123349646e+00, 5.02184792008e+01, −2.10070221912e+00, 2.05060506552e−01) |
| 6 | 1.0942085363e+01 | (−6.24652949505e+00, 4.76050485755e+01, −1.97431837106e+00, 2.53439391801e−01) |
| 7 | 1.0745912324e+01 | (−7.30567854186e+00, 4.47349782782e+01, −1.82838840429e+00, 2.95710475343e−01) |
| 8 | 1.0543714233e+01 | (−8.17915954565e+00, 4.11508108368e+01, −1.64832839207e+00, 3.30754209295e−01) |
| 9 | 1.0335506711e+01 | (−8.76509910607e+00, 3.70540225185e+01, −1.44156444675e+00, 3.54525410401e−01) |
| 10 | 1.0121474693e+01 | (−8.97432560521e+00, 3.27551946459e+01, −1.22143574046e+00, 3.63463384917e−01) |
| 11 | 9.9019153074e+00 | (−8.80015208947e+00, 2.76983915872e+01, −9.71301869386e−01, 3.57354768910e−01) |
| 12 | 9.6774814837e+00 | (−8.16195680841e+00, 2.27577796399e+01, −7.26323848357e−01, 3.32910646341e−01) |
| 13 | 9.4484234174e+00 | (−7.06575685391e+00, 1.75955941388e+01, −4.79250762317e−01, 2.90388230931e−01) |
| 14 | 9.2155352697e+00 | (−5.60762563079e+00, 1.23669861038e+01, −2.38161479064e−01, 2.33535020863e−01) |
| 15 | 8.9795679223e+00 | (−3.77052273681e+00, 7.74479647932e+00, −3.49039253972e−02, 1.61528029972e−01) |
| 16 | 8.7414772457e+00 | (−1.71785807871e+00, 2.03434555631e+00, 1.96943233655e−01, 8.03156792010e−02) |
| 17 | 8.5052543934e+00 | (5.06092325796e−01, −4.46899050442e+00, 4.42579821491e−01, −8.24699078928e−03) |
| 18 | 8.2741044454e+00 | (2.74485776680e+00, −1.16500671006e+01, 6.94421237073e−01, −9.74865696033e−02) |
| 19 | 8.0516175632e+00 | (4.82917625787e+00, −2.00110461836e+01, 9.71050333412e−01, −1.80716555601e−01) |
| 20 | 7.8468442012e+00 | (6.70860143528e+00, −5.23882372324e+01, 2.10673692237e+00, −2.58577208010e−01) |

Here, $x_0 = (-0.012153, -0.10968985, 3.0009696, 0.00)$

**Table 12** Local minima of the homotopic path $\theta_0$–$t_f$ for $\theta_0$ fixed and $\varepsilon = 10\text{N}$

| Num. | $t_f$ | $(x_0, p_0)$ |
|---|---|---|
| 1 | 2.5559195559 | x0 = (6.21066150680e−02, 8.07302549835e−02, −2.20867323166e+00, 2.03164506327e+00)<br>p0 = (−6.66443552579e+00, −6.25173710798e+00, 2.71700855321e−01, −2.16542651328e−01) |
| 2 | 2.4877948854 | x0 = (5.78471160865e−02, 8.44502708426e−02, −2.31044795603e+00, 1.91511079266e+00)<br>p0 = (−6.28986613857e+00, −6.58580267048e+00, 2.84096838074e−01, −2.05132653869e−01) |
| 3 | 2.4152338141 | x0 = (5.89589525192e−02, 8.35161943465e−02, −2.28489285585e+00, 1.94552917010e+00)<br>p0 = (−6.31179701154e+00, −6.36448410888e+00, 2.80093512573e−01, −2.05851412612e−01 ) |
| 4 | 2.3383263886 | x0 = (4.30199848124e−02, 9.48040413121e−02, −2.59371345157e+00, 1.50946004928e+00)<br>p0 = (−5.01682430960e+00, −7.53159608900e+00, 3.20380424397e−01, −1.63281437222e−01) |
| 5 | 1.2164636130 | x0 = (3.40824580273e−02, −9.94693265805e−02, 2.72134949944e+00, 1.26494111184e+00)<br>p0 = (4.16086874010e+00, −6.01023069117e+00, 2.79810763690e−01, 8.02669222580e−02 ) |
| 6 | 2.6648355866 | x0 = (−6.43377120172e−03, 1.09540654392e−01, −2.99688773664e+00, 1.56470551902e−01)<br>p0 = (1.84143605513e+00, 1.15160214759e+01, −5.22937010460e−01, −1.80219702492e−02) |
| 7 | 2.7553318369 | x0 = (−5.24985178093e−02, 1.02000508900e−01, −2.79059930716e+00, −1.10380012079e+00)<br>p0 = (−2.83255512265e+00, 8.32582148917e+00, −4.10828349203e−01, −1.79352374657e−01) |

**Table 13** Local Minima of the homotopic path $\theta_0$-$t_f$ for $\theta_0$ fixed and $\varepsilon = 5\text{N}$

| Num. | $t_f$ | $(x_0, p_0)$ |
|---|---|---|
| 1 | 3.7536181208 | x0 = (−1.94215853959e−02, 1.09448764991e−01, −2.99437376388e+00, −1.98858903633e−01)<br>p0 = (−5.82743545042e−01, −1.45939805915e+01, 5.28108712479e−01, 9.17938940083e−02 ) |
| 2 | 3.6509835654 | x0 = (−1.69972571399e−02, 1.09582834731e−01, −2.99804173504e+00, −1.32532482085e−01)<br>p0 = (−9.53230835340e−01, −1.42854614796e+01, 5.24210432656e−01, 8.11068337560e−02) |
| 3 | 3.5439391311 | x0 = (−1.68414052537e−02, 1.09589613320e−01, −2.99822718828e+00, −1.28268580164e−01)<br>p0 = (−1.00456664254e+00, −1.39836723639e+01, 5.20093415016e−01, 8.08316251597e−02 ) |
| 4 | 3.4300586117 | x0 = (−2.71066881650e−02, 1.08665779839e−01, −2.97295232347e+00, −4.09113172891e−01)<br>p0 = (3.57976926298e−01, −1.39256054397e+01, 5.21330369402e−01, 1.28696805687e−01) |
| 5 | 1.9938793490 | x0 = (4.57631694913e−02, −9.31535390469e−02, 2.54855788786e+00, 1.58450996175e+00)<br>p0 = (8.68472172157e+00, -1.16744246852e+01, 5.34314865825e−01, 2.79941139278e−01 ) |
| 6 | 5.1452276858 | x0 = (−8.18646156795e−02, -8.46885779963e−02, 2.31696772525e+00, −1.90721780228e+00)<br>p0 = (−1.46895925996e+01, −1.67744576566e+01, 7.48345214075e−01, −5.83852266304e−01) |
| 7 | 5.2618852143 | x0 = (6.37976388579e−02, −9.67713587949e−02, 2.64753666151e+00, −1.41292916054e+00)<br>p0 = (−9.33069203371e+00, −1.67310113847e+01, 7.85284331298e−01, −4.04362665917e−01 ) |
| 8 | 5.3718352579 | x0 = (−5.30765714538e−02, −1.01769965950e−01, 2.78429195632e+00, −1.11961490562e+00)<br>p0 = (−7.46146481108e+00, −1.72142885351e+01, 8.46354933519e−01, −3.20501384394e−01) |

**Table 14** Local Minima of the homotopic path $\theta_0$-$t_f$ for $\theta_0$ fixed and $\varepsilon = 1$N

| Num. | $t_f$ | $(x_0, p_0)$ |
|---|---|---|
| 1 | 1.1613772473 | x0 = (−4.94521037379e−02, 1.03153484481e−01, −2.82214319740e+00, −1.02045425234e+00)<br>p0 = (9.01995283736e+00, −3.46406158244e+01, 9.87371645985e−01, 4.85208334522e−01) |
| 2 | 1.1447550943 | x0 = (−3.19708106454e−02, 1.07884748127e−01, −2.95158432661e+00, −5.42189144473e−01)<br>p0 = (2.90230260337e+00, −3.41066967680e+01, 9.95854575771e−01, 3.05746057876e−01) |
| 3 | 1.1278080850 | x0 = (−1.51263711042e−02, 1.09649548919e−01, −2.99986694718e+00, −8.13475092700e−02)<br>p0 = (−2.50988487628e+00, −3.28125831403e+01, 9.83895951743e−01, 1.50964707129e−01) |
| 4 | 1.1105567895 | x0 = (2.95705857691e−03, 1.08644146766e−01, −2.97236047116e+00, 4.13391260989e−01)<br>p0 = (−7.69998223257e+00, −3.02620884245e+01, 9.33675575493e−01, −2.24004350034e−03) |
| 5 | 1.0929544056 | x0 = (2.04744264680e−02, 1.04724952106e−01, −2.86513647766e+00, 8.92643327742e−01)<br>p0 = (−1.21596048543e+01, −2.71809934984e+01, 8.71972929361e−01, −1.36697854580e−01) |
| 6 | 1.0750144861 | x0 = (3.83142161995e−02, 9.73905775437e−02, −2.66447766925e+00, 1.38071643052e+00)<br>p0 = (−1.59157802958e+01, −2.29695613710e+01, 7.73993351114e−01, −2.54453250197e−01) |
| 7 | 1.0566954554 | x0 = (5.44818172342e−02, 8.71301649529e−02, −2.38376632206e+00, 1.82304065745e+00)<br>p0 = (−1.86958979730e+01, −1.85218216807e+01, 6.71937688249e−01, −3.48430037725e−01) |
| 8 | 1.0379987132 | x0 = (6.88212850245e−02, 7.39934433897e−02, −2.02436295744e+00, 2.21534957152e+00)<br>p0 = (−2.02518665484e+01, −1.35149987541e+01, 5.52549152903e−01, −4.05199546746e−01) |
| 9 | 1.0188925627 | x0 = (8.03393763200e−02, 5.89663024700e−02, −1.61324021412e+00, 2.53046934823e+00)<br>p0 = (−2.08004284902e+01, −8.77291310702e+00, 4.43835886576e−01, −4.38897740719e−01) |
| 10 | 9.9936149094 | x0 = (8.95455084671e−02, 4.11008259435e−02, −1.12446435452e+00, 2.78233697387e+00)<br>p0 = (−2.00552248948e+01, -3.91048195304e+00, 3.29850374180e−01, −4.36668392516e−01) |
| 11 | 9.7938762650 | x0 = (9.53384673112e−02, 2.18506056371e−02, −5.97803732642e−01, 2.94082468252e+00)<br>p0 = (−1.84286807100e+01, 3.08770987062e−01, 2.33200787107e−01, −4.17738961747e−01) |
| 12 | 9.5893510522 | x0 = (9.75357647120e−02, 4.89297497890e−04, −1.33865337862e−02, 3.00093983950e+00)<br>p0 = (−1.55475035168e+01, 3.99090490857e+00, 1.50081331902e−01, −3.64314206604e−01) |
| 13 | 9.3800526379 | x0 = (9.55661265389e−02, −2.06991372746e−02, 5.66301077909e−01, 2.94705313854e+00)<br>p0 = (−1.21277045301e+01, 6.54661511552e+00, 9.57838813041e−02, −3.03714475156e−01) |
| 14 | 9.1654637945 | x0 = (8.88535253959e−02, −4.27731969234e−02, 1.17021821740e+00, 2.76340523030e+00)<br>p0 = (−7.71976404968e+00, 7.84981751360e+00, 7.58962825540e−02, −2.16097777716e−01) |
| 15 | 8.9466593775 | x0 = (7.66012223247e−02, −6.44558182773e−02, 1.76342612175e+00, 2.42819838839e+00)<br>p0 = (−2.49582677532e+00, 7.08434895345e+00, 1.15517251704e−01, −1.06074141800e−01) |
| 16 | 8.7241872271 | x0 = (5.72812946469e−02, −8.49160952469e−02, 2.32319229695e+00, 1.89963066482e+00)<br>p0 = (2.35166174593e+00, 3.73477013320e+00, 2.26388522820e−01, −1.23043980769e−02) |
| 17 | 8.4992141171 | x0 = (3.14063514616e−02, −1.00669992540e−01, 2.75419813551e+00, 1.19172636803e+00)<br>p0 = (5.39316031292e+00, −2.35824421954e+00, 4.16994199205e−01, 2.06174555327e−02) |
| 18 | 8.2736157050 | x0 = (3.92050344557e−05, −1.09010158391e−01, 2.98237406617e+00, 3.33562638939e−01)<br>p0 = (4.93801945083e+00, −1.09681340969e+01, 6.81788653598e−01, −5.92784509757e−02) |
| 19 | 8.0504545877 | x0 = (−3.23905959857e−02, −1.07806791098e−01, 2.94945152519e+00, −5.53673917369e−01)<br>p0 = (−5.59633476936e−01, −2.08904205668e+01, 9.86344334901e−01, −3.07897514898e−01) |
| 20 | 7.8363385435 | x0 = (−6.50028495259e−02, −9.61184576538e−02, 2.62967414796e+00, −1.44590213383e+00)<br>p0 = (−1.65966613960e+01, −3.56290598456e+01, 1.46688275370e+00, −9.16052717119e−01) |

# 6 Tables from GEO to MO Transfer Problem

**Table 15** ($\mathscr{C}_1$ class) Initial conditions for minimum-time transfer orbits for $\varepsilon = 10$N and $\theta_0 = \pi$, $x_0 = (-0.121842, 0.00, 0.00, -3.00096)$

| $k$ | $t_f$ | $p_0$ |
|---|---|---|
| 1 | 2.6295366742 | (1.06192601736e+01, 1.37248310916e+00, 5.35735968598e−02, 4.25981151967e−01) |
| 2 | 2.5591513809 | (1.03908338033e+01, 1.40394233209e+00, 5.51645610936e−02, 4.14831750559e−01) |
| 3 | 2.4845856122 | (1.01139459577e+01, 1.46253515025e+00, 5.78478667723e−02, 4.02851462389e−01) |
| 4 | 2.4046902936 | (9.78964388018e+00, 1.48536195756e+00, 5.92289931606e−02, 3.87549322564e−01) |
| 5 | 2.3188203948 | (9.43753819675e+00, 1.55135836535e+00, 6.23078750780e−02, 3.72156376196e−01) |
| 6 | 2.2258097799 | (8.94542905348e+00, 1.56715282851e+00, 6.36188551329e−02, 3.50206904265e−01) |
| 7 | 2.1236421514 | (8.46157104228e+00, 1.62768549326e+00, 6.66669404664e−02, 3.28774925899e−01) |
| 8 | 2.0109987788 | (7.64680356801e+00, 1.65788540815e+00, 6.89101321311e−02, 2.94809550939e−01) |
| 9 | 1.8831760208 | (6.90929970278e+00, 1.67399427716e+00, 7.05530473416e−02, 2.62079313820e−01) |
| 10 | 1.7368384045 | (5.39899067667e+00, 1.74013293816e+00, 7.49642789303e−02, 2.01920571395e−01) |
| 11 | 1.5620914595 | (3.92859812720e+00, 1.65447845805e+00, 7.34194614218e−02, 1.40088341785e−01) |
| 12 | 1.3450368533 | (−1.07601143153e−01, 1.77014236738e+00, 8.04240137313e−02, −1.74291709898e−02) |
| 13 | 2.7874623764 | (−6.49469099937e+00, −2.98006923087e+00, −1.14119319520e−01, −2.75907575312e−01) |
| 14 | 2.8890843383 | (−4.08758212257e+00, −2.98692269111e+00, −1.12596624508e−01, −2.15040713745e−01) |

# 7 Summary of the Cut Points on the GEO to $L_1$ Transfer

In this Appendix we summarize the results for the all the CUT points that we have found for the GEO to $L_1$ transfer problem. The initial conditions for the CUT points are summarized in Table 3. For each pair of cut points we have done the same analysis. First for each cut point we have computed the transfer trajectory, the energy variation along the trajectory ($J_c(t)$), the control along the trajectory and also the variation of

**Table 16** ($\mathscr{C}_1$ class) Initial conditions for minimum-time transfer orbits for $\varepsilon = 10\text{N}$ and $\theta_0 = 0$, $x_0 = (0.0975368, 0.00, 0.00, 3.00096)$

| $k$ | $t_f$ | $p_0$ |
|---|---|---|
| 1 | 2.6382335901 | (−8.15469444647e+00, 1.70467856003e−01, 2.36653029038e−03, −2.73809684344e−01) |
| 2 | 2.5701583528 | (−8.02693895139e+00, 2.42252173754e−01, 5.11383610870e−03, −2.71042424433e−01) |
| 3 | 2.4971963201 | (−7.80352301218e+00, 3.38330092208e−01, 8.68257649872e−03, −2.63922213562e−01) |
| 4 | 2.4206053972 | (−7.62548760869e+00, 4.10629220864e−01, 1.14039103010e−02, −2.59643712704e−01) |
| 5 | 2.3373895493 | (−7.36482326796e+00, 5.01793409601e−01, 1.47652984972e−02, −2.52022253778e−01) |
| 6 | 2.2491972124 | (−7.09747508042e+00, 5.76972246145e−01, 1.75119377896e−02, −2.44905008216e−01) |
| 7 | 2.1515314965 | (−6.76586878979e+00, 6.50789920666e−01, 2.01398087796e−02, −2.35723176492e−01) |
| 8 | 2.0463772193 | (−6.32305321005e+00, 7.36876168589e−01, 2.31369509217e−02, −2.22847227010e−01) |
| 9 | 1.9265529833 | (−5.83277580845e+00, 7.67061071412e−01, 2.38895759600e−02, −2.09001197960e−01) |
| 10 | 1.7936072565 | (−4.97749858940e+00, 8.76114686754e−01, 2.73632618304e−02, −1.82379844772e−01) |
| 11 | 1.6348670072 | (−4.03983916332e+00, 8.17496491769e−01, 2.41026868298e−02, −1.53093741527e−01) |
| 12 | 1.4489820024 | (−1.21477895977e+00, 1.08381100292e+00, 3.18058539495e−02, −5.93687133649e−02) |
| 13 | 1.6913726515 | (1.92911022593e+02, 4.09909700389e+00, 3.68559661513e−01, 6.83488813468e+00) |
| 14 | 2.7288387842 | (3.09692369616e+00, −2.60108464159e+00, −1.12213330995e−01, 2.17133715663e−02) |
| 15 | 2.8589830417 | (−2.22653391549e+00, −2.36994535022e+00, −9.99958635019e−02, −1.85205984442e−01) |

$H_{1,2}$. Moreover, for each cut solution we have integrated both optimal solutions back and forward in time ($t \in [-t_f, 2t_t]$, where $t_f$ is the transfer time). Finally, for the solutions on the homotopic curve close to them ($t_f^* \in [t_f - 0.15 : t_f + 0.15]$), for each solutions we have computed some distinctive parameters of the transfer orbits, trying to characterize their passage. In the plots that we will see, the number of turns around the Earth, and the number of times that $|(H_1, H_2)|$ comes close to zero (in particular $|(H_1, H_2)| < 0.05$).

In Fig. 70 we show the homotopic curve $t_f$ versus $\varepsilon$ and the same curve plotting $\theta_0$ versus $\varepsilon$, where $\theta_0$ in the angle that parameterizes the initial condition on the departure GEO orbit. In both plots we have highlighted in green the solutions close to the CUT pair, which are the solutions that we have analyzed. Figures 71, 72 and

**Table 17** ($\mathscr{C}_1$ class) Initial conditions for minimum-time transfer orbits for $\varepsilon = 10$N and $\theta_0 = \pi/2$, $x_0 = (-0.012153, 0.109689, -3.00096, 0.00)$

| $k$ | $t_f$ | $p_0$ |
|---|---|---|
| 1 | 2.6338132940 | (−2.58170927519e−01, −1.00417479238e+01, 4.05017074760e−01, 1.00460710975e−02) |
| 2 | 2.5653207218 | (−2.07888498661e−01, −9.89425530106e+00, 3.99790676407e−01, 7.89106089518e−03) |
| 3 | 2.4924927065 | (−1.85843554959e−01, −9.77354474323e+00, 3.97814405932e−01, 6.95648216568e−03) |
| 4 | 2.4150831353 | (−1.25626071521e−01, −9.55623627017e+00, 3.89793003207e−01, 4.31969409249e−03) |
| 5 | 2.3315729003 | (−8.39820060625e−02, −9.37472183474e+00, 3.85532601096e−01, 2.53783657098e−03) |
| 6 | 2.2419382962 | (−1.31487906225e−02, −9.04030262447e+00, 3.73024539442e−01, −6.61312444426e−04) |
| 7 | 2.1431311332 | (5.52757105262e−02, −8.74409478565e+00, 3.64176114835e−01, −3.64700523503e−03) |
| 8 | 2.0354632450 | (1.38587221198e−01, −8.18964778994e+00, 3.43706754821e−01, −7.59222655671e−03) |
| 9 | 1.9127093077 | (2.42801592628e−01, −7.64937021577e+00, 3.24724323237e−01, −1.23124693731e−02) |
| 10 | 1.7749732845 | (3.49911359159e−01, −6.58187830265e+00, 2.86744744526e−01, −1.77517569581e−02) |
| 11 | 1.6089306960 | (4.95907841498e−01, −5.40397217878e+00, 2.41166917636e−01, −2.49571014357e−02) |
| 12 | 1.4102828725 | (7.23053068198e−01, −2.59378036965e+00, 1.46469675587e−01, −3.68627889085e−02) |
| 13 | 2.4479531178 | (−3.38396896021e+00, 9.49237142101e+01, −3.37779371321e+00, 2.29212476015e−01) |
| 14 | 2.6779763301 | (1.19830766801e+00, 1.13780127873e+01, −5.12382210052e−01, −4.39638576402e−02) |
| 15 | 2.7791837819 | (2.81655473172e+00, 5.61007167226e+00, −2.93511162565e−01, −1.14270102398e−01) |

73 summarize the results for the first cut point. Similarly, Figs. 74, 75 and 76 for the second cut point, Figs. 77, 78 and 79 for the third cut point, Figs. 80, 81 and 82 for the forth cut point, Figs. 83, 84 and 85 for the fifth cut point, Figs. 86, 87 and 88 for the sixth cut point, and finally Figs. 89, 90 and 91 for the seventh cut point.

**Table 18** ($\mathscr{C}_1$ class) Initial conditions for minimum-time transfer orbits for $\varepsilon = 10$N and $\theta_0 = 3\pi/2$, $x_0 = (-0.012153, -1.096898, 3.000969, 0.00)$

| $k$ | $t_f$ | $p_0$ |
|---|---|---|
| 1 | 2.6690810760 | (−1.60148685709e+00, 9.08946766361e+00, −3.13300481199e−01, 6.61185213214e−02) |
| 2 | 2.6008623676 | (−1.54788345969e+00, 8.80209189253e+00, −2.99932270379e−01, 6.44329166721e−02) |
| 3 | 2.5287091355 | (−1.48776605296e+00, 8.47773003663e+00, −2.85028133227e−01, 6.25395502059e−02) |
| 4 | 2.4519026572 | (−1.40500394555e+00, 8.15666769590e+00, −2.70463857125e−01, 5.97473478092e−02) |
| 5 | 2.3697031266 | (−1.34275441268e+00, 7.70451974629e+00, −2.50533464519e−01, 5.79263564017e−02) |
| 6 | 2.2813887573 | (−1.22186934790e+00, 7.34625820586e+00, −2.34873041588e−01, 5.36733452623e−02) |
| 7 | 2.1850625115 | (−1.14306384179e+00, 6.70918812205e+00, −2.08000594543e−01, 5.14177402740e−02) |
| 8 | 2.0802601077 | (−9.84194129457e−01, 6.25065652108e+00, −1.89113692261e−01, 4.57761908262e−02) |
| 9 | 1.9625235243 | (−8.70139276974e−01, 5.33149659787e+00, −1.52203685456e−01, 4.24464538424e−02) |
| 10 | 1.8314874695 | (−6.71790327170e−01, 4.57367060157e+00, −1.22999765824e−01, 3.55761395655e−02) |
| 11 | 1.6770533301 | (−4.99447542856e−01, 3.10421985352e+00, −6.71298805584e−02, 3.05143051338e−02) |
| 12 | 1.4969723177 | (−2.11266620451e−01, 1.19696152516e+00, 3.54958676405e−03, 2.08238576227e−02) |
| 13 | 1.2945503501 | (5.15688868736e−01, −4.36713996233e+00, 2.05263933984e−01, −1.38119494149e−02) |
| 14 | 2.7873443220 | (1.63345406733e+00, −2.65524823134e+00, −6.44278753506e−04, −5.30504819040e−02) |
| 15 | 2.8935039376 | (2.05298326674e+00, 6.37332707982e−01, −1.25355445474e−01, −7.25661087416e−02) |

**Table 19** ($\mathscr{C}_2$ class) Initial conditions for minimum-time transfer orbits for $\varepsilon = 10$N and $\theta_0 = \pi$, $x_0 = (-0.121842, 0.00, 0.00, -3.00096)$

| $k$ | $t_f$ | $p_0$ |
|---|---|---|
| 1 | 2.5980387359 | (1.06435287264e+01, 1.38139088237e+00, 5.38698217696e−02, 4.27687562930e−01) |
| 2 | 2.5276171200 | (1.04197332870e+01, 1.41679263649e+00, 5.56093693046e−02, 4.16741367382e−01) |
| 3 | 2.4530424463 | (1.01354110133e+01, 1.47583141615e+00, 5.83202062677e−02, 4.04443207658e−01) |
| 4 | 2.3730813733 | (9.81815625576e+00, 1.50226355964e+00, 5.98360979453e−02, 3.89358210983e−01) |
| 5 | 2.2872021224 | (9.45271738688e+00, 1.56932212509e+00, 6.29723565050e−02, 3.73464035481e−01) |
| 6 | 2.1940824150 | (8.97261257687e+00, 1.58768227806e+00, 6.43744566344e−02, 3.51814760713e−01) |
| 7 | 2.0918861292 | (8.47506952249e+00, 1.65281951424e+00, 6.76178742737e−02, 3.29905135280e−01) |
| 8 | 1.9790911917 | (7.67374927169e+00, 1.67928450120e+00, 6.97076840290e−02, 2.96143235096e−01) |
| 9 | 1.8511363857 | (6.91696727321e+00, 1.70718186065e+00, 7.18325734637e−02, 2.62736457777e−01) |
| 10 | 1.7045623193 | (5.39278478726e+00, 1.77172229454e+00, 7.62018032334e−02, 2.01790101767e−01) |
| 11 | 1.5293474752 | (3.92175802120e+00, 1.69616859547e+00, 7.50545446867e−02, 1.39620793489e−01) |
| 12 | 1.3121055642 | (−4.07540989225e−01, 1.82698024096e+00, 8.24995437375e−02, −2.93268907749e−02) |
| 13 | 2.7349672541 | (−6.44652936791e+00, −3.03741643602e+00, −1.16238481485e−01, −2.70092491562e−01) |
| 14 | 2.8348007566 | (−4.08335716722e+00, −3.09857369694e+00, −1.16936332793e−01, −2.10233821136e−01) |

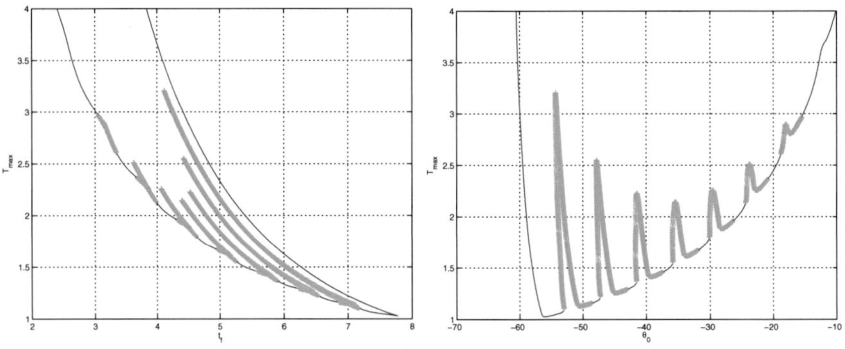


**Fig. 70** For the GEO to $L_1$ control problem, homotopic curve for $\varepsilon \in [1 : 10]$N. *Left* $t_f$ (transfer time) versus $\varepsilon$ projection. *Right* $\theta_0$ (angle defining the initial position on the departure orbit) versus $\varepsilon$

**Table 20** ($\mathscr{C}_2$ class) Initial conditions for minimum-time transfer orbits for $\varepsilon = 10\text{N}$ and $\theta_0 = 0$, $x_0 = (0.0975368, 0.00, 0.00, 3.00096)$

| $k$ | $t_f$ | $p_0$ |
|---|---|---|
| 1 | 2.6061700198 | (−8.10712545558e+00, 1.89540042126e−01, 3.02951748274e−03, −2.70993641088e−01) |
| 2 | 2.5380201140 | (−7.98832817545e+00, 2.59775949334e−01, 5.73286770389e−03, −2.68755580138e−01) |
| 3 | 2.4650494180 | (−7.75538520280e+00, 3.62884468178e−01, 9.56534705873e−03, −2.61182580565e−01) |
| 4 | 2.3883796279 | (−7.58970803264e+00, 4.31241026849e−01, 1.21532458177e−02, −2.57593525824e−01) |
| 5 | 2.3051381815 | (−7.31720208022e+00, 5.32166005246e−01, 1.58845821085e−02, −2.49451562603e−01) |
| 6 | 2.2168668949 | (−7.06641305915e+00, 5.99411681487e−01, 1.83454693321e−02, −2.43199556855e−01) |
| 7 | 2.1191349339 | (−6.71997726657e+00, 6.87257241108e−01, 2.15107998020e−02, −2.33438167469e−01) |
| 8 | 2.0139211293 | (−6.29894313030e+00, 7.57106564979e−01, 2.38986154068e−02, −2.21599131661e−01) |
| 9 | 1.8939163357 | (−5.78936380987e+00, 8.09971284648e−01, 2.55311678798e−02, −2.07142760589e−01) |
| 10 | 1.7610136117 | (−4.96411319056e+00, 8.97635598020e−01, 2.81969305386e−02, −1.81740722662e−01) |
| 11 | 1.6017081412 | (−3.99249110713e+00, 8.68838412584e−01, 2.60957559015e−02, −1.51674639469e−01) |
| 12 | 1.4167921654 | (−1.11120743455e+00, 1.13046941869e+00, 3.35818304486e−02, −5.59741017211e−02) |
| 13 | 1.6556450929 | (1.90041169951e+02, 4.11177240588e+00, 3.65721183952e−01, 6.73092385668e+00) |
| 14 | 2.6740283644 | (3.36137910646e+00, −2.72778518065e+00, −1.16970922525e−01, 3.45243380243e−02) |
| 15 | 2.8030960157 | (−2.11859368374e+00, −2.49178961344e+00, −1.04892810539e−01, −1.79419042463e−01) |

**Table 21** ($\mathscr{C}_2$ class) Initial conditions for minimum-time transfer orbits for $\varepsilon = 10$N and $\theta_0 = \pi/2$, $x_0 = (-0.012153, 0.109689, -3.00096, 0.00)$

| $k$ | $t_f$ | $p_0$ |
|---|---|---|
| 1 | 2.6019829845 | (−2.84425571457e−01, −1.00404237110e+01, 4.05458442869e−01, 1.10979295268e−02) |
| 2 | 2.5334348861 | (−2.32569565646e−01, −9.89968792431e+00, 4.00560274651e−01, 8.89110982802e−03) |
| 3 | 2.4606436274 | (−2.10541872966e−01, −9.77638734159e+00, 3.98549565013e−01, 7.95603536659e−03) |
| 4 | 2.3831582744 | (−1.47706167233e−01, −9.56627413895e+00, 3.90837330674e−01, 5.22553690464e−03) |
| 5 | 2.2996872805 | (−1.06156499893e−01, −9.38234623244e+00, 3.86599531240e−01, 3.44800514348e−03) |
| 6 | 2.2099469387 | (−3.14884828915e−02, −9.05530130469e+00, 3.74338881818e−01, 1.05119003627e−04) |
| 7 | 2.1111698233 | (3.71188462151e−02, −8.75716517386e+00, 3.65618710498e−01, −2.88387786486e−03) |
| 8 | 2.0033553989 | (1.25544557986e−01, −8.20997380954e+00, 3.45244907817e−01, −7.02816735224e−03) |
| 9 | 1.8805738016 | (2.31302088017e−01, −7.66818254447e+00, 3.26583609500e−01, −1.17992930350e−02) |
| 10 | 1.7426682066 | (3.43556047009e−01, −6.60492748194e+00, 2.88284900044e−01, −1.74520049392e−02) |
| 11 | 1.5763097264 | (4.97275584574e−01, −5.42450563786e+00, 2.43378420465e−01, −2.49405258515e−02) |
| 12 | 1.3781347238 | (7.28625942842e−01, −2.47538707912e+00, 1.43995244406e−01, −3.71575252603e−02) |
| 13 | 2.4048352258 | (−3.39252896456e+00, 9.36495680309e+01, −3.33176058888e+00, 2.28072959760e−01) |
| 14 | 2.6257251580 | (1.05189499044e+00, 1.14191299701e+01, −5.15529286934e−01, −3.80961922524e−02) |
| 15 | 2.7236432248 | (2.60098969942e+00, 6.24273168138e+00, −3.24847143004e−01, −1.04909522361e−01) |

**Table 22** ($\mathscr{C}_2$ class) Initial conditions for minimum-time transfer orbits for $\varepsilon = 10$N and $\theta_0 = 3\pi/2$, $x_0 = (-0.012153, -1.096898, 3.000969, 0.00)$

| $k$ | $t_f$ | $p_0$ |
|---|---|---|
| 1 | 2.6374327060 | (−1.62857972444e+00, 9.08391575714e+00, −3.12602494846e−01, 6.72031001922e−02) |
| 2 | 2.5691569872 | (−1.57643427418e+00, 8.78702892946e+00, −2.98800874499e−01, 6.55880396062e−02) |
| 3 | 2.4969636719 | (−1.51162908577e+00, 8.46917014342e+00, −2.84073535582e−01, 6.35051396988e−02) |
| 4 | 2.4200699217 | (−1.43149645719e+00, 8.13260148526e+00, −2.68828877698e−01, 6.08371578637e−02) |
| 5 | 2.3378362689 | (−1.36493857285e+00, 7.68625412066e+00, −2.49019602217e−01, 5.88430641217e−02) |
| 6 | 2.2493898782 | (−1.24390573655e+00, 7.31082683074e+00, −2.32688325847e−01, 5.46051794303e−02) |
| 7 | 2.1530090828 | (−1.16282817958e+00, 6.67343179150e+00, −2.05563880726e−01, 5.22675189137e−02) |
| 8 | 2.0480368975 | (−1.00003739357e+00, 6.20872801454e+00, −1.86460126643e−01, 4.64792293390e−02) |
| 9 | 1.9301495747 | (−8.83664790191e−01, 5.26814185126e+00, −1.48387158771e−01, 4.30927981545e−02) |
| 10 | 1.7989149784 | (−6.76316263624e−01, 4.52955863451e+00, −1.20054957255e−01, 3.58431970542e−02) |
| 11 | 1.6440341344 | (−4.97450251574e−01, 2.98177731120e+00, −6.06185355582e−02, 3.06225341296e−02) |
| 12 | 1.4638781887 | (−1.91345523973e−01, 1.13707535911e+00, 7.19728539102e−03, 2.00319879174e−02) |
| 13 | 1.2629462369 | (5.27114903375e−01, −4.53699320696e+00, 2.12661485766e−01, −1.45016746772e−02) |
| 14 | 2.7339401945 | (1.55300153992e+00, −2.64703138256e+00, −3.90733403113e−03, −4.97729160780e−02) |
| 15 | 2.8391341334 | (1.96486277754e+00, 6.95609002372e−01, −1.31187447456e−01, −6.91977913949e−02) |

**Table 23** ($\mathscr{C}_3$ class) Initial conditions for minimum-time transfer orbits for $\varepsilon = 10\text{N}$ and $\theta_0 = \pi$, $x_0 = (-0121842, 0.00, 0.00, -3.00096)$

| $k$ | $t_f$ | $p_0$ |
|---|---|---|
| 1 | 3.0079434320 | (1.06644572264e+01, 1.35187551484e+00, 5.26512270100e−02, 4.29500793828e−01) |
| 2 | 2.9375415112 | (1.04434578826e+01, 1.38953138641e+00, 5.44744339982e−02, 4.18722639089e−01) |
| 3 | 2.8630041962 | (1.01618362339e+01, 1.45007088633e+00, 5.72390673417e−02, 4.06603688803e−01) |
| 4 | 2.7830629302 | (9.84764443081e+00, 1.47895907175e+00, 5.88472004949e−02, 3.91667351531e−01) |
| 5 | 2.6972242903 | (9.48496629685e+00, 1.54853782815e+00, 6.20765991896e−02, 3.75979183483e−01) |
| 6 | 2.6041287654 | (9.00869563804e+00, 1.56903828964e+00, 6.35588503243e−02, 3.54472174794e−01) |
| 7 | 2.5019691384 | (8.51291261627e+00, 1.63845183385e+00, 6.69650951886e−02, 3.32749964057e−01) |
| 8 | 2.3892198510 | (7.71967340255e+00, 1.66527871957e+00, 6.90619941530e−02, 2.99253127735e−01) |
| 9 | 2.2612723039 | (6.96040246349e+00, 1.70115071916e+00, 7.14998649587e−02, 2.65876126446e−01) |
| 10 | 2.1147973125 | (5.44320887379e+00, 1.76777958465e+00, 7.59453383727e−02, 2.05235264592e−01) |
| 11 | 1.9395064592 | (3.97334759818e+00, 1.70032824652e+00, 7.51188225176e−02, 1.43112186670e−01) |
| 12 | 1.7223793899 | (−3.49402376257e−01, 1.84952575418e+00, 8.34028988787e−02, −2.52900821308e−02) |
| 13 | 3.1156574206 | (−6.51510558872e+00, −3.08751946862e+00, −1.18263364650e−01, −2.70782400660e−01) |
| 14 | 3.2127569407 | (−4.21028083233e+00, −3.19576818179e+00, −1.20901762188e−01, −2.12539884711e−01) |

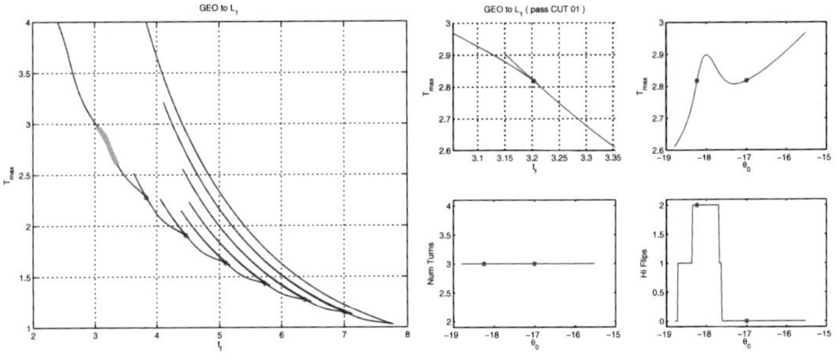


**Fig. 71** For cut point $n^o$ 1: (*left*) $t_f$ versus $\varepsilon$ homotopic curve with highlight of the cut passage in *green*; (*right*) analysis of the cut passage: (*top-left* subplot) $t_f$ versus $\varepsilon$ zoom, (*top-right* subplot) $\theta_0$ versus $\varepsilon$, (*bottom-left* subplot) $\theta_0$ versus num. turns around the Earth, (*bottom-right* subplot) $\theta_0$ versus the number of times $|(H_1, H_2)|$ passes close to zero. *Red* points are values corresponding the each cut point

**Table 24** ($\mathscr{C}_3$ class) Initial conditions for minimum-time transfer orbits for $\varepsilon = 10$N and $\theta_0 = 0$, $x_0 = (0.0975368, 0.00, 0.00, 3.00096)$

| $k$ | $t_f$ | $p_0$ |
|---|---|---|
| 1 | 3.0160686346 | (−8.07952473541e+00, 1.56009762355e−01, 1.63190503522e−03, −2.68719969249e−01) |
| 2 | 2.9478762607 | (−7.96124497160e+00, 2.28947488502e−01, 4.44172318728e−03, −2.66486754726e−01) |
| 3 | 2.8748808644 | (−7.72016520272e+00, 3.34226564842e−01, 8.34507374051e−03, −2.58467672325e−01) |
| 4 | 2.7981671236 | (−7.55602926960e+00, 4.05394083101e−01, 1.10454479194e−02, −2.54931109877e−01) |
| 5 | 2.7148931246 | (−7.27415717237e+00, 5.09903199832e−01, 1.49025953133e−02, −2.46306617667e−01) |
| 6 | 2.6265776367 | (−7.02615679895e+00, 5.80035014991e−01, 1.74807801082e−02, −2.40158650743e−01) |
| 7 | 2.5288032793 | (−6.66921179529e+00, 6.73203679187e−01, 2.08407300183e−02, −2.29902094968e−01) |
| 8 | 2.4235482061 | (−6.25292923064e+00, 7.45932195898e−01, 2.33480462145e−02, −2.18236701100e−01) |
| 9 | 2.3034863018 | (−5.73159256933e+00, 8.06189964121e−01, 2.52583831813e−02, −2.03303437857e−01) |
| 10 | 2.1710299861 | (−4.91544276746e+00, 8.96489918177e−01, 2.80205574326e−02, −1.78204644826e−01) |
| 11 | 2.0111660933 | (−3.92900285748e+00, 8.77250871764e−01, 2.63023408558e−02, −1.47666820427e−01) |
| 12 | 1.8264984515 | (−1.00305568125e+00, 1.15297499904e+00, 3.43275408112e−02, −5.02059871398e−02) |
| 13 | 2.0431049217 | (1.86578177364e+02, 4.08561828326e+00, 3.60755982581e−01, 6.60857596413e+00) |
| 14 | 3.0540126835 | (3.61881934098e+00, −2.81462056517e+00, −1.20190778166e−01, 4.54101687908e−02) |
| 15 | 3.2018637194 | (−1.96637660380e+00, −2.60107142593e+00, −1.09312588413e−01, −1.72944354724e−01) |

**Table 25** ($\mathscr{C}_3$ class) Initial conditions for minimum-time transfer orbits for $\varepsilon = 10$N and $\theta_0 = \pi/2$, $x_0 = (-0.012153, 0.109689, -3.00096, 0.00)$

| $k$ | $t_f$ | $p_0$ |
|---|---|---|
| 1 | 3.0114201851 | (−3.17790993078e−01, −9.99620628532e+00, 4.02877878386e−01, 1.23338642431e−02) |
| 2 | 2.9428784342 | (−2.66437255556e−01, −9.85807299220e+00, 3.98162529647e−01, 1.01525525586e−02) |
| 3 | 2.8700856406 | (−2.47561176375e−01, −9.73541452344e+00, 3.96250079704e−01, 9.34509289696e−03) |
| 4 | 2.7926111380 | (−1.84912105298e−01, −9.52892872608e+00, 3.88770128141e−01, 6.62912797585e−03) |
| 5 | 2.7091438748 | (−1.46456526906e−01, −9.34583623592e+00, 3.84664090025e−01, 4.97786212278e−03) |
| 6 | 2.6194200750 | (−7.15636845081e−02, −9.02376276247e+00, 3.72690032847e−01, 1.63487427495e−03) |
| 7 | 2.5206527435 | (−5.76199475645e−03, −8.72666992070e+00, 3.64151059255e−01, −1.23792031755e−03) |
| 8 | 2.4128647384 | (8.32099513855e−02, −8.18644238539e+00, 3.44124212000e−01, −5.39328168163e−03) |
| 9 | 2.2900952639 | (1.87090924654e−01, −7.64565340599e+00, 3.25712424499e−01, −1.00831811499e−02) |
| 10 | 2.1522470577 | (2.99246181934e−01, −6.59424327083e+00, 2.87877672276e−01, −1.57184440223e−02) |
| 11 | 1.9858762347 | (4.53674686372e−01, −5.41187749151e+00, 2.43244220883e−01, −2.32284106856e−02) |
| 12 | 1.7880235904 | (6.80299038636e−01, −2.46145430043e+00, 1.44255335142e−01, −3.52558523439e−02) |
| 13 | 2.7896443361 | (−3.41374362314e+00, 9.32731850142e+01, −3.31755914190e+00, 2.28456530459e−01) |
| 14 | 3.0074529459 | (9.87115821527e−01, 1.15185661024e+01, −5.20294817745e−01, −3.53915049620e−02) |
| 15 | 3.1016075052 | (2.46829014099e+00, 6.81961761072e+00, −3.51739669061e−01, −9.88979615343e−02) |

**Table 26** ($\mathscr{C}_3$ class) Initial conditions for minimum-time transfer orbits for $\varepsilon = 10$N and $\theta_0 = 3\pi/2$, $x_0 = (-0.012153, -1.096898, 3.000969, 0.00)$

| $k$ | $t_f$ | $p_0$ |
|---|---|---|
| 1 | 3.0477620262 | (−1.65436243883e+00, 9.13015536385e+00, −3.15423464048e−01, 6.81362816051e−02) |
| 2 | 2.9795368970 | (−1.60623443630e+00, 8.83016269889e+00, −3.01446202255e−01, 6.66677453622e−02) |
| 3 | 2.9073039428 | (−1.54133345202e+00, 8.51559503544e+00, −2.86806843906e−01, 6.45849249114e−02) |
| 4 | 2.8304040772 | (−1.46607509242e+00, 8.17299095306e+00, −2.71226012966e−01, 6.21184798314e−02) |
| 5 | 2.7481886105 | (−1.39982960612e+00, 7.73048160814e+00, −2.51533856789e−01, 6.01327857674e−02) |
| 6 | 2.6597101749 | (−1.28312320054e+00, 7.34735121468e+00, −2.34755001038e−01, 5.60784042063e−02) |
| 7 | 2.5633588116 | (−1.20420297838e+00, 6.70973715294e+00, −2.07551473300e−01, 5.38271396019e−02) |
| 8 | 2.4583262887 | (−1.04355838469e+00, 6.24074714632e+00, −1.88118009827e−01, 4.81332584817e−02) |
| 9 | 2.3404678233 | (−9.30788833999e−01, 5.29153019869e+00, −1.49557187330e−01, 4.49012589544e−02) |
| 10 | 2.2091494228 | (−7.23352672817e−01, 4.55685507376e+00, −1.21226959607e−01, 3.76477925265e−02) |
| 11 | 2.0542769577 | (−5.48043592799e−01, 2.98526534387e+00, −6.05924194202e−02, 3.25919877337e−02) |
| 12 | 1.8739945514 | (−2.41372435293e−01, 1.16542210106e+00, 6.40892432862e−03, 2.19566013991e−02) |
| 13 | 1.6724924789 | (4.78305672369e−01, −4.56845039900e+00, 2.15049807592e−01, −1.26469413261e−02) |
| 14 | 3.1147430656 | (1.52316731297e+00, −2.69581657727e+00, −4.16727729788e−03, −4.85564981134e−02) |
| 15 | 3.2177755237 | (1.94272857430e+00, 6.29139383819e−01, −1.31608874036e−01, −6.83226668898e−02) |

**Table 27** ($\mathscr{C}_1$ class) Local Minima of the homotopic path $\theta_0$-$t_f$ for $\theta_0$ fixed and $\varepsilon = 10$N

| Num. | $t_f$ | $(x_0, p_0)$ |
|---|---|---|
| 1 | 2.6344109813 | x0 = (−6.61692049717e−02, 9.54678694392e−02, −2.61187491303e+00, −1.47781207951e+00)<br>p0 = (5.40200904069e+00, −9.04436482091e+00, 3.81168153429e−01, 2.05269577680e−01) |
| 2 | 2.5654788135 | x0 = (−4.52828715879e−02, 1.04567089008e−01, −2.86081755165e+00, −9.06389563114e−01)<br>p0 = (3.15983116724e+00, −9.70188823099e+00, 3.98877317690e−01, 1.23233133448e−01) |
| 3 | 2.4925251859 | x0 = (−2.95411256412e−02, 1.08302897416e−01, −2.96302433931e+00, −4.75716169365e−01)<br>p0 = (1.54563324437e+00, −9.81613709586e+00, 4.01898104211e−01, 6.56372972600e−02) |
| 4 | 2.7946382317 | x0 = (−8.62590901998e−02, −8.08712057808e−02, 2.21252945945e+00, −2.02744482549e+00)<br>p0 = (−6.36693882796e−01, −4.91318631097e+00, 7.08068709438e−03, −1.47732494031e−01) |
| 5 | 2.8956929200 | x0 = (−7.16101551794e−02, −9.21776070570e−02, 2.52185767661e+00, −1.62666929644e+00)<br>p0 = (2.02907798275e+00, −2.75514593814e+00, −6.83047057606e−02, −9.50712106843e−02) |

**Table 28** ($\mathscr{C}_2$ class) Local Minima of the homotopic path $\theta_0$-$t_f$ for $\theta_0$ fixed and $\varepsilon = 10$N

| Num. | $t_f$ | $(x_0, p_0)$ |
|---|---|---|
| 1 | 2.6027273646 | x0 = (−7.08353877411e−02, 9.26727676532e−02, −2.53540461703e+00, −1.60547268183e+00)<br>p0 = (5.89933290763e+00, −8.78036555972e+00, 3.73602572575e−01, 2.24143289427e−01) |
| 2 | 2.5336945761 | x0 = (−5.28636779641e−02, 1.01855315024e−01, −2.78662699434e+00, −1.11379042071e+00)<br>p0 = (3.94605260276e+00, −9.49349757966e+00, 3.93894887009e−01, 1.51870763055e−01) |
| 3 | 2.4607414965 | x0 = (−3.58852334423e−02, 1.07091762478e−01, −2.92988928582e+00, −6.49282586092e−01)<br>p0 = (2.17392046524e+00, −9.75993925757e+00, 4.01617069536e−01, 8.86090945569e−02) |
| 4 | 2.7418546856 | x0 = (−8.84246217632e−02, −7.88321267809e−02, 2.15674294911e+00, −3.08669091108e+00)<br>p0 = (−7.80575217720e−01, −4.91458653148e+00, 2.36855239340e−03, −1.47525204726e−01) |
| 5 | 2.8414849743 | x0 = (−7.36542225280e−02, −9.08265606530e−02, 2.48489483006e+00, −1.68259228141e+00)<br>p0 = (1.96421647292e+00, −2.81151922798e+00, −7.22948807464e−02, −9.24617336107e−02) |

**Table 29** ($\mathscr{C}_3$ class) Local Minima of the homotopic path $\theta_0$-$t_f$ for $\theta_0$ fixed and $\varepsilon = 10$N

| Num. | $t_f$ | $(x_0, p_0)$ |
|---|---|---|
| 1 | 3.0122653352 | x0 = (−7.37384179257e−02, 9.07694926321e−02, −2.48333352432e+00, −1.68489575638e+00)<br>p0 = (6.16866446074e+00, −8.59208825496e+00, 3.66245485321e−01, 2.36096512184e−01) |
| 2 | 2.9431966623 | x0 = (−5.61856226705e−02, 1.00463887224e−01, −2.74855936609e+00, −1.20467444370e+00)<br>p0 = (4.25079407550e+00, −9.36060260305e+00, 3.88696084246e−01, 1.64942973814e−01) |
| 3 | 2.8702232457 | x0 = (−4.13981576422e−02, 1.05719370263e−01, −2.89234244604e+00, −8.00108916456e−01)<br>p0 = (2.69474357150e+00, −9.64529876802e+00, 3.97577540129e−01, 1.09009754379e−01) |
| 4 | 3.1227565037 | x0 = (−8.86205667692e−02, −7.86420740187e−02, 2.15154335636e+00, −2.09205170785e+00)<br>p0 = (−8.17719507113e−01, −4.95440837383e+00, 1.50403834833e−03, −1.47673574600e−01) |
| 5 | 3.2201514889 | x0 = (−7.20630362335e−02, −9.18839067611e−02, 2.51382242413e+00, −1.63905952438e+00)<br>p0 = (2.00055655527e+00, −2.78413963830e+00, −7.43272782471e−02, −9.10311210657e−02) |

**Table 30** Cut Points for GEO to MO minimum-time transfer problem for $\mathscr{C}_1$

| $N^o$ | $t_f$ | $\varepsilon$ (N) | $(x_0, p_0)$ |
|---|---|---|---|
| 1 | 5.1488625 | 1.8710653 | x0 = (−1.2183420e−01 1.3776662e−03 −3.7691130e−02 −3.0007330e+00)<br>p0 = (−2.4853749e+01 −2.5590439e+00 −1.1469955e−e−01 −7.7640343e−01) |
| | 5.1488623 | 1.8710653 | x1 = (−6.4010619e−02 9.6657394e−02 −2.6444187e+00 −1.4187560e+00)<br>p1 = (−7.6598810e+00 6.5174405e+00 −3.8985884e−01 −5.6991497e−02) |
| 2 | 5.9023179 | 1.6073723 | x0 = (−8.6446318e−02 −8.0699240e−02 2.2078247e+00 −2.0325671e+00)<br>p0 = (−2.3664669e+01 −2.9201326e+01 8.2641553e−01 −8.7846091e−01) |
| | 5.9023450 | 1.6073723 | x1 = (−1.0690133e−01 5.5268597e−02 −1.5120759e+00 −2.5921894e+00)<br>p1 = (−1.0303101e+01 1.6834274e+00 −3.1756083e−01 −2.7329434e−01) |
| 3 | 6.6599800 | 1.4056522 | x0 = (−1.8408681e−02 −1.0951133e−01 2.9960854e+00 −1.7114719e−01)<br>p0 = (−5.0074394e−01 −4.6582515e+01 1.5038084e+00 −1.6486206e−01) |
| | 6.6600036 | 1.4056522 | x1 = (−1.2149790e−01 −8.6923162e−03 2.3781030e−01 −2.9915323e+00)<br>p1 = (−9.4498793e+00 −4.5941257e+00 −1.0821845e−01 −4.0563385e−01) |
| 4 | 7.4356086 | 1.2469239 | x0 = (4.1469228e−02 −9.5689713e−02 2.6179443e+00 1.4670334e+00)<br>p0 = (2.4346187e+01 −3.8413717e+01 1.3257393e+00 6.3983535e−01) |
| | 7.4356173 | 1.2469239 | x1 = (−8.7964292e−02 −7.9274917e−02 2.1688571e+00 −2.0740969e+00)<br>p1 = (−3.6145240e+00 −9.4918222e+00 1.6199661e−01 −3.5458440e−01) |

Results obtained from Fig. 52

**Table 31** Cut Points for GEO to MO minimum-time transfer problem for $\mathscr{C}_2$

| $N^o$ | $t_f$ | $\varepsilon$ (N) | $(x_0, p_0)$ |
|---|---|---|---|
| 1 | 3.3752853 | 2.8226511 | x0 = (−3.7663428e−02 1.0668216e−01 −2.9186830e+00 −6.9793164e−01)<br>p0 = (−6.6375288e+00 1.5464483e+01 −5.2361131e−01 −1.7763009e−02) |
| | 3.3752836 | 2.8226511 | x1 = (6.1715222e−02 8.1088534e−02 −2.2184753e+00 2.0209371e+00)<br>p1 = (3.6609960e+00 8.9031244e+00 −1.9029376e−01 3.3598070e−01) |
| 2 | 4.0676620 | 2.2832275 | x0 = (−1.2136523e−01 1.0225081e−02 −2.7974471e−01 −2.9879026e+00)<br>p0 = (−4.1968548e+01 2.8323241e+00 −1.6608360e−01 −1.3456351e+00) |
| | 4.0676607 | 2.2832275 | x1 = (3.8446702e−02 9.7321810e−02 −2.6625963e+00 1.3843411e+00)<br>p1 = (2.4323326e+00 1.1587609e+01 −3.9450109e−01 3.3641398e−01) |
| 3 | 4.7758028 | 1.9114577 | x0 = (−7.6851323e−02 −8.8577601e−02 2.4233663e+00 −1.7700607e+00)<br>p0 = (−4.7402787e+01 −6.3131982e+01 2.1643608e+00 −1.5337171e+00) |
| | 4.7757718 | 1.9114577 | x1 = (−7.9147362e−03 1.0960795e−01 −2.9987287e+00 1.1595330e−01)<br>p1 = (−2.6027323e+00 1.3934522e+01 −6.2536775e−01 1.3900950e−01) |
| 4 | 5.5026835 | 1.6412384 | x0 = (2.1204286e−02 −1.0449477e−01 2.8588389e+00 9.1261133e−01)<br>p0 = (3.5025550e+01 −1.2059407e+02 4.0860053e+00 1.4312241e+00) |
| | 5.5027193 | 1.6412384 | x1 = (−6.1645451e−02 9.7889539e−02 −2.6781286e+00 −1.3540481e+00)<br>p1 = (−1.0123795e+01 1.3145099e+01 −7.2436101e−01 −2.3911818e−01) |
| 5a | 6.2531152 | 1.4359703 | x0 = (8.8714814e−02 −4.3099288e−02 1.1791396e+00 2.7596103e+00)<br>p0 = (1.6152834e+02 −7.4842600e+01 2.2694695e+00 5.8095787e+00) |
| | 6.2531294 | 1.4359703 | x1 = (−1.0551390e−01 5.7581311e−02 −1.5753487e+00 −2.5542309e+00)<br>p1 = (−1.7636018e+01 7.2559327e+00 −5.5642890e−01 −6.8757051e−01) |

(continued)

**Table 31** (continued)

| $N^o$ | $t_f$ | $\varepsilon$ (N) | $(x_0, p_0)$ |
|---|---|---|---|
| 5b | 6.4271669 | 1.3811556 | x0 = (−1.0866349e−01 5.2130500e−02 −1.4262217e+00 −2.6403997e+00)<br>p0 = (−2.2795619e+01 9.1266774e+00 −6.1238761e−01 −9.1810499e−01) |
| | 6.4255029 | 1.3811556 | x1 = (−1.0722281e−01 5.4713762e−02 −1.4968963e+00 −2.6009845e+00)<br>p1 = (−2.1556926e+01 9.0788639e+00 −6.1268640e−01 −8.5326763e−01) |
| 6 | 6.9610479 | 1.2748062 | x0 = (9.3095263e−02 3.0897695e−02 −8.4532016e−01 2.8794536e+00)<br>p0 = (2.1996284e+02 5.7639512e+01 −2.4939350e+00 7.6317637e+00) |
| | 6.9602231 | 1.2748062 | x1 = (−1.2181767e−01 2.3504855e−03 −6.4306181e−02 −3.0002806e+00)<br>p1 = (−2.0563120e+01 −2.6910797e+00 −1.3495182e−01 −9.5548578e−01) |
| 7 | 7.6312547 | 1.1477254 | x0 = (3.3347528e−02 9.9807647e−02 −2.7306055e+00 1.2448344e+00)<br>p0 = (1.3464477e+02 2.7341686e+02 −1.0034451e+01 4.2090529e+00) |
| | 7.6360553 | 1.1477254 | x1 = (−1.0627936e−01 −5.6321337e−02 1.5408774e+00 −2.5751730e+00)<br>p1 = (−1.8271336e+01 −1.4562037e+01 4.5974795e−01 −9.9004366e−01) |
| 8 | 8.3208073 | 1.0419705 | x0 = (−4.5399318e−02 1.0453012e−01 −2.8598062e+00 −9.0957538e−01)<br>p0 = (−1.0240946e+02 3.5758771e+02 −1.2592360e+01 −4.4189365e+00) |
| | 8.3210727 | 1.0419705 | x1 = (−6.0626095e−02 −9.8398290e−02 2.6920474e+00 −1.3261599e+00)<br>p1 = (−8.8941099e+00 −2.5008323e+01 1.0588586e+00 −6.4682426e−01) |

Results obtained from Fig. 53

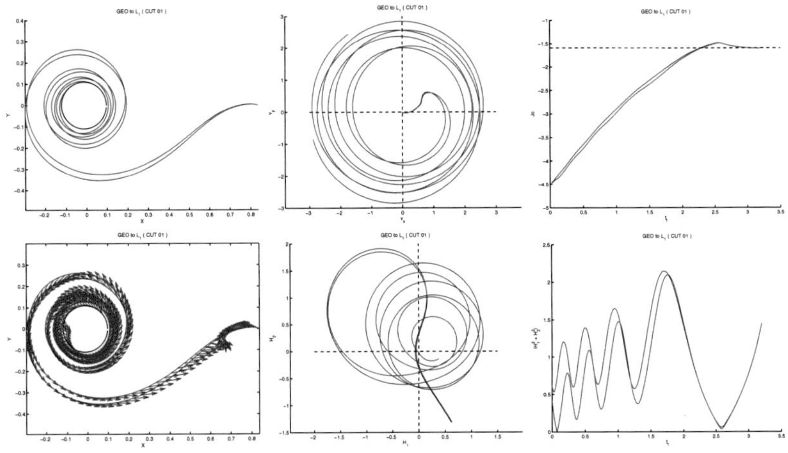

**Fig. 72** For cut point $n^o$ 1, *blue* orbits correspond to the first cut value and *red* orbits to the second cut value. (*top-left*) $\{XY\}$ projection of the transfer trajectory, (*top-center*) $\{V_x V_y\}$ projection of the transfer trajectory, (*top-right*) $t$ versus $J_c$ (energy variation along the transfer trajectory), (*bottom-left*) control along the trajectory, (*bottom-center*) $H_1$ versus $H_2$, (*bottom-right*) $t$ versus $|(H_1, H_2)|$

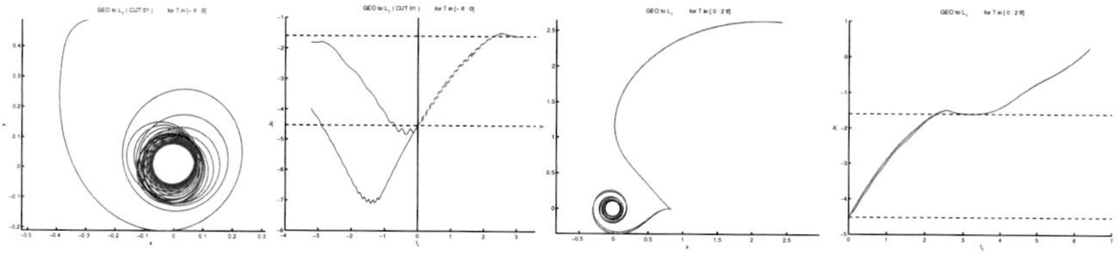

**Fig. 73** For cut point $n^o$ 1, (*left*) optimal solutions for $t \in [-t_f, 0]$ ($XY$ projection and $J_c$ variation), (*right*) optimal solutions for $t \in [0, 2t_f]$ ($XY$ projection and $J_c$ variation)

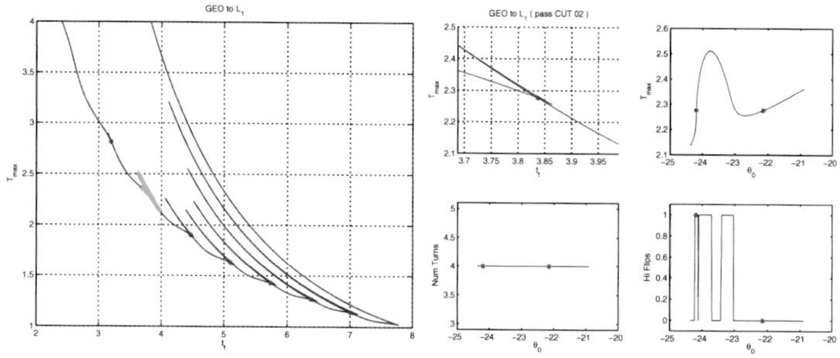


**Fig. 74** For cut point $n^o$ 2: (*left*) $t_f$ versus $\varepsilon$ homotopic curve with highlight of the cut passage in *green*; (*right*) analysis of the cut passage: (*top-left* subplot) $t_f$ versus $\varepsilon$ zoom, (*top-right* subplot) $\theta_0$ versus $\varepsilon$, (*bottom-left* subplot) $\theta_0$ versus num. turns around the Earth, (*bottom-right* subplot) $\theta_0$ versus the number of times $|(H_1, H_2)|$ passes close to zero. *Red points* are values corresponding the each cut point

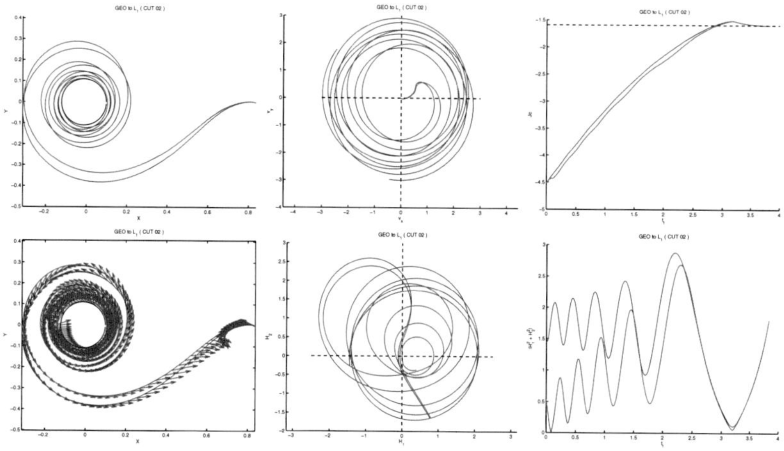

**Fig. 75** For cut point $n^o$ 2, *blue* orbits correspond to the first cut value and *red* orbits to the second cut value (*top-left*) $\{XY\}$ projection of the transfer trajectory, (*top-center*) $\{V_x V_y\}$ projection of the transfer trajectory, (*top-right*) $t$ versus $J_c$ (energy variation along the transfer trajectory), (*bottom-left*) control along the trajectory, (*bottom-center*) $H_1$ versus $H_2$, (*bottom-right*) $t$ versus $|(H_1, H_2)|$

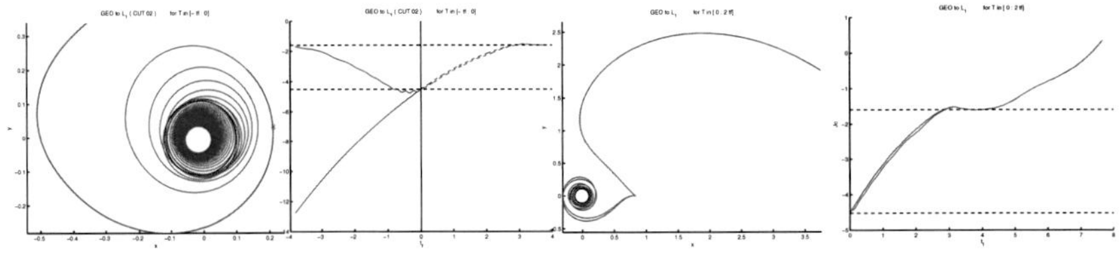

**Fig. 76** For cut point $n^o$ 2, (*left*) optimal solutions for $t \in [-t_f, 0]$ ($XY$ projection and $J_c$ variation), (*right*) optimal solutions for $t \in [0, 2t_f]$ ($XY$ projection and $J_c$ variation)

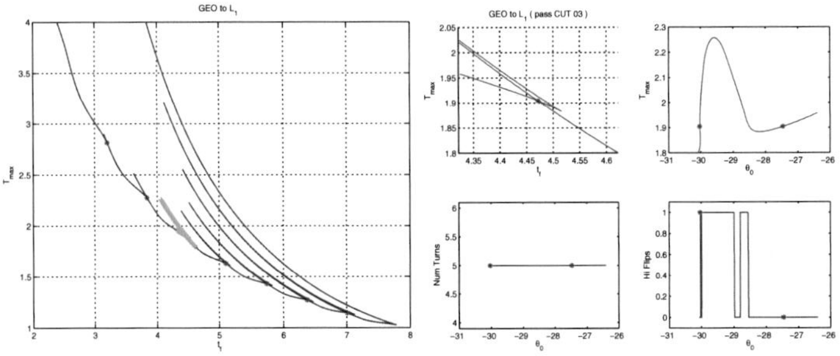


**Fig. 77** For cut point $n^o$ 3: (*left*) $t_f$ versus $\varepsilon$ homotopic curve with highlight of the cut passage in *green*; (*right*) analysis of the cut passage: (*top-left* subplot) $t_f$ versus $\varepsilon$ zoom, (*top-right* subplot) $\theta_0$ versus $\varepsilon$, (*bottom-left* subplot) $\theta_0$ versus num. turns around the Earth, (*bottom-right* subplot) $\theta_0$ versus the number of times $|(H_1, H_2)|$ passes close to zero. *Red* points are values corresponding the each cut point

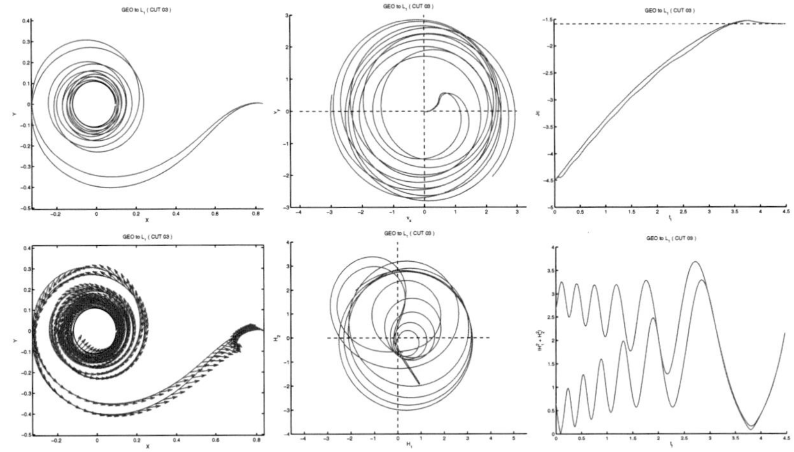

**Fig. 78** For cut point $n^o$ 3, *blue* orbits correspond to the first cut value and *red* orbits to the second cut value (*top-left*) $\{XY\}$ projection of the transfer trajectory, (*top-center*) $\{V_x V_y\}$ projection of the transfer trajectory, (*top-right*) $t$ versus $J_c$ (energy variation along the transfer trajectory), (*bottom-left*) control along the trajectory, (*bottom-center*) $H_1$ versus $H_2$, (*bottom-right*) $t$ versus $|(H_1, H_2)|$

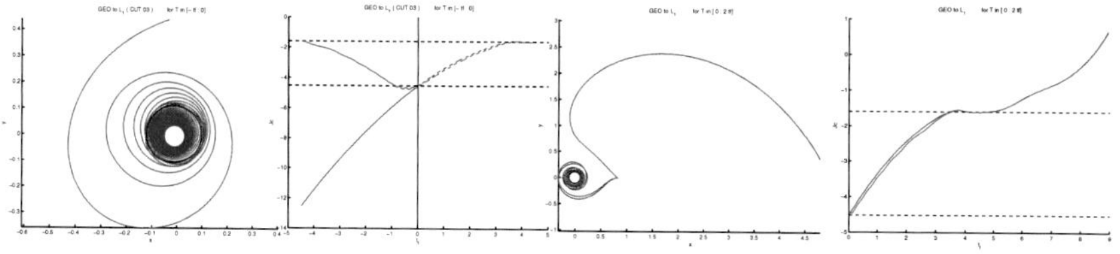

**Fig. 79** For cut point $n^o$ 3, (*left*) optimal solutions for $t \in [-t_f, 0]$ ($XY$ projection and $J_c$ variation), (*right*) optimal solutions for $t \in [0, 2t_f]$ ($XY$ projection and $J_c$ variation)

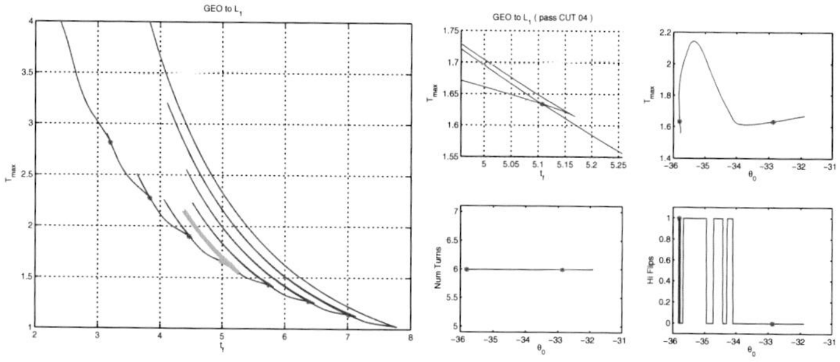


**Fig. 80** For cut point $n^o$ 4: (*left*) $t_f$ versus $\varepsilon$ homotopic curve with highlight of the cut passage in *green*; (*right*) analysis of the cut passage: (*top-left* subplot) $t_f$ versus $\varepsilon$ zoom, (*top-right* subplot) $\theta_0$ versus $\varepsilon$, (*bottom-left* subplot) $\theta_0$ versus num. turns around the Earth, (*bottom-right* subplot) $\theta_0$ versus the number of times $|(H_1, H_2)|$ passes close to zero. *Red* points are values corresponding the each cut point

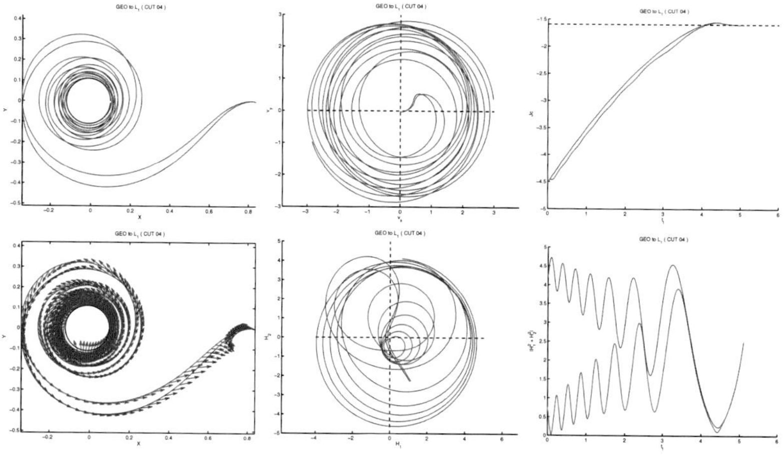

**Fig. 81** For cut point $n^o$ 4, *blue* orbits correspond to the first cut value and *red* orbits to the second cut value (*top-left*) $\{XY\}$ projection of the transfer trajectory, (*top-center*) $\{V_x V_y\}$ projection of the transfer trajectory, (*top-right*) $t$ versus $J_c$ (energy variation along the transfer trajectory), (*bottom-left*) control along the trajectory, (*bottom-center*) $H_1$ versus $H_2$, (*bottom-right*) $t$ versus $|(H_1, H_2)|$

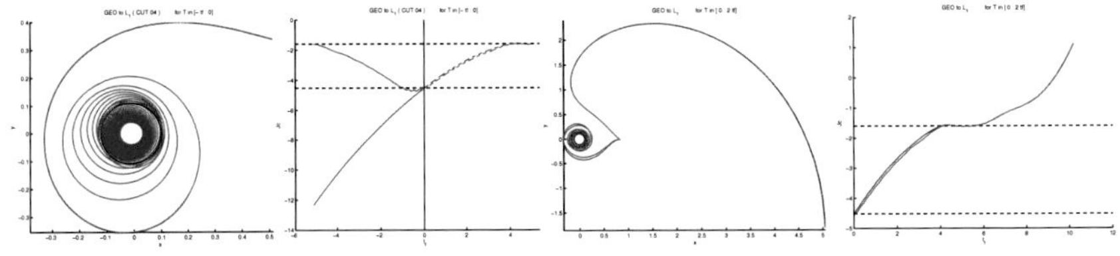

**Fig. 82** For cut point $n^o$ 4, (*left*) optimal solutions for $t \in [-t_f, 0]$ ($XY$ projection and $J_c$ variation), (*right*) optimal solutions for $t \in [0, 2t_f]$ ($XY$ projection and $J_c$ variation)

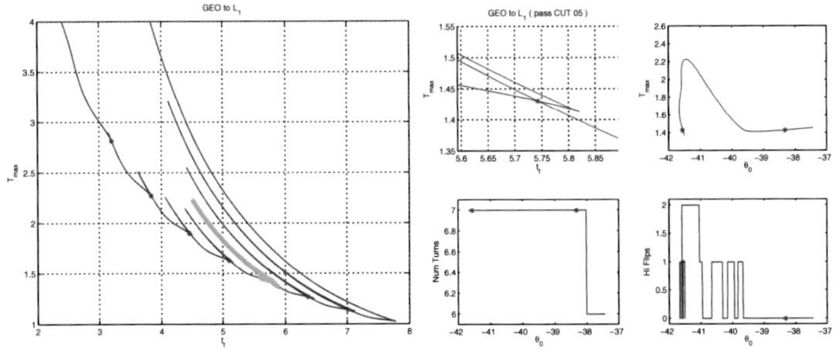


**Fig. 83** For cut point $n^o$ 5: (*left*) $t_f$ versus $\varepsilon$ homotopic curve with highlight of the cut passage in *green*; (*right*) analysis of the cut passage: (*top-left* subplot) $t_f$ versus $\varepsilon$ zoom, (*top-right* subplot) $\theta_0$ versus $\varepsilon$, (*bottom-left* subplot) $\theta_0$ versus num. turns around the Earth, (*bottom-right* subplot) $\theta_0$ versus the number of times $|(H_1, H_2)|$ passes close to zero. *Red* points are values corresponding the each cut point

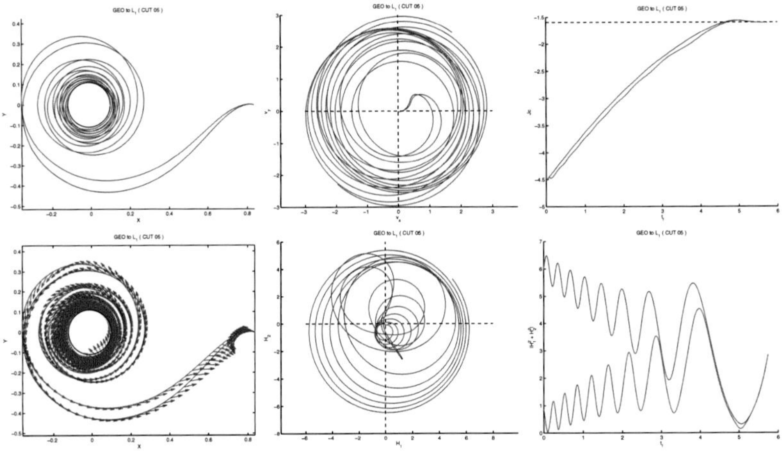

**Fig. 84** For cut point $n^o$ 5, *blue* orbits correspond to the first cut value and *red* orbits to the second cut value (*top-left*) $\{XY\}$ projection of the transfer trajectory, (*top-center*) $\{V_x V_y\}$ projection of the transfer trajectory, (*top-right*) $t$ versus $J_c$ (energy variation along the transfer trajectory), (*bottom-left*) control along the trajectory, (*bottom-center*) $H_1$ versus $H_2$, (*bottom-right*) $t$ versus $|(H_1, H_2)|$

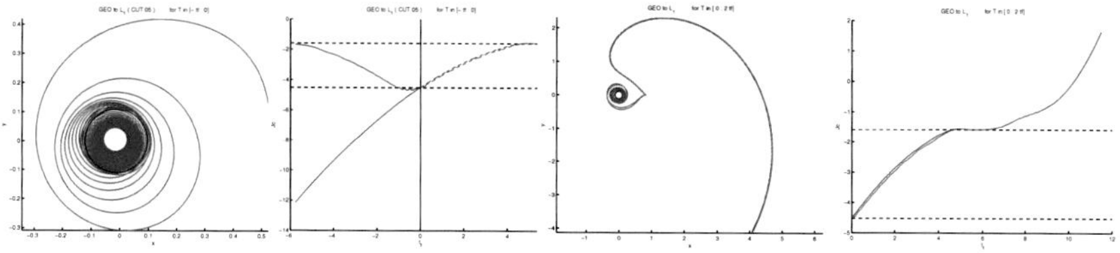

**Fig. 85** For cut point $n^o$ 5, (*left*) optimal solutions for $t \in [-t_f, 0]$ ($XY$ projection and $J_c$ variation), (*right*) optimal solutions for $t \in [0, 2t_f]$ ($XY$ projection and $J_c$ variation)

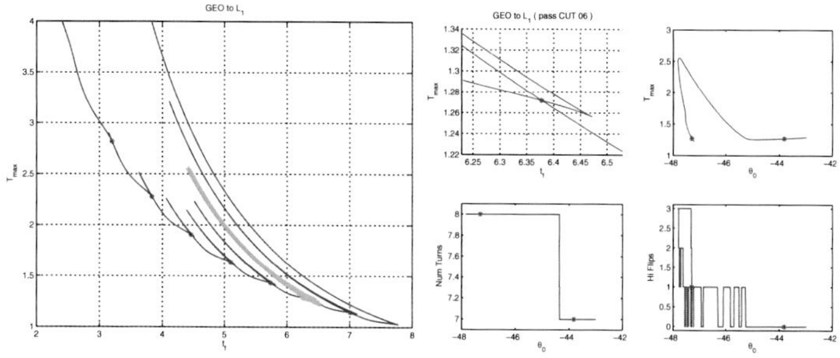


**Fig. 86** For cut point $n^o$ 6: (*left*) $t_f$ versus $\varepsilon$ homotopic curve with highlight of the cut passage in *green*; (*right*) analysis of the cut passage: (*top-left* subplot) $t_f$ versus $\varepsilon$ zoom, (*top-right* subplot) $\theta_0$ versus $\varepsilon$, (*bottom-left* subplot) $\theta_0$ versus num. turns around the Earth, (*bottom-right* subplot) $\theta_0$ versus the number of times $|(H_1, H_2)|$ passes close to zero. *Red* points are values corresponding the each cut point

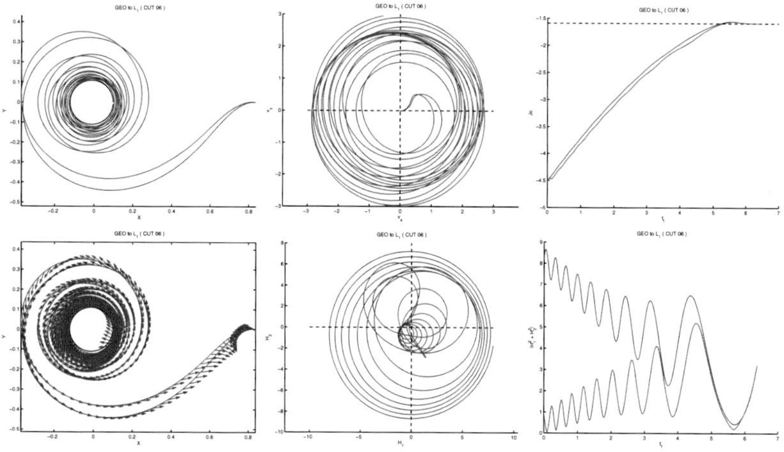

**Fig. 87** For cut point $n^o$ 6, *blue* orbits correspond to the first cut value and *red* orbits to the second cut value (*top-left*) $\{XY\}$ projection of the transfer trajectory, (*top-center*) $\{V_x V_y\}$ projection of the transfer trajectory, (*top-right*) $t$ versus $J_c$ (energy variation along the transfer trajectory), (*bottom-left*) control along the trajectory, (*bottom-center*) $H_1$ versus $H_2$, (*bottom-right*) $t$ versus $|(H_1, H_2)|$

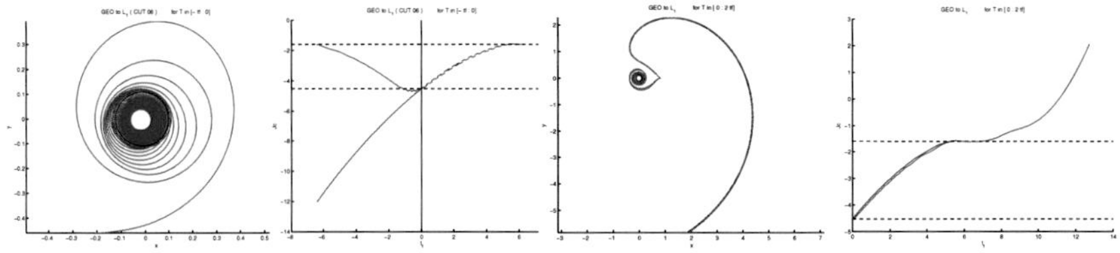

**Fig. 88** For cut point $n^o$ 6, (*left*) optimal solutions for $t \in [-t_f, 0]$ ($XY$ projection and $J_c$ variation), (*right*) optimal solutions for $t \in [0, 2t_f]$ ($XY$ projection and $J_c$ variation)

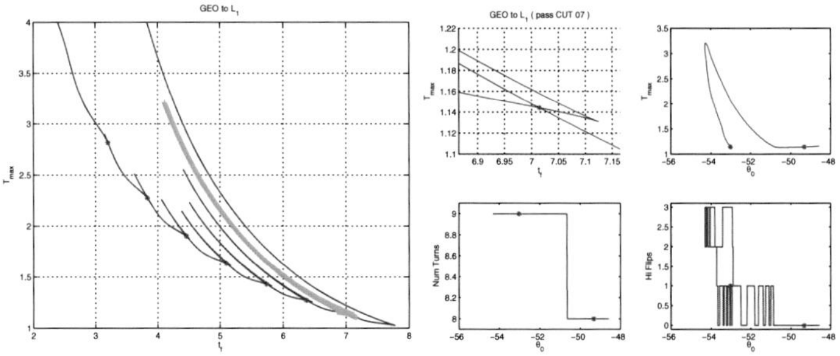


**Fig. 89** For cut point $n^o$ 7: (*left*) $t_f$ versus $\varepsilon$ homotopic curve with highlight of the cut passage in *green*; (*right*) analysis of the cut passage: (*top-left* subplot) $t_f$ versus $\varepsilon$ zoom, (*top-right* subplot) $\theta_0$ versus $\varepsilon$, (*bottom-left* subplot) $\theta_0$ versus num. turns around the Earth, (*bottom-right* subplot) $\theta_0$ versus the number of times $|(H_1, H_2)|$ passes close to zero. *Red* points are values corresponding the each cut point

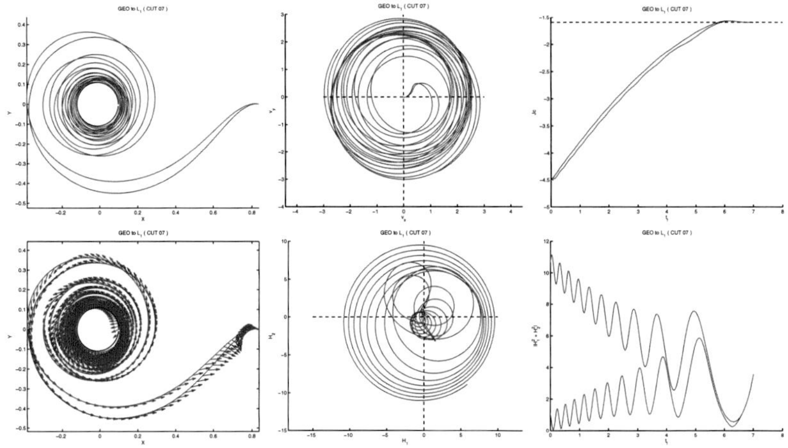

**Fig. 90** For cut point $n^o$ 7, *blue* orbits correspond to the first cut value and *red* orbits to the second cut value (*top-left*) $\{XY\}$ projection of the transfer trajectory, (*top-center*) $\{V_x V_y\}$ projection of the transfer trajectory, (*top-right*) $t$ versus $J_c$ (energy variation along the transfer trajectory), (*bottom-left*) control along the trajectory, (*bottom-center*) $H_1$ versus $H_2$, (*bottom-right*) $t$ versus $|(H_1, H_2)|$

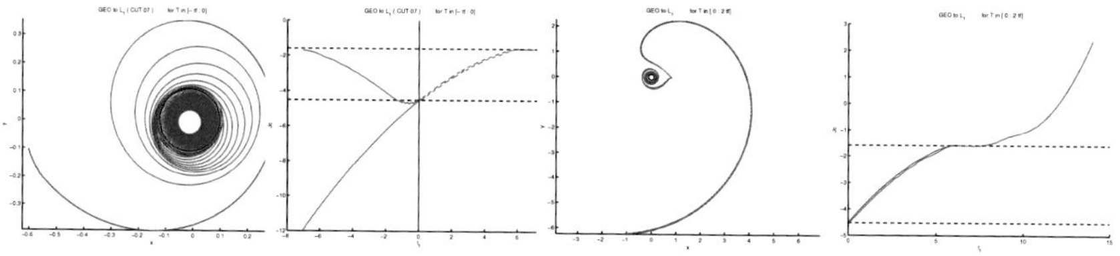

**Fig. 91** For cut point $n^o$ 7, (*left*) optimal solutions for $t \in [-t_f, 0]$ ($XY$ projection and $J_c$ variation), (*right*) optimal solutions for $t \in [0, 2t_f]$ ($XY$ projection and $J_c$ variation)

## 8 Summary of the Cut Points on the GEO to MO Transfer

In this Section we summarize the results for all the CUT points that we have found for the GEO to MO transfer problem. We recall that we have two classes of transfer trajectories, $\mathscr{C}_1$ and $\mathscr{C}_2$, and that in terms of transfer time, the solutions of type $\mathscr{C}_2$ are always better than those of type $\mathscr{C}_1$. Nevertheless, the behavior of the homotopic curve with respect to $\varepsilon$ for the $\mathscr{C}_1$ type of solutions presents a less complex structure that the $\mathscr{C}_2$ type homotopic curve. In both cases we also find CUT points, where their initial conditions are summarized in Tables 30 and 31. We have done a similar analysis as the one for the GEO to $L_1$, and for each pair of cut points we have computed the transfer trajectory, the energy variation along the transfer trajectory, $J_c(t)$, the variation of the control-law along the trajectory and the variation of $H_{1,2}$. Moreover, we have integrated the optimal solutions back and forward in time, i.e. for $t \in [-t_f, 2t_t]$, where $t_f$ is the transfer time. Finally, for the solutions along the homotopic curve close to cut point (i.e. $t_f^* \in [t_f - 0.15 : t_f + 0.15]$) we have computed transfer trajectories and some of their distinctive parameters, trying to characterize their passage. In the plots that we will see, the number of turns around the Earth, and the number of times that $|(H_1, H_2)|$ comes close to zero (in particular $|(H_1, H_2)| < 0.05$).

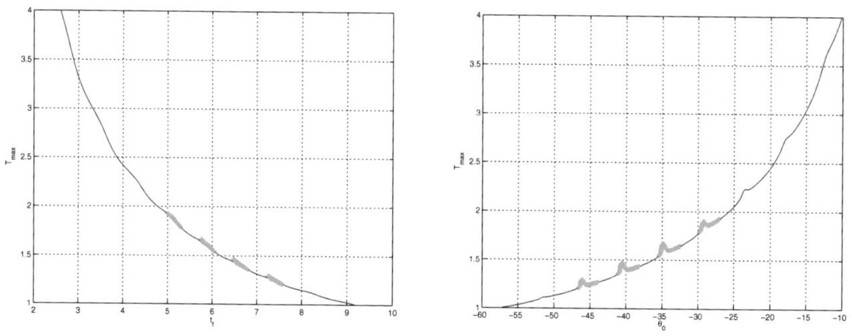

**Fig. 92** For the GEO to Mo control problem, homotopic curve for $\varepsilon \in [1 : 10]$N for the $\mathscr{C}_1$ type of solutions. *Left* $t_f$ (transfer time) versus $\varepsilon$ projection. *Right* $\theta_0$ (angle defining the initial position on the departure orbit) versus $\varepsilon$

## 8.1 $\mathscr{C}_1$ Cut Points

In Fig. 92 we show for the $\mathscr{C}_1$ type of solutions, the homotopic curve $t_f$ versus $\varepsilon$ and the same curve plotting $\theta_0$ versus $\varepsilon$, where $\theta_0$ in the angle that parameterizes the initial condition on the departure GEO orbit. In both plots we have highlighted in green the solutions close to the CUT pair, which are the solutions that we have analyzed. Moreover, Figs. 93, 94 and 95 summarize the results for the first cut point. Similarly, Figs. 96, 97 and 98 for the second cut point, Figs. 99, 100 and 101 for the third cut point, and finally Figs. 102, 103 and 104 for the fourth cut point.

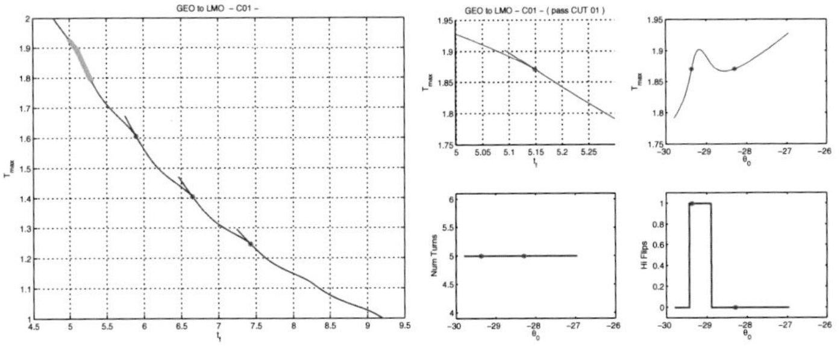


**Fig. 93** $\mathscr{C}_1$ cut point $n^o$ 1: (*left*) $t_f$ versus $\varepsilon$ homotopic curve with highlight of the cut passage in *green*; (*right*) analysis of the cut passage: (*top-left* subplot) $t_f$ versus $\varepsilon$ zoom, (*top-right* subplot) $\theta_0$ versus $\varepsilon$, (*bottom-left* subplot) $\theta_0$ versus num. turns around the Earth, (*bottom-right* subplot) $\theta_0$ versus the number of times $|(H_1, H_2)|$ passes close to zero. *Red* points are values corresponding the each cut point

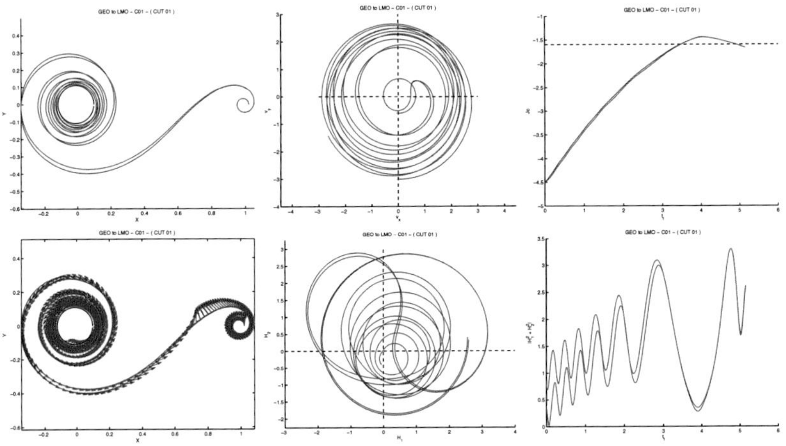


**Fig. 94** $\mathscr{C}_1$ cut point $n^o$ 1, *blue* orbits correspond to the first cut value and *red* orbits to the second cut value. (*top-left*) $\{XY\}$ projection of the transfer trajectory, (*top-center*) $\{V_x V_y\}$ projection of the transfer trajectory, (*top-right*) $t$ versus $J_c$ (energy variation along the transfer trajectory), (*bottom-left*) control along the trajectory, (*bottom-center*) $H_1$ versus $H_2$, (*bottom-right*) $t$ versus $|(H_1, H_2)|$

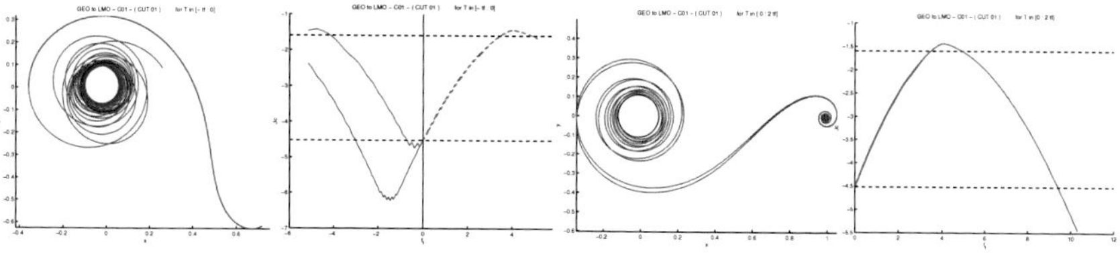

**Fig. 95** $\mathscr{C}_1$ cut point $n^o$ 1, (*left*) optimal solutions for $t \in [-t_f, 0]$ ($XY$ projection and $J_c$ variation), (*right*) optimal solutions for $t \in [0, 2t_f]$ ($XY$ projection and $J_c$ variation)

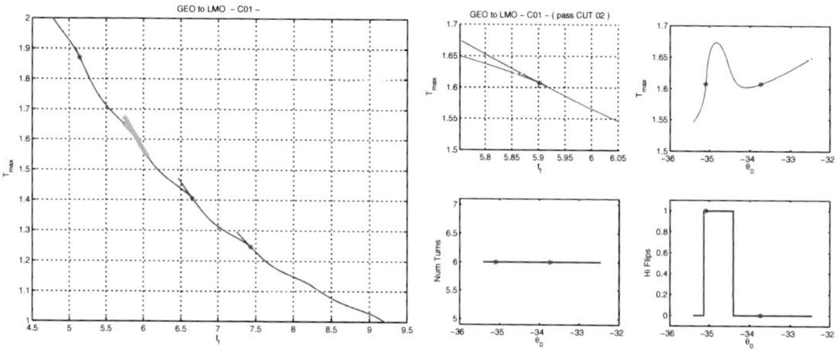


**Fig. 96** $\mathscr{C}_1$ cut point $n^o$ 2: (*left*) $t_f$ versus $\varepsilon$ homotopic curve with highlight of the cut passage in *green*; (*right*) analysis of the cut passage: (*top-left* subplot) $t_f$ versus $\varepsilon$ zoom, (*top-right* subplot) $\theta_0$ versus $\varepsilon$, (*bottom-left* subplot) $\theta_0$ versus num. turns around the Earth, (*bottom-right* subplot) $\theta_0$ versus the number of times $|(H_1, H_2)|$ passes close to zero. *Red* points are values corresponding the each cut point

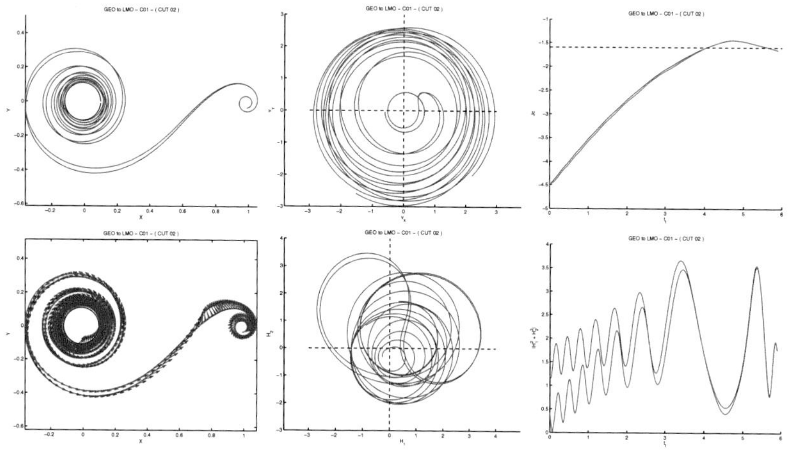

**Fig. 97** $\mathscr{C}_1$ cut point $n^o$ 2, *blue* orbits correspond to the first cut value and *red* orbits to the second cut value. (*top-left*) $\{XY\}$ projection of the transfer trajectory, (*top-center*) $\{V_x V_y\}$ projection of the transfer trajectory, (*top-right*) $t$ versus $J_c$ (energy variation along the transfer trajectory), (*bottom-left*) control along the trajectory, (*bottom-center*) $H_1$ versus $H_2$, (*bottom-right*) $t$ versus $|(H_1, H_2)|$

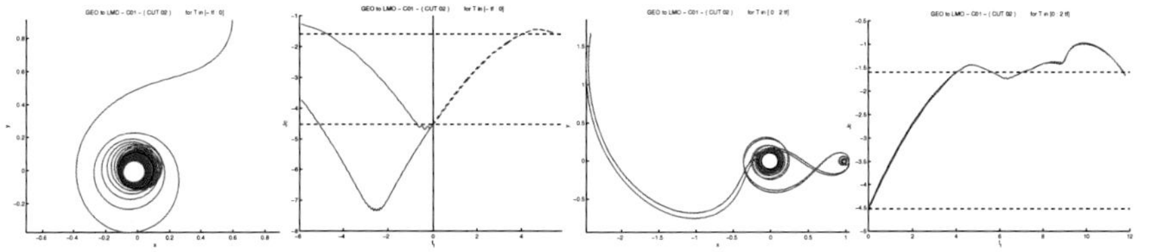

**Fig. 98** $\mathscr{C}_1$ cut point $n^o$ 2, (*left*) optimal solutions for $t \in [-t_f, 0]$ ($XY$ projection and $J_c$ variation), (*right*) optimal solutions for $t \in [0, 2t_f]$ ($XY$ projection and $J_c$ variation)

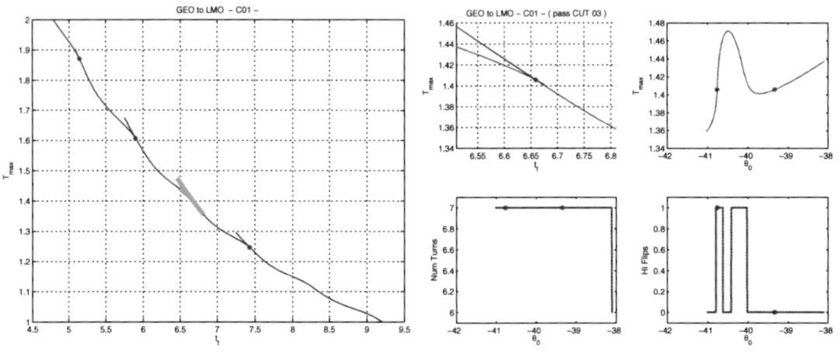


**Fig. 99** $\mathscr{C}_1$ cut point $n^o$ 3: (*left*) $t_f$ versus $\varepsilon$ homotopic curve with highlight of the cut passage in *green*; (*right*) analysis of the cut passage: (*top-left* subplot) $t_f$ versus $\varepsilon$ zoom, (*top-right* subplot) $\theta_0$ versus $\varepsilon$, (*bottom-left* subplot) $\theta_0$ versus num. turns around the Earth, (*bottom-right* subplot) $\theta_0$ versus the number of times $|(H_1, H_2)|$ passes close to zero. *Red* points are values corresponding the each cut point

## 8.2 $\mathscr{C}_2$ Cut Points

In Fig. 105 we show for the $\mathscr{C}_2$ type of solutions, the homotopic curve $t_f$ versus $\varepsilon$ and the same curve plotting $\theta_0$ versus $\varepsilon$, where $\theta_0$ in the angle that parameterizes the initial condition on the departure GEO orbit. In both plots we have highlighted in green the solutions close to the CUT pair, which are the solutions that we have analyzed. Moreover, Figs. 106, 107 and 108 summarize the results for the first cut point. Similarly, Figs. 109, 110 and 111 for the second cut point, Figs. 112, 113 and 114 for the third cut point, Figs. 115, 116 and 117 for the forth cut point, Figs. 118, 119 and 120 for the fifth cut point, Figs. 121, 122 and 123 for the sixth cut point, Figs. 124, 125 and 126 for the seventh cut point, Figs. 127, 128 and 129 for the eighth cut point, and finally Figs. 130, 131 and 132 for the ninth cut point.

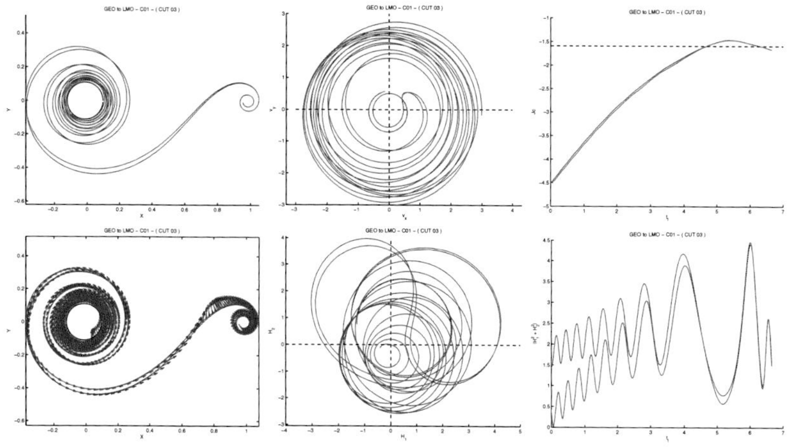


**Fig. 100** $\mathscr{C}_1$ cut point $n^o$ 3, *blue* orbits correspond to the first cut value and *red* orbits to the second cut value. (*top-left*) $\{XY\}$ projection of the transfer trajectory, (*top-center*) $\{V_x V_y\}$ projection of the transfer trajectory, (*top-right*) $t$ versus $J_c$ (energy variation along the transfer trajectory), (*bottom-left*) control along the trajectory, (*bottom-center*) $H_1$ versus $H_2$, (*bottom-right*) $t$ versus $|(H_1, H_2)|$

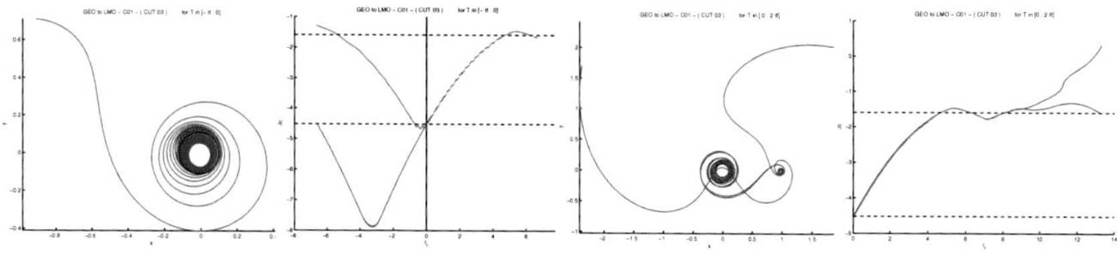

**Fig. 101** $\mathscr{C}_1$ cut point $n^o$ 3, (*left*) optimal solutions for $t \in [-t_f, 0]$ ($XY$ projection and $J_c$ variation), (*right*) optimal solutions for $t \in [0, 2t_f]$ ($XY$ projection and $J_c$ variation)

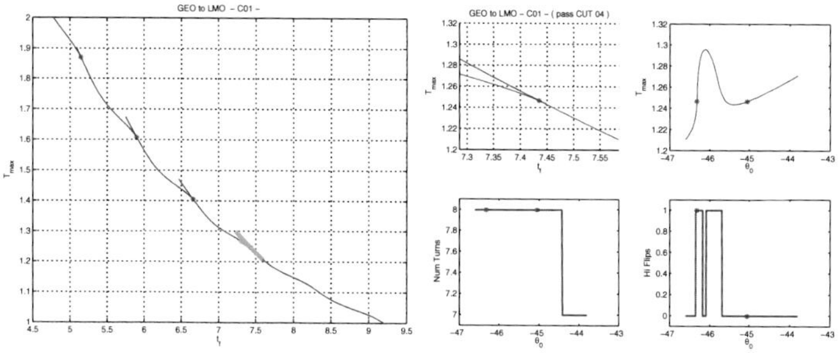


**Fig. 102** $\mathscr{C}_1$ cut point $n^o$ 4: (*left*) $t_f$ versus $\varepsilon$ homotopic curve with highlight of the cut passage in *green*; (*right*) analysis of the cut passage: (*top-left* subplot) $t_f$ versus $\varepsilon$ zoom, (*top-right* subplot) $\theta_0$ versus $\varepsilon$, (*bottom-left* subplot) $\theta_0$ versus num. turns around the Earth, (*bottom-right* subplot) $\theta_0$ versus the number of times $|(H_1, H_2)|$ passes close to zero. *Red* points are values corresponding the each cut point

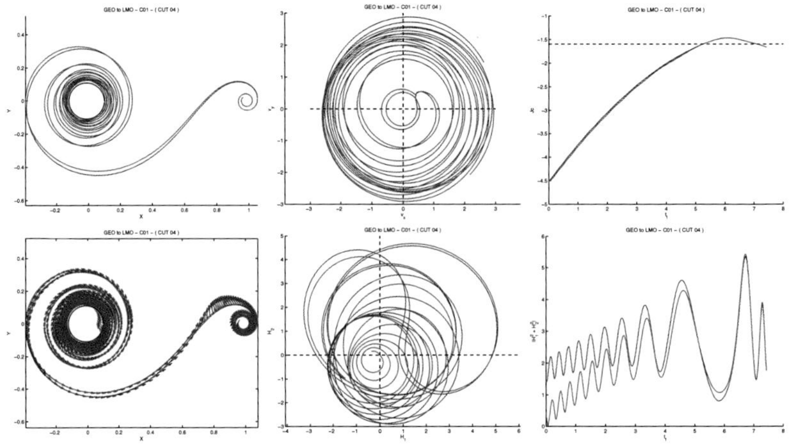


**Fig. 103** $\mathscr{C}_1$ cut point $n^o$ 4, *blue* orbits correspond to the first cut value and *red* orbits to the second cut value. (*top-left*) $\{XY\}$ projection of the transfer trajectory, (*top-center*) $\{V_x V_y\}$ projection of the transfer trajectory, (*top-right*) $t$ versus $J_c$ (energy variation along the transfer trajectory), (*bottom-left*) control along the trajectory, (*bottom-center*) $H_1$ versus $H_2$, (*bottom-right*) $t$ versus $|(H_1, H_2)|$

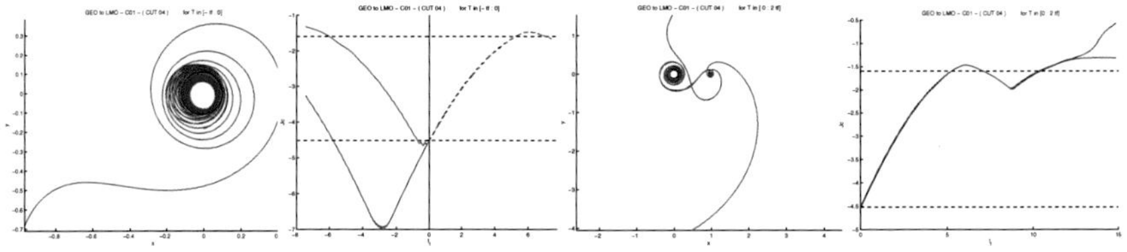

**Fig. 104** $\mathscr{C}_1$ cut point $n^o$ 4, (*left*) optimal solutions for $t \in [-t_f, 0]$ ($XY$ projection and $J_c$ variation), (*right*) optimal solutions for $t \in [0, 2t_f]$ ($XY$ projection and $J_c$ variation)

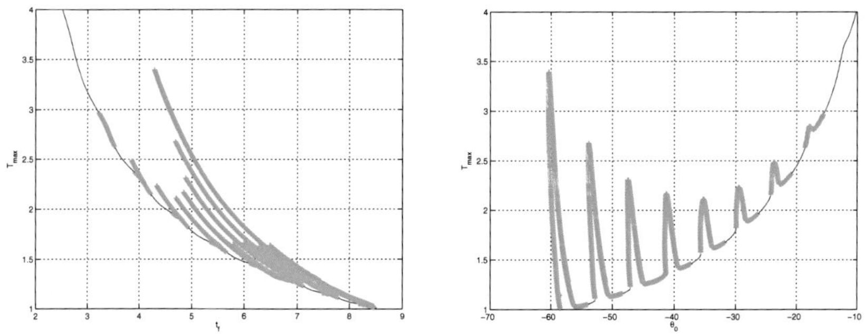

**Fig. 105** For the GEO to Mo control problem, homotopic curve for $\varepsilon \in [1 : 10]$N for the $\mathscr{C}_2$ type of solutions. *Left* $t_f$ (transfer time) versus $\varepsilon$ projection. *Right* $\theta_0$ (angle defining the initial position on the departure orbit) versus $\varepsilon$

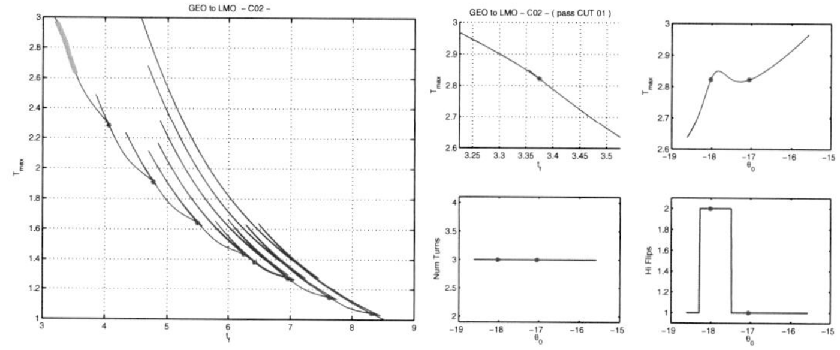


**Fig. 106** $\mathscr{C}_2$ cut point $n^o$ 1: (*left*) $t_f$ versus $\varepsilon$ homotopic curve with highlight of the cut passage in *green*; (*right*) analysis of the cut passage: (*top-left* subplot) $t_f$ versus $\varepsilon$ zoom, (*top-right* subplot) $\theta_0$ versus $\varepsilon$, (*bottom-left* subplot) $\theta_0$ versus num. turns around the Earth, (*bottom-right* subplot) $\theta_0$ versus the number of times $|(H_1, H_2)|$ passes close to zero. *Red* points are values corresponding the each cut point

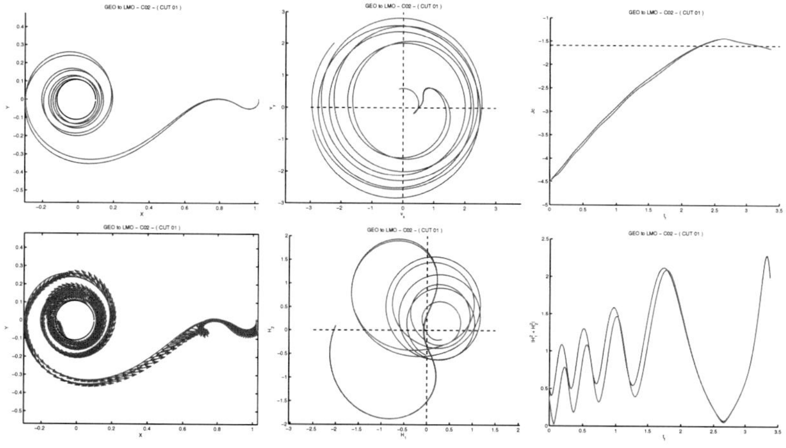

**Fig. 107** $\mathscr{C}_2$ cut point $n^o$ 1, *blue* orbits correspond to the first cut value and *red* orbits to the second cut value. (*top-left*) $\{XY\}$ projection of the transfer trajectory, (*top-center*) $\{V_x V_y\}$ projection of the transfer trajectory, (*top-right*) $t$ versus $J_c$ (energy variation along the transfer trajectory), (*bottom-left*) control along the trajectory, (*bottom-center*) $H_1$ versus $H_2$, (*bottom-right*) $t$ versus $|(H_1, H_2)|$

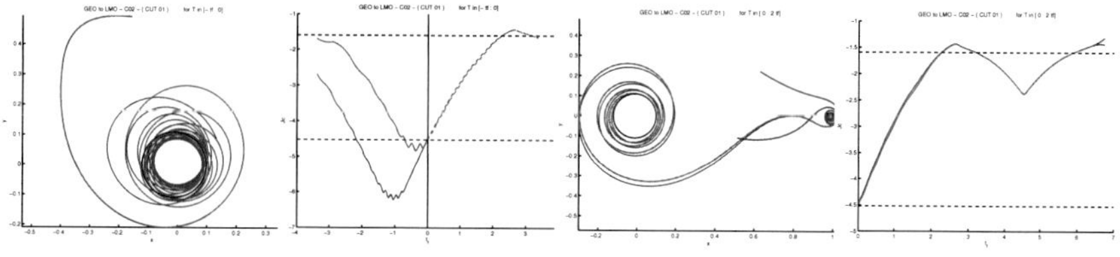

**Fig. 108** $\mathscr{C}_2$ cut point $n^o$ 1, (*left*) optimal solutions for $t \in [-t_f, 0]$ ($XY$ projection and $J_c$ variation), (*right*) optimal solutions for $t \in [0, 2t_f]$ ($XY$ projection and $J_c$ variation)

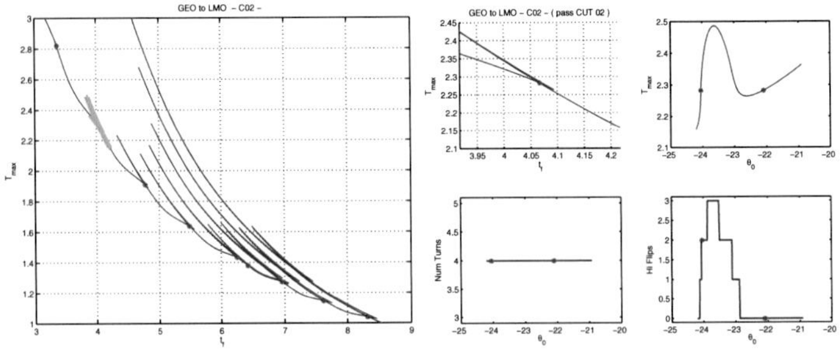


**Fig. 109** $\mathscr{C}_2$ cut point $n^o$ 2: (*left*) $t_f$ versus $\varepsilon$ homotopic curve with highlight of the cut passage in *green*; (*right*) analysis of the cut passage: (*top-left* subplot) $t_f$ versus $\varepsilon$ zoom, (*top-right* subplot) $\theta_0$ versus $\varepsilon$, (*bottom-left* subplot) $\theta_0$ versus num. turns around the Earth, (*bottom-right* subplot) $\theta_0$ versus the number of times $|(H_1, H_2)|$ passes close to zero. *Red* points are values corresponding the each cut point

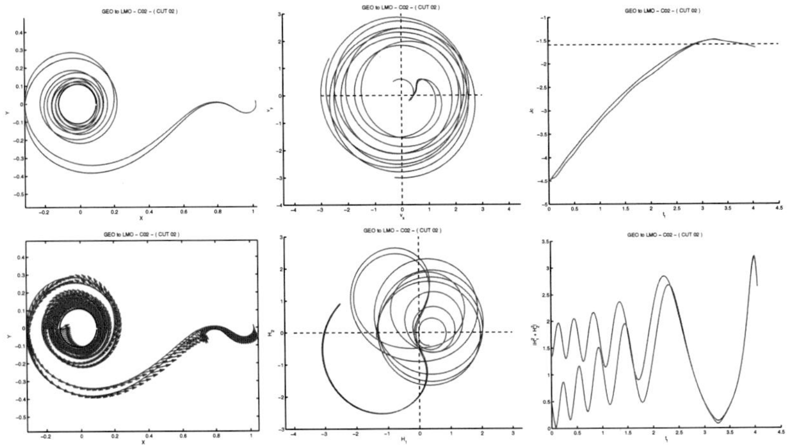

**Fig. 110** $\mathscr{C}_2$ cut point $n^o$ 2, *blue* orbits correspond to the first cut value and *red* orbits to the second cut value. (*top-left*) $\{XY\}$ projection of the transfer trajectory, (*top-center*) $\{V_x V_y\}$ projection of the transfer trajectory, (*top-right*) $t$ versus $J_c$ (energy variation along the transfer trajectory), (*bottom-left*) control along the trajectory, (*bottom-center*) $H_1$ versus $H_2$, (*bottom-right*) $t$ versus $|(H_1, H_2)|$

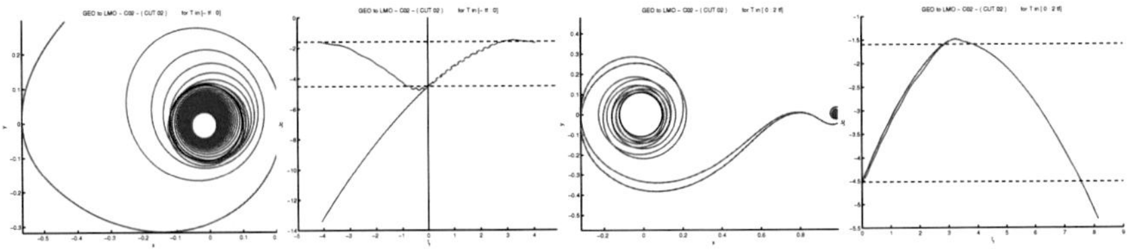

**Fig. 111** $\mathscr{C}_2$ cut point $n^o$ 2, (*left*) optimal solutions for $t \in [-t_f, 0]$ ($XY$ projection and $J_c$ variation), (*right*) optimal solutions for $t \in [0, 2t_f]$ ($XY$ projection and $J_c$ variation)

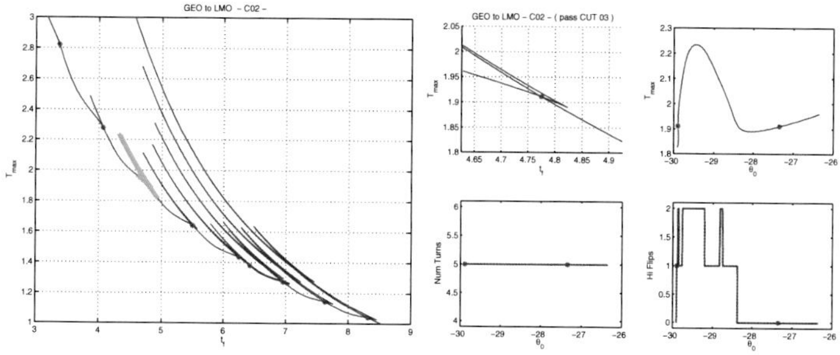


**Fig. 112** $\mathscr{C}_2$ cut point $n^o$ 3: (*left*) $t_f$ versus $\varepsilon$ homotopic curve with highlight of the cut passage in *green*; (*right*) analysis of the cut passage: (*top-left* subplot) $t_f$ versus $\varepsilon$ zoom, (*top-right* subplot) $\theta_0$ versus $\varepsilon$, (*bottom-left* subplot) $\theta_0$ versus num. turns around the Earth, (*bottom-right* subplot) $\theta_0$ versus the number of times $|(H_1, H_2)|$ passes close to zero. *Red* points are values corresponding the each cut point

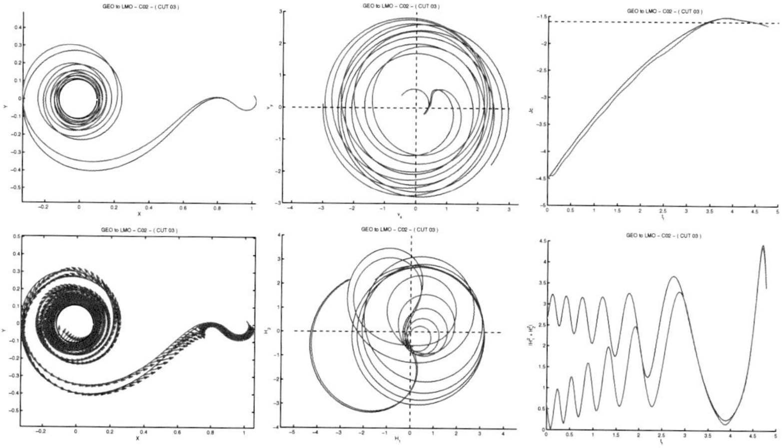


**Fig. 113** $\mathscr{C}_2$ cut point $n^o$ 3, *blue* orbits correspond to the first cut value and *red* orbits to the second cut value. (*top-left*) $\{XY\}$ projection of the transfer trajectory, (*top-center*) $\{V_x V_y\}$ projection of the transfer trajectory, (*top-right*) $t$ versus $J_c$ (energy variation along the transfer trajectory), (*bottom-left*) control along the trajectory, (*bottom-center*) $H_1$ versus $H_2$, (*bottom-right*) $t$ versus $|(H_1, H_2)|$

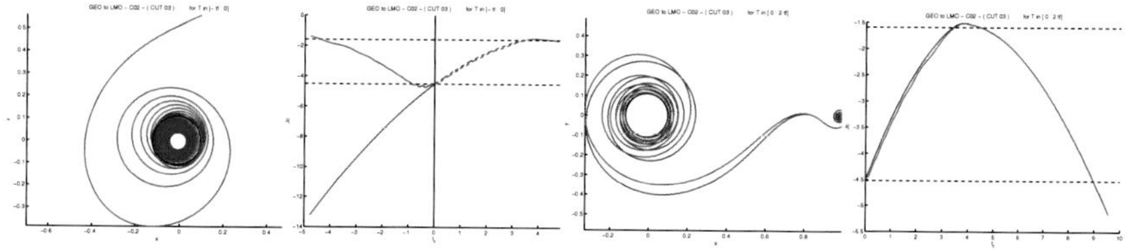

**Fig. 114** $\mathscr{C}_2$ cut point $n^o$ 3, (*left*) optimal solutions for $t \in [-t_f, 0]$ ($XY$ projection and $J_c$ variation), (*right*) optimal solutions for $t \in [0, 2t_f]$ ($XY$ projection and $J_c$ variation)

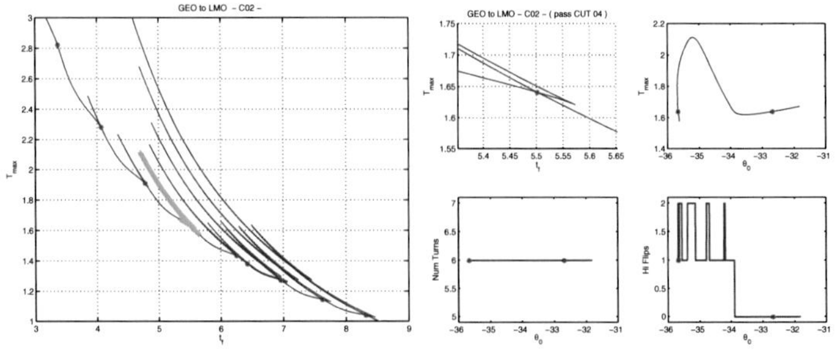


**Fig. 115** $\mathscr{C}_2$ cut point $n^o$ 4: (*left*) $t_f$ versus $\varepsilon$ homotopic curve with highlight of the cut passage in *green*; (*right*) analysis of the cut passage: (*top-left* subplot) $t_f$ versus $\varepsilon$ zoom, (*top-right* subplot) $\theta_0$ versus $\varepsilon$, (*bottom-left* subplot) $\theta_0$ versus num. turns around the Earth, (*bottom-right* subplot) $\theta_0$ versus the number of times $|(H_1, H_2)|$ passes close to zero. *Red* points are values corresponding the each cut point

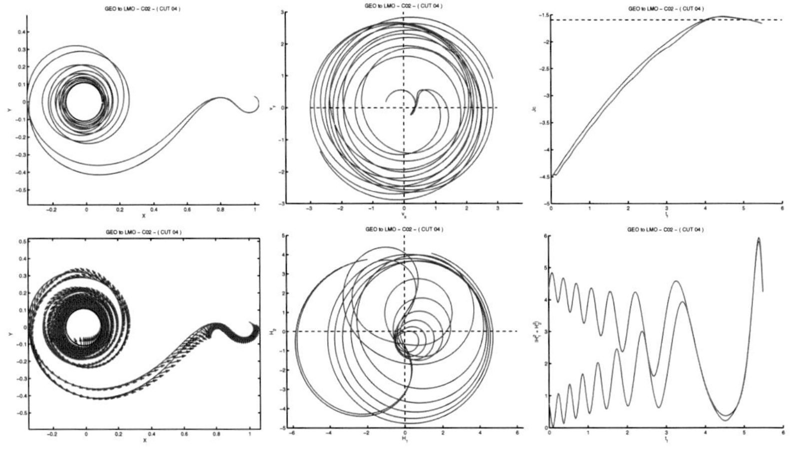


**Fig. 116** $\mathscr{C}_2$ cut point $n^o$ 4, *blue* orbits correspond to the first cut value and *red* orbits to the second cut value. (*top-left*) $\{XY\}$ projection of the transfer trajectory, (*top-center*) $\{V_x V_y\}$ projection of the transfer trajectory, (*top-right*) $t$ versus $J_c$ (energy variation along the transfer trajectory), (*bottom-left*) control along the trajectory, (*bottom-center*) $H_1$ versus $H_2$, (*bottom-right*) $t$ versus $|(H_1, H_2)|$

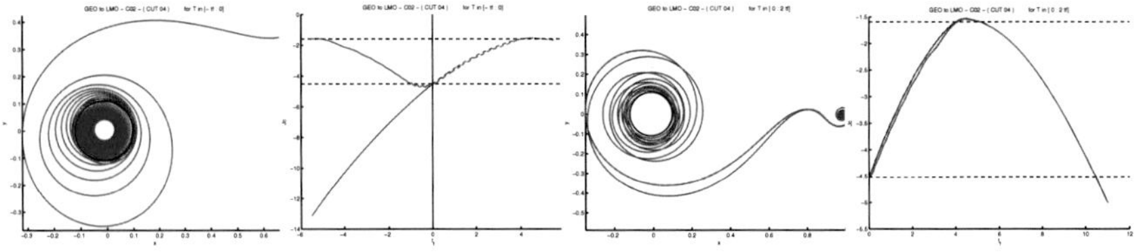

**Fig. 117** $\mathscr{C}_2$ cut point $n^o$ 4, (*left*) optimal solutions for $t \in [-t_f, 0]$ ($XY$ projection and $J_c$ variation), (*right*) optimal solutions for $t \in [0, 2t_f]$ ($XY$ projection and $J_c$ variation)

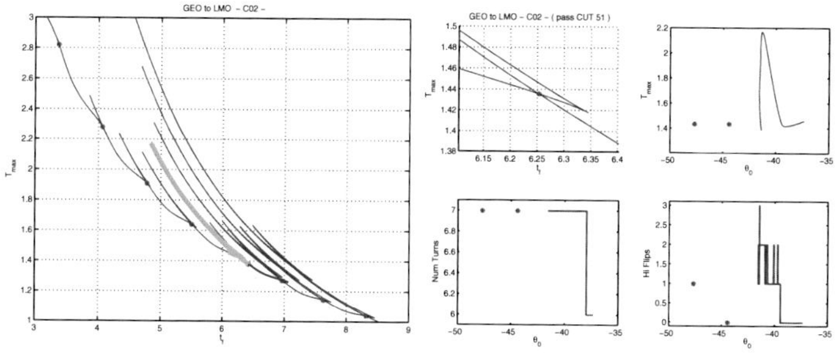


**Fig. 118** $\mathscr{C}_2$ cut point $n^o$ $5(a)$: (*left*) $t_f$ versus $\varepsilon$ homotopic curve with highlight of the cut passage in *green*; (*right*) analysis of the cut passage: (*top-left* subplot) $t_f$ versus $\varepsilon$ zoom, (*top-right* subplot) $\theta_0$ versus $\varepsilon$, (*bottom-left* subplot) $\theta_0$ versus num. turns around the Earth, (*bottom-right* subplot) $\theta_0$ versus the number of times $|(H_1, H_2)|$ passes close to zero. *Red* points are values corresponding the each cut point

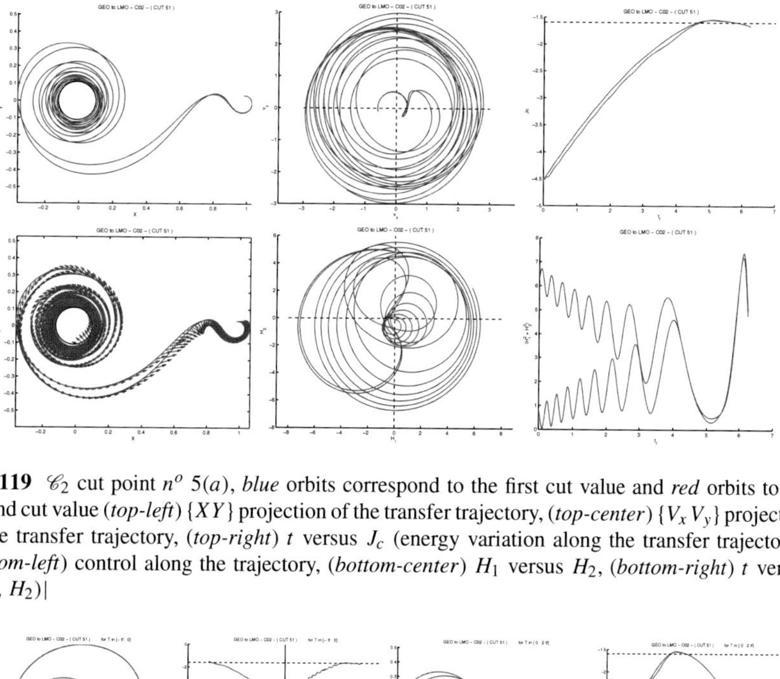

**Fig. 119** $\mathscr{C}_2$ cut point $n^o$ $5(a)$, *blue* orbits correspond to the first cut value and *red* orbits to the second cut value (*top-left*) $\{XY\}$ projection of the transfer trajectory, (*top-center*) $\{V_x V_y\}$ projection of the transfer trajectory, (*top-right*) $t$ versus $J_c$ (energy variation along the transfer trajectory), (*bottom-left*) control along the trajectory, (*bottom-center*) $H_1$ versus $H_2$, (*bottom-right*) $t$ versus $|(H_1, H_2)|$

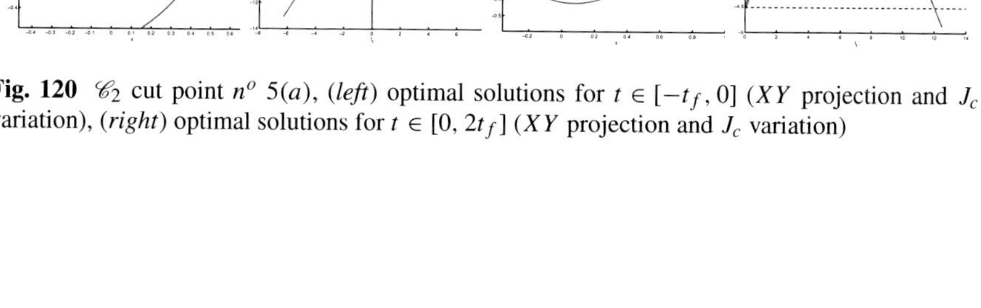

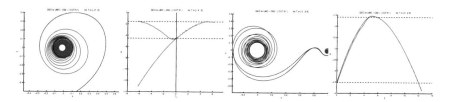

**Fig. 120** $\mathscr{C}_2$ cut point $n^o$ $5(a)$, (*left*) optimal solutions for $t \in [-t_f, 0]$ ($XY$ projection and $J_c$ variation), (*right*) optimal solutions for $t \in [0, 2t_f]$ ($XY$ projection and $J_c$ variation)

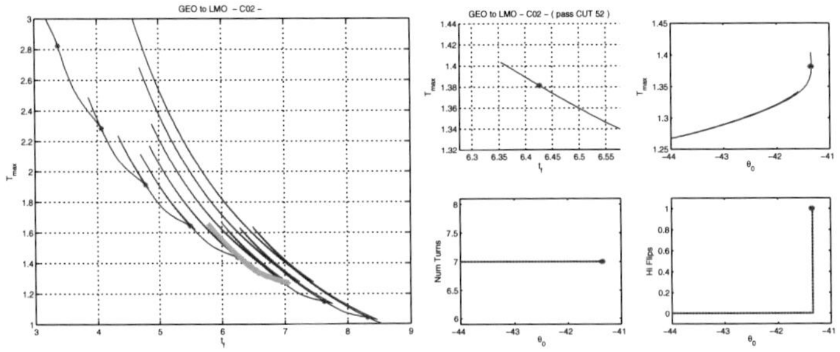


**Fig. 121** $\mathscr{C}_2$ cut point $n^o$ $5(b)$: (*left*) $t_f$ versus $\varepsilon$ homotopic curve with highlight of the cut passage in *green*; (*right*) analysis of the cut passage: (*top-left* subplot) $t_f$ versus $\varepsilon$ zoom, (*top-right* subplot) $\theta_0$ versus $\varepsilon$, (*bottom-left* subplot) $\theta_0$ versus num. turns around the Earth, (*bottom-right* subplot) $\theta_0$ versus the number of times $|(H_1, H_2)|$ passes close to zero. *Red* points are values corresponding the each cut point

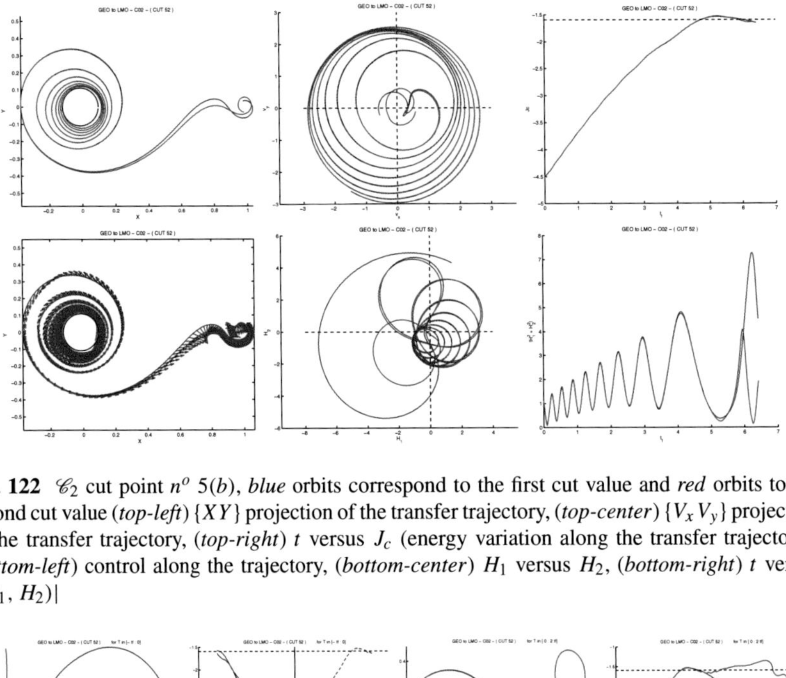

**Fig. 122** $\mathscr{C}_2$ cut point $n^o$ $5(b)$, *blue* orbits correspond to the first cut value and *red* orbits to the second cut value (*top-left*) $\{XY\}$ projection of the transfer trajectory, (*top-center*) $\{V_x V_y\}$ projection of the transfer trajectory, (*top-right*) $t$ versus $J_c$ (energy variation along the transfer trajectory), (*bottom-left*) control along the trajectory, (*bottom-center*) $H_1$ versus $H_2$, (*bottom-right*) $t$ versus $|(H_1, H_2)|$

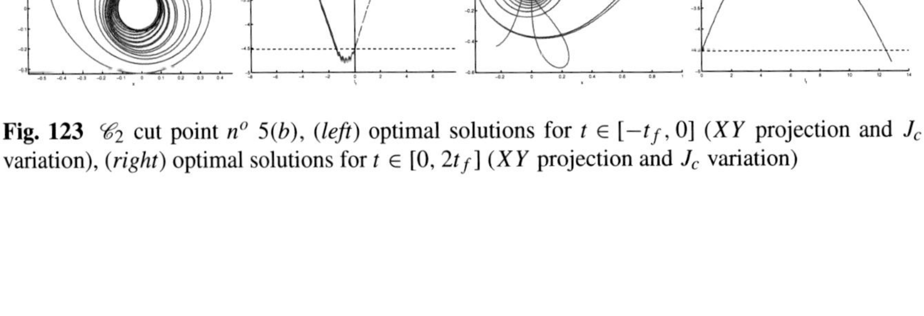

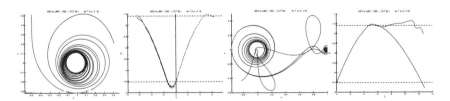

**Fig. 123** $\mathscr{C}_2$ cut point $n^o$ $5(b)$, (*left*) optimal solutions for $t \in [-t_f, 0]$ ($XY$ projection and $J_c$ variation), (*right*) optimal solutions for $t \in [0, 2t_f]$ ($XY$ projection and $J_c$ variation)

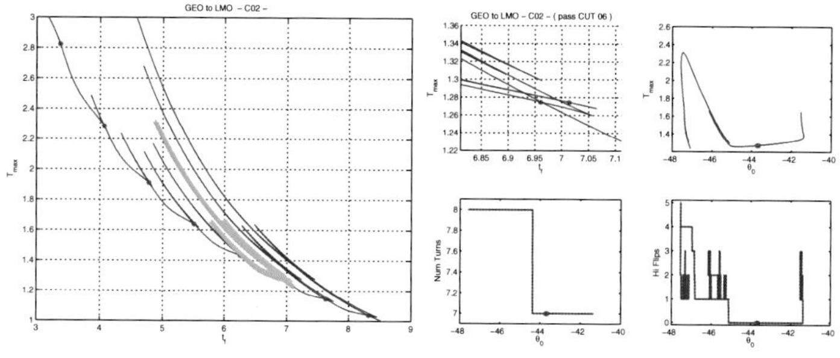


**Fig. 124** $\mathscr{C}_2$ cut point $n^o$ 6: (*left*) $t_f$ versus $\varepsilon$ homotopic curve with highlight of the cut passage in *green*; (*right*) analysis of the cut passage: (*top-left* subplot) $t_f$ versus $\varepsilon$ zoom, (*top-right* subplot) $\theta_0$ versus $\varepsilon$, (*bottom-left* subplot) $\theta_0$ versus num. turns around the Earth, (*bottom-right* subplot) $\theta_0$ versus the number of times $|(H_1, H_2)|$ passes close to zero. *Red* points are values corresponding the each cut point

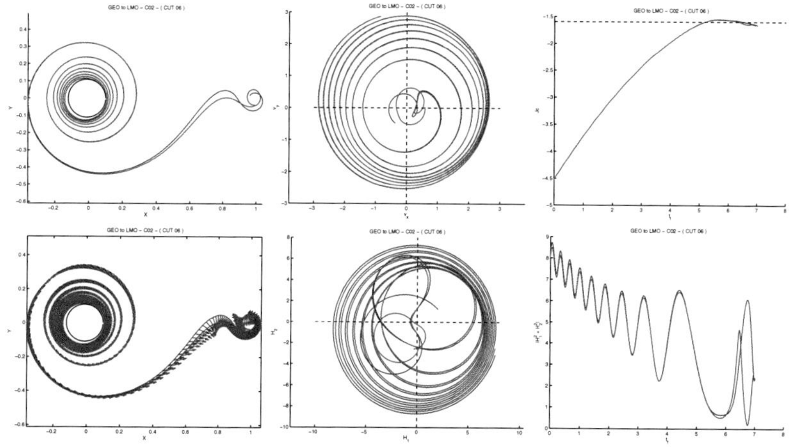

**Fig. 125** $\mathscr{C}_2$ cut point $n^o$ 6, *blue* orbits correspond to the first cut value and *red* orbits to the second cut value. (*top-left*) $\{XY\}$ projection of the transfer trajectory, (*top-center*) $\{V_x V_y\}$ projection of the transfer trajectory, (*top-right*) $t$ versus $J_c$ (energy variation along the transfer trajectory), (*bottom-left*) control along the trajectory, (*bottom-center*) $H_1$ versus $H_2$, (*bottom-right*) $t$ versus $|(H_1, H_2)|$

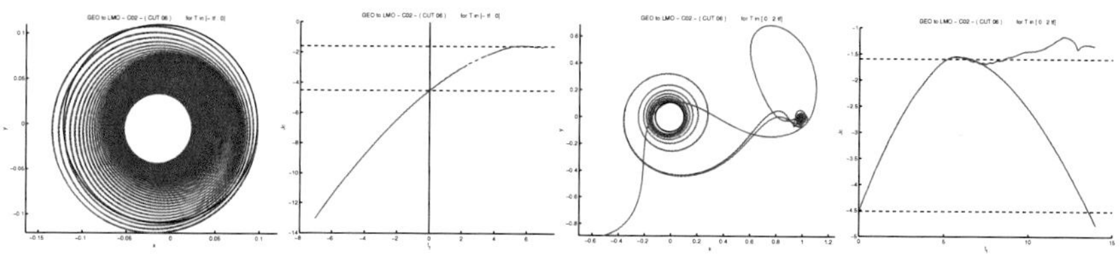

**Fig. 126** $\mathscr{C}_2$ cut point $n^o$ 6, (*left*) optimal solutions for $t \in [-t_f, 0]$ ($XY$ projection and $J_c$ variation), (*right*) optimal solutions for $t \in [0, 2t_f]$ ($XY$ projection and $J_c$ variation)

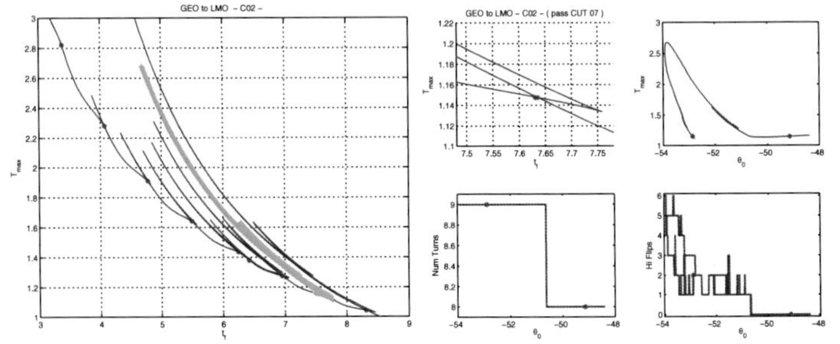


**Fig. 127** $\mathscr{C}_2$ cut point $n^o$ 7: (*left*) $t_f$ verus $\varepsilon$ homotopic curve with highlight of the cut passage in *green*; (*right*) analysis of the cut passage: (*top-left* subplot) $t_f$ versus $\varepsilon$ zoom, (*top-right* subplot) $\theta_0$ versus $\varepsilon$, (*bottom-left* subplot) $\theta_0$ versus num. turns around the Earth, (*bottom-right* subplot) $\theta_0$ versus the number of times $|(H_1, H_2)|$ passes close to zero. *Red* points are values corresponding the each cut point

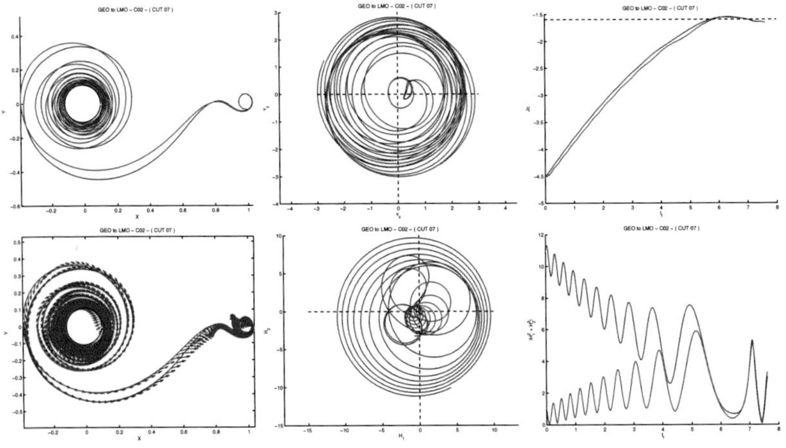

**Fig. 128** $\mathscr{C}_2$ cut point $n^o$ 7, *blue* orbits correspond to the first cut value and *red* orbits to the second cut value. (*top-left*) $\{XY\}$ projection of the transfer trajectory, (*top-center*) $\{V_x V_y\}$ projection of the transfer trajectory, (*top-right*) $t$ versus $J_c$ (energy variation along the transfer trajectory), (*bottom-left*) control along the trajectory, (*bottom-center*) $H_1$ versus $H_2$, (*bottom-right*) $t$ versus $|(H_1, H_2)|$

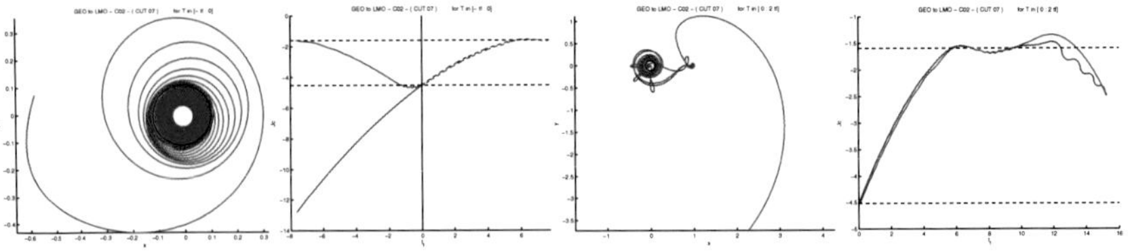

**Fig. 129** $\mathscr{C}_2$ cut point $n^o$ 7, (*left*) optimal solutions for $t \in [-t_f, 0]$ ($XY$ projection and $J_c$ variation), (*right*) optimal solutions for $t \in [0, 2t_f]$ ($XY$ projection and $J_c$ variation)

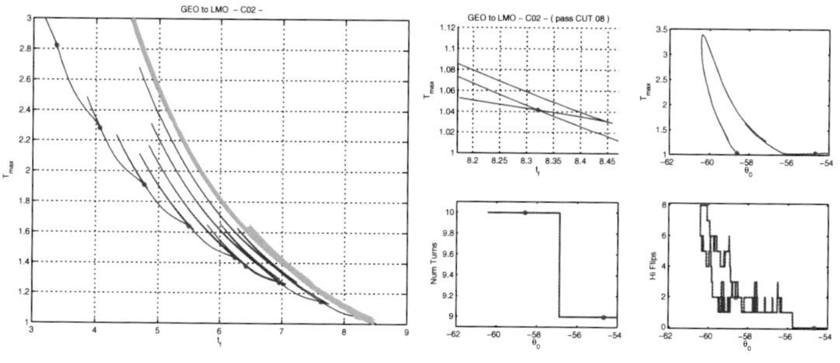


**Fig. 130** $\mathscr{C}_2$ cut point $n^o$ 8: (*left*) $t_f$ versus $\varepsilon$ homotopic curve with highlight of the cut passage in *green*; (*right*) analysis of the cut passage: (*top-left* subplot) $t_f$ versus $\varepsilon$ zoom, (*top-right* subplot) $\theta_0$ versus $\varepsilon$, (*bottom-left* subplot) $\theta_0$ versus num. turns around the Earth, (*bottom-right* subplot) $\theta_0$ versus the number of times $|(H_1, H_2)|$ passes close to zero. *Red* points are values corresponding the each cut point

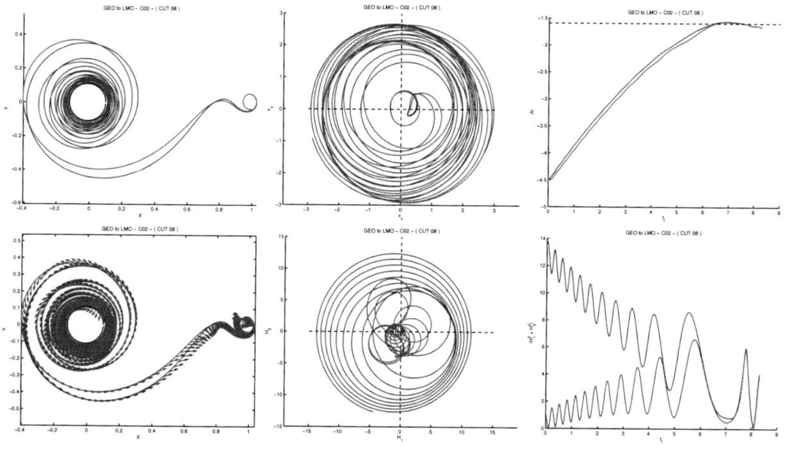

**Fig. 131** $\mathscr{C}_2$ cut point $n^o$ 8, *blue* orbits correspond to the first cut value and *red* orbits to the second cut value. (*top-left*) $\{XY\}$ projection of the transfer trajectory, (*top-center*) $\{V_x V_y\}$ projection of the transfer trajectory, (*top-right*) $t$ versus $J_c$ (energy variation along the transfer trajectory), (*bottom-left*) control along the trajectory, (*bottom-center*) $H_1$ versus $H_2$, (*bottom-right*) $t$ versus $|(H_1, H_2)|$

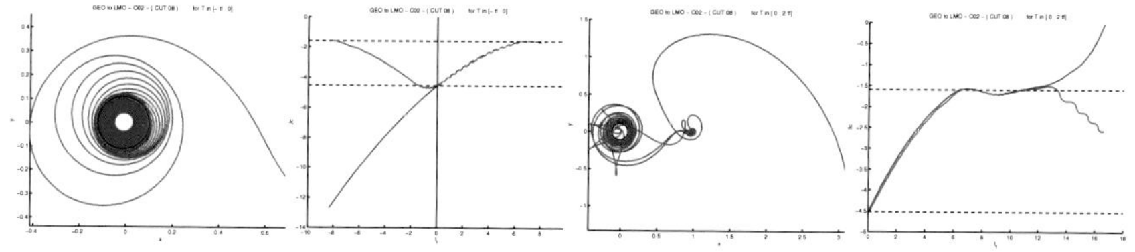

**Fig. 132** $\mathscr{C}_2$ cut point $n^o$ 8, (*left*) optimal solutions for $t \in [-t_f, 0]$ ($XY$ projection and $J_c$ variation), (*right*) optimal solutions for $t \in [0, 2t_f]$ ($XY$ projection and $J_c$ variation)

**Acknowledgments** This work has been supported by the Conseil Régional de Bourgogne grant no. 9201AAO049S0273.

## References

1. Arnold VI (2004) Catastrophe theory. Springer, New York
2. BepiColombo mission: http://sci.esa.int/bepicolombo
3. Caillau JB, Chitour Y, Zheng Y (2015) $L^1$-Minimization for mechanical systems, pp 1–22
4. Caillau J-B, Cots O, Gergaud J (2012) Differential pathfollowing for regular optimal control problems. Optim Methods Softw 27(2):177–196
5. Caillau JB, Daoud B (2012) Minimum time control of the restricted three-body problem. SIAM J Control Optim 50(6):3178–3202 (2012). http://apo.enseeiht.fr/hampath
6. Caillau JB, Daoud B, Gergaud J (2011) Discrete and differential homotopy in circular restricted three-body control. Discrete Contin Dyn Syst Suppl:229–239 (2011) (Proceedings of 8th AIMS Conference on Dynamical Systems, Differential Equations and Applications, Dresden, May 2010)
7. Caillau J-B, Daoud B, Gergaud J (2012) Minimum fuel control of the planar circular restricted three-body problem. Celestial Mech Dyn Astron 114(1):137–150
8. Caillau J-B, Noailles J (2001) Coplanar control of a satellite around the Earth. ESAIM Control Optim Calc Var 6:239–258
9. Lisa Pathfinder mission: http://sci.esa.int/lisa-pathfinder
10. Zhang C, Topputo F, Bernelli-Zazzera F, Zhao Y-S (2015) Low-thrust minimum-fuel optimization in the circular restricted three-body problem. J Guid Control Dyn 38:1501–1510